New Perspectives in Teaching and Learning With ICTs in Global Higher Education Systems

Madalina Armie
University of Almería, Spain

Germán Asensio
University of Almería, Spain

Maria Enriqueta Cortés de los Ríos
University of Almería, Spain

Angeles Jordán Soriano
University of Almería, Spain

A volume in the Advances in Educational Technologies and Instructional Design (AETID) Book Series

IGI Global
PUBLISHER of TIMELY KNOWLEDGE

Published in the United States of America by
IGI Global
Information Science Reference (an imprint of IGI Global)
701 E. Chocolate Avenue
Hershey PA, USA 17033
Tel: 717-533-8845
Fax: 717-533-8661
E-mail: cust@igi-global.com
Web site: http://www.igi-global.com

Library of Congress Cataloging-in-Publication Data

Names: Armie, Madalina, 1991- editor.
Title: New perspectives in teaching and learning with ICTs in global higher education systems / edited by Madalina Armie, Germán Asensio, Maria Enriqueta Cortés de los Ríos, and Angeles Jordán Soriano.
Other titles: New perspectives in teaching and learning with information and communications technologies in global higher education systems
Description: Hershey, PA : Information Science Reference, [2023] | Includes bibliographical references and index. | Summary: "The book's research focuses on new pedagogical methodologies and approaches that can be utilized to engage students and improve their learning outcomes. It also highlights the role of the modern lecturer in new teaching and learning contexts that utilize ICTs and emphasizes the need for educators to adapt their teaching approaches to meet the changing needs of today's learners. This book is an essential resource for educators, policymakers, and researchers seeking to stay up to date with the latest trends and approaches in higher education and ICTs"-- Provided by publisher.
Identifiers: LCCN 2023019731 (print) | LCCN 2023019732 (ebook) | ISBN 9781668488614 (h/c) | ISBN 9781668488652 (s/c) | ISBN 9781668488621 (ebk)
Subjects: LCSH: Education, Higher--Effect of technological innovations on. | Education, Higher--Computer-assisted instruction. | Internet in higher education. | Educational technology.
Classification: LCC LB2395.7 .N495 2023 (print) | LCC LB2395.7 (ebook) | DDC 378.1/7344678--dc23/eng/20230602
LC record available at https://lccn.loc.gov/2023019731
LC ebook record available at https://lccn.loc.gov/2023019732

This book is published in the IGI Global book series Advances in Educational Technologies and Instructional Design (AE-TID) (ISSN: 2326-8905; eISSN: 2326-8913)

British Cataloguing in Publication Data
A Cataloguing in Publication record for this book is available from the British Library.

For electronic access to this publication, please contact: eresources@igi-global.com.

Advances in Educational Technologies and Instructional Design (AETID) Book Series

Lawrence A. Tomei
Robert Morris University, USA

ISSN:2326-8905
EISSN:2326-8913

Mission

Education has undergone, and continues to undergo, immense changes in the way it is enacted and distributed to both child and adult learners. In modern education, the traditional classroom learning experience has evolved to include technological resources and to provide online classroom opportunities to students of all ages regardless of their geographical locations. From distance education, Massive-Open-Online-Courses (MOOCs), and electronic tablets in the classroom, technology is now an integral part of learning and is also affecting the way educators communicate information to students.

The **Advances in Educational Technologies & Instructional Design (AETID) Book Series** explores new research and theories for facilitating learning and improving educational performance utilizing technological processes and resources. The series examines technologies that can be integrated into K-12 classrooms to improve skills and learning abilities in all subjects including STEM education and language learning. Additionally, it studies the emergence of fully online classrooms for young and adult learners alike, and the communication and accountability challenges that can arise. Trending topics that are covered include adaptive learning, game-based learning, virtual school environments, and social media effects. School administrators, educators, academicians, researchers, and students will find this series to be an excellent resource for the effective design and implementation of learning technologies in their classes.

Coverage

- Virtual School Environments
- Digital Divide in Education
- Hybrid Learning
- Instructional Design
- Social Media Effects on Education
- Bring-Your-Own-Device
- Adaptive Learning
- Higher Education Technologies
- Collaboration Tools
- K-12 Educational Technologies

IGI Global is currently accepting manuscripts for publication within this series. To submit a proposal for a volume in this series, please contact our Acquisition Editors at Acquisitions@igi-global.com or visit: http://www.igi-global.com/publish/.

The Advances in Educational Technologies and Instructional Design (AETID) Book Series (ISSN 2326-8905) is published by IGI Global, 701 E. Chocolate Avenue, Hershey, PA 17033-1240, USA, www.igi-global.com. This series is composed of titles available for purchase individually; each title is edited to be contextually exclusive from any other title within the series. For pricing and ordering information please visit http://www.igi-global.com/book-series/advances-educational-technologies-instructional-design/73678. Postmaster: Send all address changes to above address.

Titles in this Series

For a list of additional titles in this series, please visit: www.igi-global.com/book-series

Developing Skills and Competencies for Digital and Green Transitions
Patricia Ordóñez de Pablos (The University of Oviedo, Spain) Mohammad Nabil Almunawar (School of Business and Economics, University of Brunei Darussalam, Brunei) and Muhammad Anshari (School of Business and Economics, University of Brunei Darussalam, Bunei)
Business Science Reference • © 2023 • 367pp • H/C (ISBN: 9781668490891) • US $250.00

Artificial Intelligence Applications Using ChatGPT in Education Case Studies and Practices
Muhammad Shahzad Aslam (School of Traditional Chinese Medicine, Xiamen University Malaysia, Malaysia) and Saima Nisar (School of Computing, UUM College of Arts and Sciences, University Utara Malaysia, Sintok, Malysia)
Information Science Reference • © 2023 • 234pp • H/C (ISBN: 9781668493007) • US $215.00

Using Assistive Technology for Inclusive Learning in K-12 Classrooms
Jeremy Bell (University of Louisiana, Monroe, USA) and Timothy Gifford (Movia Robotics Inc., USA)
Information Science Reference • © 2023 • 280pp • H/C (ISBN: 9781668464243) • US $215.00

The Impact and Importance of Instructional Design in the Educational Landscape
Christie F. Calhoun (Jacksonville State University, USA)
Information Science Reference • © 2023 • 295pp • H/C (ISBN: 9781668482087) • US $215.00

Global Perspectives on Effective Assessment in English Language Teaching
Miguel Fernández Álvarez (Universidad Politécnica de Madrid, Spain) and Amanda Lira Gordenstein Montes (Northeastern Illinois University, USA)
Information Science Reference • © 2023 • 287pp • H/C (ISBN: 9781668482131) • US $215.00

Handbook of Research on Innovative Frameworks and Inclusive Models for Online Learning
Jared Keengwe (University of North Dakota, USA)
Information Science Reference • © 2023 • 412pp • H/C (ISBN: 9781668490723) • US $270.00

Technology Management and Its Social Impact on Education
P.C. Lai (University of Malaya, Malaysia)
Information Science Reference • © 2023 • 316pp • H/C (ISBN: 9781668491034) • US $215.00

IGI Global
PUBLISHER of TIMELY KNOWLEDGE

701 East Chocolate Avenue, Hershey, PA 17033, USA
Tel: 717-533-8845 x100 • Fax: 717-533-8661
E-Mail: cust@igi-global.com • www.igi-global.com

Table of Contents

Detailed Table of Contents

This chapter presents the results of a study that examines the effects of an online gamification of the end-of-unit tests of a course in a master's degree. Learners completed a survey about the effects of the online gamified experience on their motivation and their progress and results in the course. Participants (n = 41) were in- or pre-service teachers, so the survey examined their teachers' sense of self-efficacy in terms of the application of active methodologies and ICTs in their classes after the gamified experience. Results show that the gamification had positive effects in terms of increasing motivation for the subject and, therefore, their perceived academic results. Also, participants felt more confident to use gamification in their classes. Results also showed significant differences in terms of motivation between nontraditional and traditional learners.

The teaching paradigm is undergoing a real revolution in addition the pandemic situation suffered from COVID-19. Therefore, both teachers and faculty must adapt and train so as to give an adequate response that improves the results and favors the motivation for learning. They are putting their focus on the permanent training of teachers and on educational innovation. One of the most current and innovative teaching innovation tools is gamification. Therefore, this study aims to analyze the existing scientific literature, carrying out a systematic review of the empirical studies that show experiences carried out in university classrooms, especially in careers related to teaching in different specialties. Discarding the texts that did not meet the eligibility criteria, six articles were selected that are committed to teaching innovation and the practice of educational strategies such as gamification in university classrooms to promote meaningful learning as well as the involvement and motivation of students towards learning.

The European project entitled "Empowering Nurses and Healthcare Professionals to Promote Vaccination and Tackle Vaccine Hesitancy. IENE 11 PROVAC" is an initiative that has been developed by three European Universities (from Spain, Cyprus, and Greece) and an educational organization from Romania. Its goal is to train healthcare professionals to become reliable sources of information as well as address the doubts of the general population. Furthermore, this program aims to create a curriculum, information tools, and a massive online open course (MOOC) targeting healthcare professionals, individuals from disadvantaged groups, and those who are more vaccine hesitant to help them become better informed, overcome their fears, and ultimately strengthen confidence in vaccines. This MOOC aspires to be an example for innovative online courses for healthcare professionals around the world.

This study aimed to examine the predictive role of digital literacy skills on mobile English learning attitudes. It was conducted with 135 university students studying in the school of foreign languages of a state university in Turkey, and the data were gathered using digital literacy scale and English mobile learning attitude scale. This study showed that university students' digital literacy skills and mobile English learning attitudes were high and their digital literacy skills have a moderate and positive relationship with their mobile English learning attitudes. Besides, it was concluded that university students' digital literacy skills significantly predicted their mobile English learning attitudes and they together explained 50% of the total variance on their mobile English learning attitudes. Also, university students' scores on the attitude subscale of the digital literacy scale presented the largest effect on their mobile English learning attitudes.

The present study deals with a systematic review on virtual education in higher education during the COVID-19 health crisis to know the digital competence of teachers and student perception of online learning. The scientific literature was analyzed during the years 2020-2022 based on empirical and theoretical studies in the Spanish university environment. The results show a certain weakness of telematic teaching with respect to face-to-face teaching. Technology is a powerful tool that requires planning, an aspect that was not adequately developed in times of pandemic. COVID-19 has revealed technological deficiencies in teachers, which implies that we are still developing in the digital framework and their use is reduced to the minimum expression of all the potential they have. In general, university students feel satisfied with virtual teaching. They have recognized the potential that ICT has in the "learning to learn" competition, but have emphasized an increase in academic stress, backed by the psychological impact of confinement.

The ephemeral nature of life is such that what was present in our lives yesterday may undergo a transformation or cease to exist altogether, as if it had never been. Technology has emerged as a significant factor in mitigating the challenges posed by various crises, particularly during recent global pandemics, natural calamities, and mass migrations. Undoubtedly, the incorporation of technology in the realm of education is a significant phenomenon. Currently, attention is directed towards academic institutions that offer programs for teacher education. Numerous responsibilities are assigned to the faculty members who provide instruction to individuals pursuing a career in teaching. The education of pre-service teachers regarding technology necessitates a distinct emphasis and instruction. The primary subject matter of this chapter pertains to the instruction of technology integration for pre-service English teachers, with a particular emphasis on the digital divide.

The aim of this research is to display how project-based learning (PBL), one of the most innovative methodologies in the contemporary educational panorama, can be beneficial in teaching culture in English language classes at the tertiary level. Students will be due to produce a final project that examines cultural elements in several English-speaking countries combining various disciplines, such as geography, history, literature, and economics. The results will be drawn from two questionnaires: a pre-task survey that will assess the expectations and motivation levels of tertiary students in connection to this pedagogical approach and a post-task survey which will—in this case—assess the effects of these final results in students' motivation and the potential changes concerning their views on PBL.

The objective of the chapter is to propose the incorporation of active methodologies for the development of the competencies of postgraduate students in order to promote quality professional training. This comparative study comes from research whose results emanate from the use of active and innovative teaching strategies, put into practice in the classroom, with the aim of defeating the problems that frequently arise such as academic lapsing, lack of interest and communication, work team up. Within the solution alternatives, strategies and materials were used to promote learning related to the needs and in the context of the students, to develop skills and knowledge for solving administrative situations. Therefore, a proposal for successful training practices in project-based learning and the flipped classroom is presented to promote quality academic training in postgraduate students.

This chapter explores students' opinions and experiences with distance education (DE) from a multinational perspective, focusing on Portugal, Ukraine, and the United Arab Emirates. The research questions address students' previous experiences with online courses, their opinions about DE, reasons for enrolling or not enrolling in DE courses, perceived preparedness for DE, receptiveness to DE courses after the pandemic, preferences for DE formats, and differences based on age, gender, and country. By examining these aspects, the study aims to provide research-informed support for future DE models and contribute to understanding students' preferences, needs, and opinions regarding DE. The findings can potentially inform IHEs' digitalization strategies, recruitment efforts, learning approaches, and the development of purposefully designed DE and blended learning programs. The multinational approach strengthens the analysis and enhances understanding of the impact of DE on students' learning experiences in different countries.

Stigma towards mental health disorders is a powerful obstacle to the recovery and social inclusion. This is why social awareness and empathy must be addressed as soon as possible. Among the possibilities for intervention, new technologies, and more specifically virtual reality (VR), are a very attractive resource for young people who spend a large part of their free time playing video games. The aim of this chapter is to present a proposal that can be applied to university students to address stigma in mental health whose central axis is a virtual reality programme, Inclúyete VR. This programme fosters empathy towards those who suffer from a mental health disorder, allowing them to "experience" how a person feels when they have hallucinations and what the possibilities are for a psychosocial approach to the disorder. The session should be led by a mental health professional familiar with the software. The total duration of the VR programme is approximately 10 minutes, which is very appropriate to initiate a discussion-reflection about mental health disorders.

The ubiquity of new portable devices with internet access and recent events such as the global pandemic caused by COVID-19 and the Russian invasion of Ukraine has highlighted the importance of media and information literacy. In parallel, developing tools based on artificial intelligence will allow the developing of sophisticated disinformative content (deepfakes). This media context demands new and refined informational competences to identify and verify information. This research presents a proposal to develop media and information literacy using AI tools to create audiovisual content. Strategies such as flipped classrooms, thematic debates, and playful activities are used. A selection of evaluation instruments is presented, which, through a pre-and post-experimental design, will allow the evaluation of this proposal and the learning developed by the students.

This pilot study aimed to investigate the effectiveness of game-based learning in teaching executive functions (EF) to preservice teachers. Various games were employed as training tools to improve EF skills, and students' satisfaction and perceptions of the training were evaluated. The results revealed a highly positive response to the utilized methodology, highlighting its efficacy in stimulating students with special educational needs. Participants valued the adaptability of the games and considered the experience enjoyable and innovative. These findings emphasize the importance of providing future educators with practical exposure to active learning methodologies.

Foreword

In a time defined by rapid change, interconnectedness and the transformative impact of the COVID-19 pandemic, the field of higher education stands as a crossroads of dynamic innovation and steadfast tradition. *New Perspectives in Teaching and Learning With ICTs in Global Higher Education Systems* emerges as a beacon of insightful discourse on the multifaceted landscape of learning and teaching within higher education systems across the globe. The carefully selected contributions bring together students' and teachers' perspectives and experiences from Turkey, México, the United Arab Emirates, Ukraine, Portugal and Spain examining the new problems faced in the classrooms of the twenty-first century, the new challenges posed by a society in constant change, the programs and pedagogical methodologies employed, as well as the new approaches and instruments aimed at improving teaching and learning.

The breadth of subjects covered in this volume is nothing short of inspiring, encompassing themes that range from harnessing the power of mobile gamification tools for foreign language education to the pivotal role of digital literacy and the innovative use of project-based learning to enrich the teaching of English language and culture. Additionally, as the world grapples with the consequences of a global pandemic, the exploration of online learning and its intersection with the pedagogical landscape provides invaluable insights. The authors' and editors' commitment to inclusivity recognizes the diverse audience that this publication serves. Policymakers, pre-service and in-service educators, students, and researchers alike will find within the following pages a rich array of experiences that resonate across the educational spectrum.

In the so-called digital age or the era of information (Pratt-Johnson, 2019), where ideas, trends and innovations transcend borders with unprecedented ease, the narratives shared within the contributions included in this volume reflect the richness of experiences from education spanning diverse cultures and contexts. Research in higher education is woven not only with the threads of pedagogy and methodology, but also with the intricate fibers of culture, communication, and global awareness (Lütge et al., 2023). As the authors and editors of this publication embark on the pursuit of excellence, they encounter a landscape defined by its diversity and the challenges that arise from it. In fact, the traditional model of classroom and teaching methodologies seem to have stopped working (Lee & Hammer, 2011). Readers will find that the chapters resonate with recent shifts in history, such as globalization, digitalization, and the most recent COVID-19 pandemic, that have shaken the very foundations of education systems and have compelled educators to rethink, adapt, and innovate like never before (Neuwirth et al., 2021). The pandemic has accelerated the adoption of online learning methodologies, thrusting educators into uncharted territories where digital platforms become the conduits of knowledge dissemination (Vergara-Rodríguez et al., 2022; Guppy et al., 2022; Wieland & Kollias, 2020). The exploration of online active methodologies in higher education, as discussed in these pages, rises as a testament to the resilience and adaptability of educators in embracing technology to facilitate meaningful and successful learning experiences.

In line with the above, bearing in mind that the task of continuous training and updating is found at the core of the twenty-first-century educator, the ground-breaking chapters included in this volume disrupt the notion of education as a static field. Some contributions delve into the emergence of mobile gamification tools for foreign language teaching, as well as the creative integration of game-based learning, attesting to the contemporary fusion of pedagogy and technology, which fosters a greater motivation and engagement of the learners in the dynamic lessons that consolidate the content through play (Armie & Membrive, 2022). This also reflects a powerful recognition of the evolving learning preferences of the digital-native generations (Çimen & Hangül, 2021). In parallel, the pandemic has unveiled the intricate tapestry of mental health stigma among students, adding a poignant layer to the academic discourse, which is also explored in this volume. Recent studies (Abi-Jaoude et al., 2020) observe how student bullying, depression and suicide rates have disturbingly risen. As educators and researchers, we are entrusted with nurturing not only intellectual growth but also the emotional well-being of those under our guidance. The two very timely and comprehensive systematic reviews included in the following pages—the first on the existing literature on innovative approaches that promote meaningful learning, and the second on the impact of the COVID-19 health crisis on virtual learning in higher education—answer to the very nature of education as a reflection of the societies it serves, and as the tides of globalization ebb and flow, so do the landscapes of learning and teaching.

We stand at the crossroads of tradition and transformation, where pedagogical methodologies are reimagined to accommodate the needs of a diverse world. The collective wisdom shared within these pages serves as a guidepost for educators, researchers, students, and policymakers alike navigating the complexities of higher education in a time defined by change and diversity. As we strive to shape the minds of future generations, let the new contexts and new ways discussed in this volume inspire you to embrace the challenges and opportunities presented by our rapidly evolving and interconnected world.

J. Javier Torres-Fernández
University of Almería, Spain

J. Javier Torres-Fernández *has a BA and MA in English Studies with Honorary Distinction. He is currently registered for a Ph.D. and works at the University of Almería as a predoctoral researcher funded by the Spanish Ministry of Science, Innovation and Universities (FPU21/01232) in the Department of Philology in the Women, Literature and Society (HUM-874) research group and the Communication and Society I+D Research Centre. His fields of interest include women, gender, sexuality studies, popular culture, Irish literature, and North American literature. His thesis examines the literary production that emerged from the HIV/AIDS crisis in Ireland and the United States and conducts a comparative study of the genre. It tackles stigma, shame, sexuality, identity, trauma, and cultural politics. Most recently, he has completed a master's in Secondary Education, Baccalaureate, Vocational Training, and Language Teaching at the University of Almería, where he specialized in teaching English as a foreign language through graphic novels in the classroom. He also examined the transversal inclusion of LGBTQIA+ and mental health topics in the Heartstopper books and Netflix series.*

REFERENCES

Abi-Jaoude, E., Naylor, K. T., & Pignatiello, A. (2020). Smartphones, social media use and youth mental health. *Canadian Medical Association Journal, 192*(6), 136–141. doi:10.1503/cmaj.190434 PMID:32041697

Armie, M., & Membrive Pérez, V. (2022). *Escaping boredom in the classroom: Breakouts, breakout boxes and escape rooms*. Cambridge Scholars Publishing.

Çimen, B., & Hangül, Ş. (2022). Digital immigrant teachers' perceptions about digital native students: An investigation into Turkish school context. *European Journal of Education and Psychology*, *14*(2), 1–21. doi:10.32457/ejep.v14i2.1576

Guppy, N., Verpoorten, D., Boud, D., Lin, L., Tai, J., & Bartolic, S. (2022). The post-COVID-19 future of digital learning in higher education: Views from educators, students, and other professionals in six countries. *British Journal of Educational Technology*, *53*(6), 1750–1765. doi:10.1111/bjet.13212

Lee, J. J., & Hammer, J. (2011). Gamification in education: What, how, why bother? Definitions and uses. *Exchange Organizational Behavior Teaching Journal*, *15*(2), 1–5.

Lütge, C., Merse, T., & Rauschert, P. (2023). *Global Citizenship in Foreign Language Education: Concepts, Practices, Connections*. Routledge.

Neuwirth, L. S., Jović, S., & Mukherji, B. R. (2021). Reimagining higher education during and post-COVID-19: Challenges and opportunities. *Journal of Adult and Continuing Education*, *27*(2), 141–156. doi:10.1177/1477971420947738

Pratt-Johnson, Y. (2019) The fourth industrial revolution and its impact on teacher education. *Tạp chí Nghiên cứu dân tộc,* *8*(2), 95-97. . doi:10.25073/0866-773X/307

Vergara-Rodríguez, D., Antón-Sancho, Á., & Fernández-Arias, P. (2022). Variables influencing professors' adaptation to digital learning environments during the COVID-19 Pandemic. *International Journal of Environmental Research and Public Health*, *19*(6), 3732. doi:10.3390/ijerph19063732 PMID:35329422

Wieland, N., & Kollias, L. (2020). Online learning before, during and after COVID-19: Observations over 20 years. *International Journal of Advanced Corporate Learning*, *13*(2), 84–92. doi:10.3991/ijac.v13i2.16779

Preface:
New Contexts, New Ways?

Globalization, the frenetic evolution of science, and the recent COVID-19 global pandemic have triggered dramatic social and cultural transformations. In the midst of all of this disruption of traditional identities and sociocultural practices, we have learned how to relate, socialize, communicate, learn, work, and, to an extent, live differently. It is not odd, therefore, that in the context of the 21st century, education too finds itself under continuous pressure and in constant need of innovation. The reasons for this imperative, as the following pages of this volume will reveal, are manifold, although they all stem from the changes taking place in human lifestyles in the context of the growing "technoculture" (Hand, 2016, p. 4).[1]

To begin with, the ever-growing body of research in the field of education emphasizes how, regardless of the level under consideration, traditional classroom teaching is no longer working (Lee & Hammer, 2011). Evidence is provided by data on learners' intrinsic and extrinsic motivation and demotivation, which also reinforces how certain subjects may become sources of frustration or boredom. Frustration and boredom may subsequently lead to increased cheating, disengagement from school, and high attrition rates, all of which pose major problems for present-day educational systems. These contributions from research also highlight how teaching and learning, though they may appear to be universal experiences, in fact, are not; simultaneously, they stress the need to understand the impressive variety of educational approaches to the art and science of teaching that explain how knowledge is transmitted and stored in the human brain. An awareness of how teaching and learning work is what makes teaching effective (Shastri, 2010, p. 18), and the discipline of educational psychology—which explains how human beings learn—sets the basis for this awareness. Educational psychological theories, which stem from multidisciplinary collaboration, can inform successful approaches to teaching while also serving as a foundation for developing lesson plans. One of these approaches, which complements many others—such as cognitivism, constructivism, behaviorism, and humanism—and which lies at the very heart of this book, is connectivism, also known as the learning theory of the digital era. Connectivism, additionally referred to as "networked learning" or "technologically mediated learning" (AlDahdouh et al., 2018), has achieved prominence lately due to the evolution of new information and communication technologies (ICTs) and artificial intelligence (AI), which have propelled a dizzying digital transformation, conducting our analog world into a new stage known under many different names, such as "the digital age," "the information age," "the era of information," or "the Fourth Industrial Revolution" (Pratt-Johnson, 2019), to mention just a few.

Connectivism was introduced and modeled as a concept in 2005 by George Siemens and Stephen Downes ("Connectivism Learning Theory," 2021).[2] Although Siemens focuses on the social aspects of connectivism—that is, the way humans can learn from one another—and Downes on machine-based learning—that is, the way humans can use machines and ICTs to widen their knowledge—each perspective constructs an additional layer in the understanding of education. Connectivism aims to bridge the gaps between various traditional ideas of teaching while promoting the creation of new methodologies and methods to maintain the rapid processes of metamorphosis taking place in society and the current demands of the labor market. According to the premises of connectivism, learning is more than our own internal construction of knowledge, as external networks such as social media, online networks, blogs, and information databases can widen the horizons of knowledge. Based on this idea, therefore, connectivism starts from the foundational premise that "[k]nowledge has many authors, knowledge has many facets, it looks different to each person, and it changes moment to moment. A piece of knowledge isn't a description of something, it is a way of relating to something" (Downes, quoted in "Connectivism Learning Theory," 2021). Technology becomes a major agent in the learning process of relating to this "something" Downes mentions, and our constant connectedness provides an opportunity to make choices about the ways in which we learn and what we want to learn. People are all "nodes" in this network or system, that is, "elements" that can be connected to other "elements," be they other human beings, printed materials, or online resources. Connectivist learning, therefore, occurs when links are established—and, more importantly, nurtured—between various nodes of information. Consequently, connectivism promotes group collaboration and discussion and encourages learning that draws on different positions and perspectives.

Understanding connectivism and its pillars also implies an understanding of how people have coped with the new reality imposed by the COVID-19 pandemic. Well before the onset of the pandemic, new and hybrid forms of education such as e-learning, blended learning, flipped learning, and mobile learning had already been used either to supplement or to replace more traditional forms of knowledge acquisition. New methods of teaching and learning such as teaching through cinema or song, project-based learning, or applying methodologies such as gamification or flipped classrooms have been proved to work effectively in tandem with new technologies and hybrid educational forms. Traditional resources are now easily created, personalized, adapted to more or less interactive formats, duplicated, shared, and stored. Computers and ICTs have become useful for research, recordkeeping, review, communication, evaluation, and more. This digital boom was drastically accelerated by the pandemic from 2019 onward, triggering at the same time a series of undesired responses. According to the LEGO Foundation (n.d.), many students' recent experiences of schooling have included extended periods of isolation where "collaboration and socialisation have been deprioritised to ensure academic progression during a stressful and disruptive period." Of course, what little collaboration and socialization students have been able to practice over the last four years has been restricted to the digital dimension, and this has had a negative impact on students' abilities to communicate, negotiate, interact socially, and cooperate. Moreover, it has been claimed that as a result, many students are now more "dependent on adults supporting and leading their learning. They are therefore less able or willing to take risks and show leadership. They also lack some of the reflective skills necessary for their cognitive development" (The LEGO Foundation, n.d.). That this is in fact a generational issue that existed even prior to the pandemic is suggested by the findings of the researcher and psychologist Jean M. Twenge (2017), who claims in her study that "the allure of independence, so powerful to previous generations, holds less sway over today's teens." Troublingly, as economies become more globally integrated, the demand for workers possessing soft skills—critical thinking, problem-solving, public speaking, teamwork, leadership, career management, and intercultural fluency, among others—continues to rise (Chadid, 2023, p. 1).

In a world affected in a positive way by globalization and multiculturalism, which have also transformed the once homogeneous composition of the classroom, these soft skills are rendered absolutely essential. Educators are now more aware than ever of the increasing diversity of their students not only in terms of geographical, religious, cultural, racial, or ethnic background but also in terms of capacities, learning styles, and special needs. This is because "today, the demands of a modern productive economy and a modern inclusive society compel our schooling system to develop the special talent of every child. We also need to prepare each member of the first genome generation to play his or her part in a complex and knowledge-rich society" (Hopkins, 2003, p. XI). The challenges faced by teaching and learning in the 21st century are significantly different to those of the past, as teaching nowadays

Embraces not only the transmission of knowledge, but also learning how to learn, horizon-broadening and teacher development. The vision is of all students being engaged in appropriately challenging curriculum tasks, created by skillful teachers. As students master information and skills, the result of each learning experience is not only the content they learn but also the greater ability they acquire to approach future learning tasks in and out of [the educational space]. (Hopkins, 2003, p. xi)

In this "age of insecurity" (Hopkins, 2003), both "the knowledge economy [and] the Knowledge Society" (Hargreaves, 2003) are rapidly evolving, exerting unparalleled pressure on education. In the words of Andy Hargreaves (2003),

A knowledge society is really a learning society. . . . knowledge societies process information and knowledge in ways that maximize learning, stimulate ingenuity and invention and develop the capacity to initiate and cope with change. In the knowledge economy, wealth and prosperity depend on people's capacity to out-invent and outwit their competitors, to tune into the desires and demands of the consumer market, and to change jobs or develop new skills as economic fluctuations and downturns require. (p. xiii)

With expectations now taking a central role, it is more than obvious that the profile of the learner, and that of the future graduate, has also undergone a transformation, as today's students require the necessary training to function as adults in tomorrow's society and, by extension, in a highly skilled and competitive job market. For Prensky (2001), today's students, these "digital natives," "have not just changed *incrementally* from those of the past, nor simply changed their slang, clothes, body adornments, or styles, as happened between generations previously. A really big *discontinuity* has taken place" (p. 1). Prensky considers the contrast between this generation of students and those of the past as being that the former "*think and process information fundamentally differently* from their predecessors" (p. 1) and that

[They] are used to receiving information really fast. They like to parallel process and multi-task. They prefer their graphics before their text rather than the opposite. They prefer random access (like hypertext). They function best when networked. They thrive on instant gratification and frequent rewards. They prefer games to serious work. (p. 2)

The generation to which Prensky refers is that of the last three decades, made up of learners who "have spent their entire lives surrounded by and using computers, video games, digital music players, video cams, cell phones, and all the other toys and tools of the digital age" (Prensky, 2001, p. 1). However, exposure to information does not mean that learning takes place: at times, the opposite is the case.

Students are exposed to excessive and sometimes unreliable information, and digital competence and its subskills—information and data literacy, communication and collaboration, digital content creation, an awareness of safety, and problem-solving—may fail to materialize. According to information provided by the Central Statistics Office and also reported by Yahoo, *The Journal*, RTE, Head Topics, the *Irish Times*, and the *Irish Examiner*, the numbers could hardly be more concerning: at least 62% of all information found on the internet in 2022 was unreliable (Yaqub, 2023).

Another characteristic of these modern students is their overuse—and, by extension, misuse—of technological devices. Cross-sectional, longitudinal, and empirical studies highlight concerning figures on depression, suicide rates, and the prevalence of bullying (Abi-Jaoude et al., 2020). In the same vein, Twenge (2017) affirms that "rates of teen depression and suicide have skyrocketed since 2011. It's not an exaggeration to describe iGen as being on the brink of the worst mental-health crisis in decades. Much of this deterioration can be traced to their phones"—and definitely to their social media.

The roles and obligations of educators have diversified in this new context. Professors are now motivators, sources of information, facilitators, guides, observers, planners, counselors, mentors, tutors, and participants, to mention but a few roles. Following the transformation of society and education, teachers are expected to learn, plan, publish, share, and collaborate with colleagues (Heick, 2022). Expectations have varied, too, with pressure coming at different times from both the educational community and society, as educators are subjected to increased levels of scrutiny seeking to ensure that they are doing everything possible to guarantee each student's success while also proving their "teacherpreneurship" by acting as groundbreaking figures that find novel ways to create and share materials and connect with students and communities (Heick, 2022).

In light of all of this, education is a highly valued profession that requires continuous training and updating. In higher education in particular, perceptions of the field's static nature are counterbalanced by the need to respond to the aforementioned changing contexts (Brennan, 2008). With that point in mind, the chapters of this book aim to provide examples of this changing context, with a particular focus on higher education. For example, in the first chapter, titled "Mobile Gamification Tools for Foreign Language Teaching in Higher Education," Muhammet Demirbilek offers an overview of the literature on mobile gamification tools for foreign language teaching in higher education by evaluating not just some of the tools available but also the features that distinguish each tool from the rest. In addition, the chapter presents instances where mobile gamification tools have been successfully implemented in language classrooms, while also delving into the pedagogical implications of implementing these tools in a higher education context.

Chapter 2, titled "The Essence of Learning Management Systems Acceptance: A Phenomenological Study of Pre-Service Teachers' Experiences," investigates the experiences of pre-service teachers in the use of Learning Management Systems (LMS) from a phenomenological perspective. By analyzing the responses given by 73 pre-service teachers across a wide variety of academic fields, author Tugra Karademir Coşkun concludes that pre-service teachers consider the implementation of LMS necessary. Her discussion of the results is organized under six major themes: "effective learning processes," "professional development," "assessment and evaluation/feedback," "flexibility in access," "time management and organization," and "content management."

In Chapter 3, "Use of Online Active Methodologies in Higher Education," Patricia Bárcena-Toyos presents the results of a study involving 41 participants, including both in- and pre-service teachers, and their experiences in the application of active methodologies and ICTs in their classrooms following online gamification of end-of-unit tests in a Master's degree–level course. By examining the responses obtained by survey, Bárcena-Toyos arrives at the conclusion that participants felt predisposed and motivated to introduce gamification in their current or future classes as a means of promoting meaningful learning and the engagement of learners.

Chapter 4, "Educational Innovation in Teacher Classrooms: A Systematic Review," is co-authored by María del Pilar Díaz-López, Ana María Martínez-Martínez, Nuria Torres López and Noelia Navarro Gómez. The authors analyze existing scientific literature on the introduction of innovative approaches that promote meaningful learning, such as gamification. Despite the acknowledged limitations of the studies considered, their selection of six articles on experiences related to teaching innovation and gamified learning environments reveals that use of these strategies is reported to yield significant benefits in terms of learning and motivation.

Chapter 5, "A MOOC on Promoting Vaccination for Healthcare Professionals in Higher Education: Project IENE11 PROVAC," authored by Remedios López Liria, Patricia Rocamora Pérez, María de los Ángeles Valverde Martínez, and María Jesús Benzo Iglesias, describes the creation of a massive online open course (MOOC) intended to promote and spread information and confidence about vaccination among the general public, particularly in the case of disadvantaged individuals and the vaccine-hesitant. The MOOC, currently in development for a minimum of 120 pilot participants and with an expected launch window of 2024, is based on the four constructs of the Papadopoulos, Tilki, and Taylor Model: cultural awareness, cultural knowledge, cultural sensitivity, and cultural competence.

Chapter 6, titled "Predicting University Students' Mobile Learning Attitudes in Language Classrooms: The Role of Digital Literacy," addresses the role of digital literacy in using mobile technology devices in the learning process (m-learning), regarded among scholars as a challenging issue. Ali Orhan explores how the digital literacy levels of students from Turkish universities contribute to predicting their attitudes towards m-learning in language classrooms. The implementation of this predictive method shows that the efficacy of m-learning implementation is closely related to the willingness of learners and teachers to adopt m-learning, along with their respective attitudes toward this educational approach.

In Chapter 7, "Online Learning in Times of Pandemic: A Systematic Review," the objective is to examine the impact of the COVID-19 health crisis on virtual learning in higher education. Co-authors Ana María Martínez-Martínez, María del Pilar Díaz-López, Christian Roith, and Noelia Navarro Gómez conduct a systematic review to analyze a number of scientific studies on this topic. Their analysis focuses on various aspects, including the digital competence of teachers, the perception of online learning by students, and the new challenges faced in higher education during the pandemic. Their conclusions also highlight issues such as the limited digital training of new students and the weaknesses of online learning as a mode of education.

Chapter 8, "Becoming a Successful Pre-Service English as a Foreign Language Lecturer in the Digital Divide," presents an evaluation of pre-service teacher education programs. Gülin Zeybek investigates whether these programs adequately equip English as a foreign language (EFL) instructors with the essential knowledge and skills to effectively incorporate technological innovations into their teaching methods. Through a thorough analysis of existing research in this area, the author explores the advantages of incorporating various approaches in teacher training. Additionally, this chapter aims to demonstrate that in order to successfully integrate technology into teaching, teacher education programs should provide comprehensive and well-structured training that includes experiential learning and reflective practices.

Chapter 9, titled "Using Project-Based Learning to Teach English Language and Culture at the Tertiary Level," displays the results of applying a project-based learning (PBL) methodology in courses at a tertiary level with the aim of incorporating culture into tertiary English language instruction. In this study, co-authors Madalina Armie, María Enriqueta Cortés de los Ríos, Ángeles Jordán Soriano, Nuria del Mar Torres López and María del Mar Sánchez Pérez, present a PBL implementation which was reviewed by 110 undergraduates—both local and international—studying Infant Education, Primary Education, Humanities, and Spanish Philology. Based on students' participation in an anonymous online survey, the results show that the majority of the students agreed that the implementation of PBL enhanced their skills in the classroom.

Chapter 10, "Innovative Active Methodologies That Promote Learning in Postgraduate Students in the 21st Century: Project-Based Learning and Flipped Classroom," discusses the post-pandemic panorama and explores the need for new methodological approaches to develop competencies in postgraduate students. The chapter focuses on documenting an educational experience involving the use of PBL, flipped learning, and gamification in groups of students. The authors adopt a comparative approach to assess the effectiveness of each active methodology used. As a result of the study, co-authors Edgar Oliver Cardoso Espinosa, María Elena Zepeda Hurtado, and Jésica Alhelí Cortés Ruiz provide recommendations for the implementation of PBL and the flipped classroom.

Chapter 11, "The Future of Distance Education: An International Comparative Study," examines students' perspectives and experiences with distance education (DE), considering a multinational perspective with a specific focus on Portugal, Ukraine, and the United Arab Emirates. The aim of authors Patricia Fidalgo, Joan Thormann, Oleksandr Kulyk, José Alberto Lencastre, and Maria Joao Figueiras is to gather evidence-based data that can assist institutions of higher education (IHEs) in expanding and diversifying their learning methodologies. Examining various educational models, they suggest, will enable countries to gain a deeper understanding of how digital tools can support student learning. The authors of this chapter have modified and adjusted research questions from a study conducted before the pandemic in order to suit the current pandemic situation. These research questions prompt the investigation of various aspects related to online courses, such as students' past experiences with such courses, their opinions on DE, factors influencing their decision to enroll in DE courses or not, their perceived level of preparedness for DE, their willingness to continue with DE courses after the pandemic, and their preferences for different DE formats; the authors further examine the responses based on age, gender, and country to identify similarities and differences. The study's findings have the potential to inform IHEs' digitalization strategies, recruitment efforts, and learning approaches, and to inform the development of purposefully designed DE and blended learning programs. The conclusions show that given the impact of the pandemic, it is crucial to consider long-term educational prospects.

Chapter 12, titled "*Inclúyete* Virtual Reality. A Teaching Innovation Proposal to Work on Mental Health Stigma in University Students – An Innovative Software for the Intervention on Mental Health Stigma," explores the existence of stigma associated with mental health disorders, emphasizing its role as a major obstacle to recovery and social integration. Written by Noelia Navarro Gómez, Anabel Corral Ganados, Ana María Martínez-Martínez, María del Pilar Díaz-López, and Remedios López Liria, the chapter argues that promptly addressing social awareness and empathy is of utmost importance. Furthermore, the chapter aims to showcase the potential of new technologies, particularly virtual reality (VR), as a promising intervention strategy, especially among young individuals who spend a significant amount of their leisure time playing video games. With this objective in mind, the authors propose a solution specifically designed for university students, which tackles mental health stigma through a VR program called *Inclúyete VR*. This program aims to foster empathy toward individuals with mental health disorders by allowing users to "experience" hallucinations and explore available psychosocial approaches for managing such conditions. This research provides further evidence to support the use of innovative interventions to fight stigma.

In Chapter 13, "Creative Writing of Disinformation Narratives: Teaching Innovation With the Use of Artificial Intelligence," Anastasio García Roca, Ana Manzano León, José Miguel Rodríguez Ferrer, and Rubén Camacho Sánchez propose the development of media and information literacy through the utilization of Artificial Intelligence (AI) tools for creating audiovisual content. The authors emphasize that integrating media literacy education into primary education, especially in subjects such as Didactics of Language and Literature, can equip children with the essential skills needed to critically assess and interact with media content. In addition, they highlight the increasing significance of media literacy due to the advancement of AI technologies and the potential risks associated with deepfake creation in today's evolving technological landscape. The resulting proposal incorporates flipped classes, thematic debates, and engaging activities. Additionally, a set of evaluation instruments is presented to measure the effectiveness of this approach and the learning outcomes achieved by students using a pre- and post-experimental design.

Finally, Chapter 14, "Game-Based Learning to Teach Executive Functions (EFs): A Pilot Study With Master's Degree Students in Special Education," presents a pilot study carried out to explore the effectiveness of game-based learning in the Master's degree in Special Education at the University of Almería (Spain). The study, carried out by José M. Rodríguez Ferrer, Rubén Camacho Sánchez, Anastasio García Roca, and Ana Manzano León, introduced this approach for the teaching of students with specific learning needs, focusing on enhancing students' executive functions in both their daily lives and their academic activities. To achieve this, a theoretical–practical workshop was conducted to teach the future educators how to design specific activities and adapt them to their students' needs using games. Participants valued the use of the games and found the experience enjoyable and innovative. The study's findings emphasize the potential advantages and challenges of integrating game-based learning into educational settings, showcasing games as an effective pedagogical tool.

Madalina Armie
University of Almería, Spain

Germán Asensio
University of Almería, Spain

María Enriqueta Cortés de los Ríos
University of Almería, Spain

Ángeles Jordán Soriano
University of Almería, Spain

REFERENCES

Abi-Jaoude, E., Naylor, K. T., & Pignatiello, A. (2020). Smartphones, social media use and youth mental health. *Canadian Medical Association Journal*, *192*(6), 136–141. doi:10.1503/cmaj.190434 PMID:32041697

Aldahdouh, T. Z., Nokelainen, P., & Korhonen, V. (2020). Technology and social media usage in higher education: The influence of individual innovativeness. *SAGE Open*, *10*(1). Advance online publication. doi:10.1177/2158244019899441

Brennan, J. (2008). Higher education and social change. *Higher Education*, *56*(3), 381–393. doi:10.100710734-008-9126-4

Chadid, O. (2023). *The importance of soft skills for graduate students: University of Almeria as a case study* [Unpublished master's dissertation]. University of Almeria, Almeria, Spain.

Connectivism learning theory. (2021, May 27). Western Governors University. Retrieved July 19, 2023, from https://www.wgu.edu/blog/connectivism-learning-theory2105.html

Darmawan, R. (2011). Web 2.0 and idiosyncrasy of cultural heritage: A perspective from Indonesia. In G. Styliaras, D. Koukopoulos, & F. Lazarinis (Eds.), *Handbook of research on technologies and cultural heritage: Applications and environments* (pp. 481–494). IGI Global. doi:10.4018/978-1-60960-044-0.ch024

Downes, S. (2008). An introduction to connective knowledge. In T. Hug (Ed.), *Media, knowledge and education: Exploring new spaces, relations and dynamics in digital media ecologies* (pp. 77–102). Innsbruck University Press.

Hand, M. (2016). *Making digital cultures: Access, interactivity, and authenticity*. Routledge. doi:10.4324/9781315593289

Hargreaves, A. (2004). *Teaching in the knowledge society: Education in the age of insecurity*. Open University Press.

Heick, T. (2023, March 13). *How technology changed teaching and learning*. TeachThought. Retrieved July 19, 2023, from https://www.teachthought.com/the-future-of-learning/how-technology-changed-teaching/

Hopkins, D. (2004). Foreword. In A. Hargreaves (Author), Teaching in the knowledge society: Education in the age of insecurity (pp. x-xii). Open University Press.

Lee, J. J., & Hammer, J. (2011). Gamification in education: What, how, why bother? Definitions and uses. *Exchange Organizational Behavior Teaching Journal*, *15*(2), 1–5.

Pratt-Johnson, Y. (2019) The fourth industrial revolution and its impact on teacher education. *Tạp chí Nghiên cứu dân tộc*, *8*(2), 95–97. . doi:10.25073/0866-773X/307

Prensky, M. (2001). Digital natives, digital immigrants. *MCB University Press*, *9*(5), 1-6. https://www.marcprensky.com/writing/Prensky%20-%20Digital%20Natives,%20Digital%20Immigrants%20-%20Part1.pdf

Shastri, P. D. (2010). *Communicative approach to the teaching of English as a second language*. Himalaya Pub. House.

Siemens, G. (2005). Connectivism: A learning theory for the digital age. *International Journal of Instructional Technology and Distance Learning*, *2*(1), 1–6. http://www.itdl.org/Journal/Jan_05/article01.htm

The Lego Foundation. (n. d.). *Social learning and collaboration in school: Learning to thrive through play*. Future Learn. Retrieved July 19, 2023, from https://www.futurelearn.com/courses/social-learning-collaboration-in-school

Twenge, J. M. (2017, September). Have smartphones destroyed a generation? *The Atlantic*. Retrieved July 19, 2023, from https://www.theatlantic.com/magazine/archive/2017/09/has-the-smartphone-destroyed-a-generation/534198/

Yaqub, M. (2023, April 17). *62% percent of unreliable information on the internet in 2022*. BusinessDIT. Retrieved July 19, 2023, from https://www.businessdit.com/fake-news-statistics/

KEY TERMS AND DEFINITIONS

Connectivism: A new approach that aims to explain how learning takes place in the digital age.

Node: A term that is relevant to the understanding of connectivism. Nodes can be inanimate elements, such as books or webpages, that are connected to animate ones, such as human learners. Knowledge transfer takes place through interactions between these nodes or connections between the different inanimate and animate elements.

Teacherpreneurship: A word meant to designate educators' assertiveness and leadership in the world of teaching.

Technoculture: A neologism meant to describe the results or repercussions visible in culture from the interaction between people and technology or digital tools.

ENDNOTES

1. Darmawan (2011) defines technoculture as "[a] dimension of culture as well as cultural discourse which focus[es] on the experience, consciousness, and meaning that is derived from mutual interaction between people and tools and/or technology" (p. 494).
2. For more information, consult "Connectivism: Learning as a Network Creation" (Siemens, 2004) and "An Introduction to Connective Knowledge" (Downes, 2005).

Acknowledgment

The research involved in this project has been supported by the Ministry of Universities of Spain (FPU grant 2021), CEI Patrimonio (University of Almería,) CySOC (University of Almería,) and the Teaching and Innovation Group 21_22_1_10C, La cultura inglesa a través del aprendizaje basado en proyectos en el aula universitaria de lengua extranjera (University of Almería.) We would also like to thank the Lindisfarne research group (HUM-807) of the University of Almería for its encouragement and counsel.

Likewise, we would like to extend our most sincere gratitude to our families and all of our contributors for their invaluable work.

Chapter 1
Mobile Gamification Tools for Foreign Language Teaching in Higher Education

Muhammet Demirbilek
Suleyman Demirel University, Turkey

ABSTRACT

Gamification has gained popularity in recent decades as an effective approach to enhancing student engagement and motivation in various fields, including foreign language teaching. Mobile gamification tools have been identified as a promising avenue for incorporating game elements into language learning, as they enable students to access learning materials and participate in interactive activities anytime and anywhere. This chapter provides an overview of the literature on mobile gamification tools for foreign language teaching in higher education. The review focuses on the benefits and challenges of using mobile gamification tools in language learning, the types of tools available, and the factors that influence their effectiveness. In addition, the paper presents examples of successful implementations of mobile gamification tools in language classrooms and discusses the pedagogical implications of using these tools.

INTRODUCTION

Information and Communication Technologies offer new possibilities to create engaging contexts for learning. Especially, mobile technologies have increasingly become a mainstream tool among students. Educational institutions have endorsed mobile applications in order to attract young learners who are more accustomed to, or even expect this type of interactivity (Pachler et al., 2010). According to Klopfer (2008), accessing updated information, promoting active and participatory learning, personalized learning, promoting mobility in education, increasing distribution of educational content, and increased interaction during learning are the befits of using mobile devises in education. Due to ubiquitous feature, smart phones may support a range of pedagogical approaches (Pegrum et al., 2013). They are particularly suitable for the application of constructivist approaches, centered on the active and collaborative

DOI: 10.4018/978-1-6684-8861-4.ch001

student (Sharples et al., 2009). As Klopfer (2008) highlights, when used to their full potential, smart phones can be transformed into language learning laboratories. Therefore, mobile devices such as smart phones, tablets, and personal computers are powerful devices for learning anytime and anywhere due to their portability and connectivity features in addition to social interactivity (Klopfer, 2008). These characteristics of the mobile devices may allow teachers to use them in their teaching context. Mobile devices can be utilized to employ gamified applications for teaching and learning process. Therefore, gamification on mobile devices has received much attention especially in both educational institutions and marketing industry. Gamification can be defined as the utilization of game design techniques, game thinking and game mechanics to enhance non-game context (Nicholson, 2015).

The term "gamification" was first coined in 2008 by Nick Pelling, a British computer programmer and inventor. However, the concept of using game-like elements in non-game contexts dates back to the 1980s when marketers started using loyalty programs and reward systems to incentivize customer behavior (Kapp, 2012). In the early 2000s, researchers and practitioners began exploring the potential of gamification as a strategy to enhance user engagement and motivation in various domains, including education, healthcare, and workplace management. The emergence of mobile devices, social media, and big data also contributed to the rapid growth of gamification as a popular and effective approach to behavior change.

Gamification involves applying game mechanics and dynamics to non-game situations, even in contexts that are not traditionally associated with playfulness, such as education (Deterding, Dixon, Khaled, & Nacke, 2011; Innocenti, 2021). It involves using game design elements, mechanics, and principles in non-game contexts to motivate and engage students to achieve desired behaviors, tasks, or goals (Deterding, Dixon, Khaled, & Nacke, 2011). In general, gamification is seen as the application of game(s), allowing game players (e.g., users, learners) to understand a situation, acquire knowledge, and learn skills through the interplay of games. Studies show that gamification can increase motivation and learning. (Deterding, Dixon, Khaled & Nacke, 2011; Hamari, Koivisto, & Sarsa, 2014; Werbach, 2014). It is not just about playing games. But it is the use of game thinking and mechanics in a non-game context to engage employees and students in the learning process (Landers et al. 2015).

The gamification of learning is an evolving educational approach aimed at motivating students and enhancing their engagement within learning environments. It incorporates elements of videogame design and game mechanics to achieve this objective. Gamification has the potential to increase learner engagement, reduce the effective filter and create better learning opportunities (Hamari et al., 2014). Gamification often involves the use of rewards, badges, points, leaderboards, and challenges, which trigger the learner's competitive and social instincts and increase their sense of achievement and progress (Landers & Armstrong, 2017). One of the key benefits of gamification is its ability to foster engagement and learning in educational settings. Studies have shown that gamification can increase students' motivation, interest, and academic performance by providing immediate feedback, clear goals, and a sense of progress (Figueroa Flores, 2015, Dichev & Dicheva, 2017; Hew, et al. 2016; Severengiz, et al., 2018). Moreover, gamification can be used to enhance learning outcomes in foreign language learning by creating an immersive and interactive learning experience. Therefore, in foreign language education, gamification has the potential to foster student motivation and active participation within the classroom setting. It equips teachers with effective tools to guide and incentivize students, encouraging them to fully immerse themselves in the learning process. By demonstrating that education can be an enjoyable experience, gamification blurs the lines between informal and formal learning, inspiring students to embrace lifelong and holistic learning approaches, both within and beyond the classroom (Lee & Hammer, 2011).

The adoption of gamification in in foreign language teaching has been widespread, serving as a means to enhance foreign language learning across various contexts and subjects. It addresses various attitudes, activities, and behaviors related to education, such as promoting participatory approaches, fostering collaboration, encouraging self-guided study, facilitating assignment completion, improving the effectiveness of assessments, integrating exploratory learning approaches, and boosting student creativity and retention (Dichev & Dicheva, 2017).

In the realm of learning, gamification encompasses several key aspects, namely game elements, the educational context, desired learning outcomes, learner profiles, and the gamified environment. Thus, gamification can be considered a constructivist model that places the student at the core of the educational process. It enables learners to construct their knowledge through hands-on experiences, fostering self-motivated learning. This approach facilitates better retention of notions and concepts over the long term, going beyond mere focus on grades (Innocenti, 2021; Pappas, 2014).

Educational gamification revolves around the concept of "customization" or personalization of the learning path. It involves dividing the learning process into successive phases, with a strong focus on promoting socialization and bridging the gap between theory and practice. By employing game-like rule systems, player experiences, and cultural roles, this approach aims to shape learners' behavior in a constructive manner.

The rationale behind gamifying learning is rooted in the idea that integrating game elements into educational activities can generate a sense of immersion similar to that experienced in games (Codish & Ravid, 2015). This belief leads to the notion that by incorporating game mechanics into the design of a learning process, learners can become actively engaged in a productive learning experience and, in a broader sense, positively transform their behavior in desirable ways (Holman et al., 2013).

Among the educators, there is a confusion between the concept of gamification and game-based learning. Gamification should be distinguished from "game-based learning". The latter involves using video games in a direct manner to facilitate learning. On the other hand, gamification adopts the dynamics and mechanics found in video games but does not necessarily rely on actual video games. It does not aim to turn the entire learning process into a game but rather integrates game elements and tools into educational contexts, making use of game-based approaches in non-game settings. Contrary to game-based learning, Gamification involves applying game mechanisms to non-game contexts with the aim of encouraging prosocial behaviors and facilitating complex learning outcomes within conventional tasks. It is a less familiar approach compared to game-based learning because it is more intricate and structured, making it particularly suitable for older students who can effectively employ elaborate strategies. Furthermore, gamification may reduce feedback loops, provide learners with low-stakes methods of evaluating their skills, and provide an environment where effort, rather than expertise, is rewarded. Therefore, an effective gamification system can assist students in assuming roles that are beneficial to their learning. We can influence pupils to rethink their potential in school and what school could entail for them by making the building of a new identity entertaining and rewarding it properly. Studies shows that gamification can increase student motivation (Subhash & Cudney, 2018; Dicheva et al., 2015, Azzouz Boudadi & Gutiérrez-Colón, 2020, Chen et al., 2015, Kapp, 2012). Gamification activities can serve particular psychological functions, which create a motivational pull toward engaging in certain sustained behaviors. Gamification has been argued to be a fun and enjoyable method to support learning in foreign language classrooms and to ease the gap between students' learning and educational practice.

THE USE OF GAMIFICATION IN EDUCATION

The use of gamification in education has been an increasingly popular and promising approach to engage students and enhance their motivation, learning outcomes, and overall experience. The literature has shown that gamification leverages several psychological and behavioral theories, such as self-determination theory, goal-setting theory, and operant conditioning, to promote intrinsic motivation, mastery, and autonomy (O'Donovan et al., 2013, Landers, 2015, Werbach & Hunter, 2015). One of the key benefits of gamification in education is its ability to increase student engagement and motivation. Several studies have shown that gamification can enhance students' interest, enjoyment, and motivation in various subjects and activities by providing immediate feedback, clear goals, and a sense of progress. Gamification can also increase students' sense of autonomy and competence by providing them with choices and challenges that match their abilities and interests (Landers, 2015).

Studies show that gamification improve students' learning outcomes and academic performance. For instance, a meta-analysis of 68 studies on gamification in education found a small but significant positive effect on student achievement and retention compared to traditional instruction (Nair & Mathew, 2019). Gamification can also enhance students' problem-solving, critical thinking, and collaboration skills by providing them with opportunities to practice and apply these skills in a game-like context (Abramovich, et al, 2013). The effectiveness of gamification in education depends on several factors, such as the quality of game design, the alignment of game mechanics with learning objectives, and the context and culture of the educational setting (Nair & Mathew, 2019).

According to Joey Lee and Jessica Hammer, gamification can be employed as an intervention in three main areas: cognitive, emotional, and social (Lee & Hammer, 2011). Moreover, Subbash and Cudey's research in 2013, gamification, whether implemented in face-to-face or online settings, has the potential to elevate student engagement and motivation. Nevertheless, it is imperative to ensure that "the game" is well-suited to encourage participants to concentrate on the content and the educational process itself.

In the cognitive domain, games offer intricate systems of rules that encourage players to actively experiment and discover. They also facilitate players' mastery process, keeping them engaged even when faced with challenging tasks (Koster, 2004). Moreover, games present multiple paths to success, enabling students to select sub-goals within the overall task.

Secondly, games evoke a wide range of powerful emotions, including curiosity, frustration, and joy (Morschheuser et al. 2018). They maintain a positive relationship with failure by ensuring rapid feedback cycles and keeping the stakes low. Gamification offers the potential to foster resilience in the face of failure by redefining it as an essential part of the learning process. Through gamification, feedback cycles can be shortened, and learners are provided with low-stakes opportunities to assess their capabilities. This creates an environment where effort, rather than immediate mastery, is rewarded (Lee & Hammer, 2011). Consequently, students can learn to view failure as an opportunity for growth, rather than feeling fearful or overwhelmed by it.

Lastly, game-like elements in education enable students to explore and adopt new identities and roles. Additionally, these elements can offer social validation and recognition for academic accomplishments that might otherwise go unnoticed or be undervalued by their peers. While teachers can provide recognition, gamification also empowers students to acknowledge each other's achievements through in-game currency and rewards. This further enhances a positive and motivating learning environment where students can support and appreciate one another's efforts.

In foreign language learning, the elements of game design employed in crafting gamification scenarios can be categorized into three main groups: dynamics, mechanics, and components (Werbach & Hunter, 2012). Mechanics constitute a set of rules that govern the results of interactions within the system, while dynamics are the users' reactions to the interplay of those mechanics. Game mechanics, specifically, refer to the elements that propel the action forward. These encompass challenges, chance, competition, cooperation, feedback, resource acquisition, and rewards. At the fundamental level of the gamification process, components encompass specific instances of mechanics and dynamics. These components consist of achievements, avatars, badges, collections, content unlocking, gifting, leaderboards, levels, points, virtual goods, and more. For example, points serve as components that offer rewards as mechanics and contribute to creating a sense of progression as dynamics within the system. Involvement, feedback, risk related to certain actions, and setting up point systems are the core elements of gamification in the classroom.

According to Yu-kai Chou there are eight principles, defined as core drives of gamification to motivate learners (Chou, 2015). These principles are:

- Epic meaning and calling, concerning, for example, completing a mission alone or in a group.
- Development and accomplishment, through progress, developing skills, conquering, and overcoming challenges, gaining points, badges, credits, bonuses, etc.
- Empowerment of creativity and feedback
- Ownership and possession
- Social pressure and relatedness, through challenges, bonus exchanges, and requests for help
- Scarcity and impatience, through certain mechanisms such as countdown, time-on-task, etc.
- Unpredictability and curiosity
- Loss and avoidance

Today, the challenge that education encounters may involve harnessing these principles (while not necessarily adhering to all of them) to captivate learners, spark their curiosity, and grasp their focus. At the same time, gamification facilitates the fulfillment of three fundamental psychological needs: autonomy, relatedness, and competence. Furthermore, by incorporating challenges, progress indicators, feedback mechanisms, achievement levels, and other elements, Gamification effectively stimulates and enhances students' performance. This is achieved through their heightened commitment to the goals, the valuable feedback they receive, the complexity of the activities, and the situational constraints that shape their learning experience. To implement gamification successfully in foreign language education, it necessitates clear and comprehensible goals, timely feedback, achievement markers, and a proper balance between challenges, student skills, and the perceived value of the activities. By achieving an optimal psychophysical state, both the entertainment value and student commitment can be maximized.

Gamification in Foreign Language Teaching

The increasing prevalence and accessibility of smartphones has created new opportunities for the use of these mobile devices in foreign language education. Combining traditional teaching methods with (online) games or gamified activities is a good method for teaching foreign languages. Gamification has been identified as a promising approach to improve foreign language education by engaging learners and enhancing their motivation, participation, and language skills. The theoretical background for gamification in foreign language education is based on several psychological and pedagogical theories, such as self-determination theory, cognitive load theory, and social constructivism (Nair & Mathew, 2019, Dehghanzadeh, et al, 2019):

- The self-determination theory posits that learners are more motivated and engaged when they have a sense of autonomy, competence, and relatedness. Gamification in foreign language education can provide learners with choices, challenges, and feedback that match their abilities and interests, and create a supportive and social learning environment that promotes interaction and collaboration among learners (Nair & Mathew, 2019, Sun et al, 2018).
- The cognitive load theory emphasizes the importance of reducing extraneous cognitive load and enhancing germane cognitive load to optimize learning. Gamification in foreign language education can reduce extraneous cognitive load by providing clear goals, instructions, and feedback, and enhance germane cognitive load by providing opportunities for practice, feedback, and reflection in a game-like context (Nair & Mathew, 2019, Werbach & Hunter, 2015).
- Social constructivism emphasizes the importance of social interaction and collaboration in learning. Gamification in foreign language education can facilitate social interaction and collaboration by providing opportunities for learners to share their progress, achievements, and challenges, and to collaborate on tasks and projects that require communication and language use (Nair & Mathew, 2019).

Moreover, gamification in foreign language teaching can leverage several game mechanics and principles, such as points, badges, leaderboards, quests, and narrative, to enhance learners' engagement and motivation. These game principles can provide learners with immediate feedback, clear goals, and a sense of progress, and create a more immersive and enjoyable learning experience.

In essence, with the gamification, the learning experience is transformed into an educational game using missions, levels, goals, rewards. and point systems. Knowing fundamental aspects of gamification may help foreign language educators to implement gamification effectively in their teaching and learning process. As per Kapp's findings in 2012, gamification relies on the following fundamental aspects:

- Rules: In the realm of gaming, rules play a crucial role, whether explicitly stated or implicitly understood. Often, students seek to challenge not only the imposed rules but also create their own set of rules to conquer.
- Competition: The primary aim is to foster a spirit of healthy competition among students, emphasizing collaborative strategies as the foundation. This encourages both competition and cooperation, promoting a positive and enriching gaming experience.
- Reward and feedback: Rewards and feedback are integral components of the system, ensuring that students consistently receive outcomes for their efforts, whether positive or negative. Various types of rewards and reinforcement mechanisms are employed to motivate students and reinforce their continued engagement in the learning process.
- Levels of difficulty: Incorporating different levels of difficulty in the assigned tasks serves as an additional motivation for students to enhance their performance. The challenge of progressing through various difficulty levels inspires them to strive for continuous improvement.
- Storytelling: Storytelling holds great motivational power, especially for students who find themselves captivated and connected to specific character arcs or plots. The incorporation of avatars

allows them to immerse themselves in the narrative, sparking their interest and enthusiasm for the learning experience.

Gamification activities can provide foreign language learners with challenging and rewarding tasks that are aligned with their interests and preferences, and that promote their sense of autonomy and competence. Mobile gamification tools can also enable learners to access and use language learning resources and tools in a flexible and convenient manner, and to receive immediate feedback and support from instructors and peers. Game-based activities can provide learners with opportunities to practice and apply their language skills in a variety of contexts and situations, and to receive feedback and correction on their language use from instructors and peers. Mobile gamification tools can also enable learners to interact and communicate with native speakers of the target language and with other learners from different cultural backgrounds, thus enhancing their intercultural awareness and communication skills. However, the proper design and implementation of gamification in foreign language education is crucial to ensure its effectiveness and avoid potential drawbacks. Gamification should be aligned with the learning objectives, curriculum, and assessment criteria of the foreign language course, and should be based on a thorough needs analysis and learner-centered approach. Furthermore, gamification should be used judiciously and ethically, without undermining intrinsic motivation or creating excessive competition or pressure.

Using Mobile Gamification Tools in Foreign Language Education?

Mobile gamification tools have been increasingly used in foreign language education to enhance learners' motivation, participation, and language skills. These tools can take various forms, such as mobile applications, games, quizzes, and simulations, and can leverage several game mechanics and principles, such as points, badges, leaderboards, quests, and narrative, to create an immersive and enjoyable learning experience. Foreign language teachers need to know how to relate knowledge of the concept of motivation to their classroom and their role as instructors in the teaching-learning. Gamification is not only entertaining but also an important instructional material. mobile gamification tools have many useful features that foreign language teachers can utilize. One of the key benefits of mobile gamification tools in foreign language education is their accessibility and flexibility (Subhash & Cudney, 2018). Learners can access these tools anytime and anywhere, using their mobile devices, and can engage in self-directed and personalized learning activities that match their interests and needs. Mobile gamification tools can also provide immediate feedback and progress tracking, which can enhance learners' motivation and sense of achievement.

Mobile gamification tools in foreign language education can also enhance learners' language skills and communicative competence. For example, mobile games and simulations can provide learners with opportunities to practice and apply their language skills in a contextually relevant and engaging way. Mobile quizzes and exercises can also help learners to develop their vocabulary, grammar, and pronunciation skills in a fun and interactive way. Additionally, mobile gamification tools in foreign language education can enhance learners' social and cultural awareness and understanding. Mobile applications and games can expose learners to different cultures, customs, and perspectives, and promote intercultural competence and empathy. Mobile collaborative tools, such as language exchange platforms, can also facilitate language and cultural exchange among learners from different countries and backgrounds.

Among the advantages of gamification in the foreign language classroom, it has been found that:

- it changes the mood in the classroom,
- increases the students' sense of happiness,
- provides breaks when students are tired,
- increases motivation and attention,
- increases student participation in classroom activities,
- encourages goal-oriented activities,
- makes learning fun.

Because of these advantages, gamification is often used in foreign language teaching as a source of motivation and as an incentive to practice the language. However, the effectiveness of mobile gamification tools in foreign language education depends on several factors, such as the quality of game design, the alignment of game mechanics with learning objectives, and the integration with the foreign language curriculum and assessment. Mobile gamification tools should also be designed and implemented ethically and inclusively, without promoting stereotypes or discriminating against certain groups. However, there are also several constraints and challenges that need to be considered when using mobile gamification tools in foreign language teaching in higher education. One constraint is the technical and pedagogical feasibility of game-based activities. Mobile gamification tools need to be designed and implemented in a way that is compatible with the learners' devices, operating systems, and internet connectivity, as well as with the course objectives, syllabus, and assessment. In addition, game-based activities need to be pedagogically sound, meaning that they need to be aligned with the learners' needs and goals, and that they need to promote meaningful and authentic language use. Another constraint is the potential for distraction and disengagement. Mobile gamification tools need to be designed and implemented in a way that balances the intrinsic motivation and enjoyment of game-based activities with the extrinsic motivation and learning objectives of the course. In addition, mobile gamification tools need to provide learners with clear instructions, goals, and feedback, and to scaffold their learning process in a way that promotes their sense of progress and achievement.

A third constraint is the potential for unequal access and participation. Mobile gamification tools need to be designed and implemented in a way that takes into account the learners' diverse backgrounds, needs, and abilities, and that promotes equitable and inclusive access and participation. In addition, mobile gamification tools need to respect learners' privacy and data protection, and to provide them with informed consent and choice in the use of their personal information.

As stated before, gamified learning has been gaining momentum and acceptance in the field of foreign language education. Educators have been borrowing various game mechanics to enhance learning experiences in a wide array of subjects, including science, language, and communication (Subhash & Cudney, 2018).

Contrary to common assumptions, incorporating individual game elements into apps does not automatically ensure a truly game-like experience (Ferrara, 2013). Consequently, several studies have aimed to evaluate gamified mobile apps based on core game elements and shed light on their effectiveness in particular contexts, such as healthcare, fitness, and the energy sector (Lister, West, Cannon, Sax, & Brodegard, 2014; Beck, Chitalia, & Rai, 2019; Edwards et al., 2016).

Sun and Hsieh (2018) conducted a study with the objective of examining the impact of a gamified interactive response system on the motivation, engagement, and attention of EFL learners. The results of their research indicated that the gamified learning tool successfully heightened interest, intrinsic motivation, attention, and emotional engagement among the participants. However, no significant differences were observed in terms of behavioral and cognitive engagement. The underlying reasons for these outcomes warrant further investigation to gain deeper insights into the effects of gamification on language learning.

In their experimental study, Dindar, Ren, and Järvenoja (2020) investigated the influence of gamified cooperation and competition elements on English vocabulary learning. Surprisingly, they found no significant differences in behavioral engagement (task effort), academic achievement, and motivation among the participants. However, it's important to note that the study lacked an un-gamified control group, which means the effectiveness of the results should be interpreted with caution. Further research and a more comprehensive experimental design would be necessary to draw definitive conclusions about the impact of gamification on English vocabulary learning.

In their study, Reynolds and Taylor (2020) presented varied outcomes while assessing the effects of a gamified EFL tool (Kahoot!) on user experiences and vocabulary knowledge. During their investigation of the gamified learning tool's weaknesses, they identified a crucial issue: the tool's incompatibility with certain students. While most students responded positively to the gamified approach, some individuals experienced demotivation during the experiment. Additionally, while instructors generally held a positive attitude toward the gamified learning tool, they were not entirely convinced of its effectiveness in terms of instructional aspects (Reynolds & Taylor, 2020).

Chen, Li, and Chen (2020) conducted a study to examine the influence of a web-based collaborative reading annotation system on reading performance. The findings revealed that while the experimental group produced more annotations, there was no discernible difference in reading comprehension performance. This suggests that merely increasing behavioral investment does not necessarily lead to improvements in academic achievement.

According to Dehganzadeh and Dehganzadeh's (2020) research, Duolingo emerged as the most commonly utilized gamified platform for language learning in recent times. This platform offers playful and engaging opportunities for language learners to enhance their language skills.

Future Trends and Issues

The use of mobile gamification tools in foreign language teaching in higher education is a relatively new and evolving field, with several emerging trends and issues.

One trend is the integration of mobile gamification tools with other emerging technologies, such as augmented reality and virtual reality. This can create immersive and interactive language learning environments that simulate real-world contexts and provide learners with authentic language input and output opportunities.

Another trend is the use of personalized and adaptive gamification, which tailors the game-based activities to the individual needs, preferences, and progress of each learner. This can increase learners' engagement and motivation, as well as their language learning outcomes, by providing them with challenging and relevant tasks that match their abilities and interests.

A third trend is the integration of social and collaborative features in mobile gamification tools, which enable learners to interact and communicate with each other, as well as with instructors and native speakers of the target language. This can enhance learners' language proficiency, as well as their intercultural and communication skills, by providing them with opportunities to practice and receive feedback on their language use in a social and authentic context.

However, there are also several issues and challenges that need to be addressed in the use of mobile gamification tools in foreign language teaching in higher education. One issue is the lack of standardization and quality assurance in the design and evaluation of game-based activities. This can lead to variability in the effectiveness and relevance of the game-based activities, as well as in the learners' engagement and motivation.

Another issue is the need for greater alignment and integration of mobile gamification tools with the curriculum and assessment of the course. This can ensure that the game-based activities are relevant and meaningful for the learners, as well as that the learners' language proficiency and progress are appropriately assessed and evaluated.

A third issue is the need for greater awareness and attention to the ethical and social implications of the use of mobile gamification tools in foreign language teaching in higher education. This includes concerns related to data privacy and security, as well as to the potential impact of game-based activities on learners' social and cultural identities and values.

Finally, there is a need for greater collaboration and partnership between language instructors, educational technologists, and game designers in the design, implementation, and evaluation of mobile gamification tools in foreign language teaching in higher education. This can ensure that the game-based activities are pedagogically sound, technically feasible, and socially responsible, as well as that they reflect the needs and expectations of the learners and the broader educational context.

In conclusion, the use of mobile gamification tools in foreign language teaching in higher education is a rapidly evolving field with several emerging trends and issues. By addressing the issues and challenges related to standardization and quality assurance, alignment with the curriculum and assessment, ethical and social implications, and collaboration and partnership, the use of mobile gamification tools can enhance and support the language learning process in higher education, as well as promote the development of learners' digital literacy and 21st-century skills.

Mobile Gamification Tools

Mobile gamification tools are effective in foreign language teaching with no age limit. Gamification allows teachers making foreign language learning fun and motivational for learners (Arnold, 2014). Research shows that learning a foreign language is more effective when gamification facilitated (Purgina& Blake, 2019; Zarzycka-Piskorz, 2016). Trivia can be incorporated into the classroom in an enjoyable and interactive manner, encouraging students to participate in team competitions while learning about specific subjects. This approach not only promotes teamwork but also nurtures skills like quick thinking and strategic planning, all without the fear of failure due to the context.

One engaging way to use Trivia is by structuring it as a game show, similar to Jeopardy. This format allows students to select categories aligned with the current subjects being taught, adding an element of choice and excitement. Moreover, trivia can serve as a rewarding incentive after complex and information-packed lessons, enabling students to decompress while reinforcing their learning. Additionally, trivia can be employed to celebrate class milestones, consolidating knowledge acquired over a certain period and concluding chapters with a delightful and enjoyable activity. It serves as a supplementary tool for educators, who can utilize existing apps and resources to simplify the process.

The playful context of trivia fosters high levels of intrinsic motivation among students. However, educators can also incorporate occasional extrinsic motivators like badges or small prizes if they choose to do so. Overall, trivia proves to be an effective and versatile educational tool that enhances both learning and enjoyment in the classroom. Riddles and puzzles serve as fantastic brain teaser tools, offering a variety of activities to engage learners and recapture their focus. While riddles are a specific type of puzzle, they present an enigmatic and mysterious format. This makes them perfect for creating escape games, a rising trend in education, especially in the digital realm.

Escape rooms provide a refreshing break from routine learning, allowing students to work together in teams or individually, tackling riddles, and honing their reflective and problem-solving skills. This interactive learning approach can be seamlessly integrated into lesson plans, thanks to the availability of supporting tools like Genially, which aid in creating captivating escape games.

In the realm of digital learning, puzzles offer a plethora of opportunities for creating gamified content. Among them, crosswords stand out as a timeless and relevant option. Crosswords not only promote individual engagement but also provide an excellent platform for small group activities to enhance language and memory skills. To ensure a positive learning experience, crosswords can be thoughtfully designed with helpful clues or multiple-choice options, avoiding unnecessary frustration for the learners.

Another captivating approach is to incorporate scavenger hunts using online resources relevant to the lesson. Educators can provide clear guidelines and let learners independently explore the internet to find the required information, fostering a sense of discovery and autonomy in the learning process.

To encourage comprehensive coverage of the material, educators can employ the strategy of hiding puzzle pieces within the course module. This method ensures that learners delve into the entire content while simultaneously piquing their interest and attention in the provided reading materials. By incorporating puzzles in a digital learning environment, educators can create an engaging and immersive experience that enhances students' learning outcomes.

Similar to crosswords, bingos represent another widely used format in the digital classroom to gamify general learning materials. Educators can utilize this format creatively by reading word definitions, and students then search for the corresponding words on their bingo sheets. Additionally, bingos can be applied in mathematics lessons to teach concepts like fractions, with educators calling out decimal numbers for students to identify. In the context of mathematics, educators can take it a step further by incorporating random representations, such as a slot machine, to teach probability. This approach enables learners to engage in hands-on experiments and exercises, fostering a deeper understanding of the subject matter. By employing bingos in the digital classroom, educators can add an element of fun and interactivity to their lessons, making the learning experience both enjoyable and effective.

In the digital classroom context, numerous class activities can be employed to prevent boredom and combat online learning fatigue. These engaging activities serve as a means for learners to rediscover their motivation and interact with the content in a more dynamic manner. The beauty of these activities lies in their high adaptability and customizability, enabling educators to cater to the unique needs of their students. Since various aspects of gamification impact motivation differently, these activities can be tailored to suit different types of learners, fostering a more inclusive and stimulating learning environment.

1. Busuu

A language-learning platform allows users learn languages at their own pace. It uses gamification to assist you in practicing a new language. Learners can connect with native speakers and have one-on-one conversations in targeted language providing digital flashcards, a convenient tool for developing users' vocabulary. The flashcards have example sentences, audio clips in native tongue, and meaning in target language. Lessons start with a vocabulary drill and cover reading, writing, speaking and spelling practices through different activities. Busuu has three phases; namely, "onboarding", "scaffolding", and "endgame".

- Onboarding: In this phase, the users become familiar with the rules of the game, the options, the mechanics, and the win-states. In this phase, the user goes through all the first Desired Actions through an interactive step-by-step tutorial rewarding user with small High-Fives once they accomplish it. This technique will help users to get to know the app better and to use it as much as possible.
- Scaffolding: During the scaffolding phase of the experience, users are familiar with all the rules and options they learned during onboarding to try to achieve as many Win-States as possible. Ideally this is where they come back on a regular basis to commit to Desired Actions.
- Endgame: This phase starts when players believe that they have done everything there is to do at least once and they start to feel like there are no longer any unexplored Win-States.

2. Duolingo

Duolingo works through a language skills tree. It starts with the very basics and progressing to more advanced lessons organized by topics. Duolingo focuses on vocabulary learning in a variety of contexts that range from simple to more complex. Most of lessons are powered by artificial intelligence (AI). Duolingo's language learning method also focuses only on the language itself and not on the cultural tips and contexts in which the language is used. Duolingo utilizes gamification by incorporating elements such as streaks, levels, and rewards. Therefore, it makes the learning experience more enjoyable but also helps users stay on track and see measurable progress over time.

3. Memrise

Learn Languages with Memrise was developed in 2017. Memrise Learn Language is a very comprehensive application that includes not only English but also different languages. The application can be accessed from Android devices, iOS devices and the website. There are more than 200 language courses prepared by expert linguists. With the Learn Language with Memrise application, the user can learn the sentence patterns used in daily life, how these sentences are pronounced and how they should be used in sentences. Learner can also watch videos made by people from the country that uses the language the user wants to learn and learn how they speak or pronounce the words. The app includes chat and linguistics skills from local accents, more than 20,000 videos of native speakers and new chat bots. With offline mode without an internet connection, the user can continue learning anytime, anywhere, even in the most remote places. Gamification techniques was used by Memrise to keep users coming back to learn more. For instance, gamification lies in overall learning process. Memrise allows users "plant" the words, similar to how a gardener would plant the seeds of a new beautiful flower. This learning system involves choosing the correct translation several times and entering the word for correct spelling.

4. Mondly

Learning languages gives people the freedom to travel safely on unexpected paths, and Mondly speaks new languages fast. With Mondly, mastering a new language is fun, fast, and easy. It is an app developed by ATi Studios that helps in learning many languages. You can use this app from the website, your android device, and your iOS device. You can access the app for free or you can buy the lessons sold within the app to use it as Premium and start learning. Mondly offers those who want to use this app on android devices to learn 30 languages for free. These languages are: English, Spanish, French, German, Portuguese, Russian, Italian, Italian, American English, Norwegian, Swedish, Korean, Chinese, Arabic, Thai, Greek, Romanian, Hindi, Hebrew, Polish, Bulgarian, Croatian, Persian (Farsi), Japanese and Turkish. For those who want to use it on iOS devices, it offers the opportunity to learn 33 languages for free. These languages are Spanish, American English, British English, French, German, Portuguese, Italian, Russian, Norwegian, Danish, Dutch, Dutch, Swedish, Arabic, Korean, Japanese, Chinese, Greek, Romanian, Vietnamese, Indonesian, Hindi, Hebrew, Polish, Bulgarian, Ukrainian, Hungarian, Czech, Croatian, Finnish, Persian (Farsi), Afrikaans, Vietnamese and Turkish.

5. HelloTalk

HelloTalk is a language exchange mobile app that provides opportunities to learn and build on your language skills. Language learning is primarily done by chatting with native speakers of the language you are trying to learn. HelloTalk also provides access to a newsfeed functionality called Moments, which allows you to learn more about the lives of those who live in the country of your desired language. Teachers could reward users who reach certain number of messages with their partner over the span of a week. Instructors could also utilize a reward system to try and incentivize engagement on the application. The success of such loyalty system can be observed by:

- Analysing the number of messages sent in conversations that are A/B testing this gamification compared to the number of messages sent in conversations without this gamification.
- Analysing the length of overall text in conversations that are A/B testing this gamification vs. those that are not testing it.

6. Babbel

Babbel allows language learners who want to have a conversation in their target language. The content of the app is focused on real-life application—like knowing your coffee order, or how to greet people in different settings. It uses gamified learning to reinforce, track and teach learners' lesson progress. Babbel immerses learner in a conversation using what was learned, showing real-life examples of how to apply phrases and expressions in actual situations with native speakers. A combination of matching, fill-in-the-blank and speaking exercises are mixed in with cultural and grammar tips that bring the language into context. It uses interactive dialogs and a speech recognition system with more than 60,000 language lessons. Babbel supports 14 languages with short courses for beginners and intermediate-level students.

7. MindSnacks

Mindsnacks brings gamification to a new level. Each of the 7 languages they offer comes with eight or nine small games intended to help you learn vocabulary, grammar, and listening. There are quick, easy lessons explaining concepts which are then learned or evaluated in games before they are perfected. Mindsnacks tracks your progress so that you can easily see how much further you need to practice in order to gain competence in every area. MindSnacks is an app that includes nothing but games. The organization touts that not a single boring quiz, class, or lesson is involved in their learning process. MindSnacks breaks learning into a series of games, each designed to teach vocabulary and grammar. Each language includes over 1000 words and phrases available to learn, as well as around 50 lessons.

DISCUSSIONS

The use of mobile gamification tools in higher education foreign language teaching offers a range of benefits and challenges. The literature suggests that the affordances of these tools include enhancing learner motivation and engagement, promoting language proficiency and intercultural competence, and enabling flexible and convenient access to learning resources and tools. However, the constraints of using these tools include technical and pedagogical feasibility, potential distraction and disengagement, and unequal access and participation.

One of the main advantages of using mobile gamification tools in foreign language teaching is their ability to enhance learner motivation and engagement. By incorporating game-based activities into language learning, instructors can provide learners with a more enjoyable and challenging learning experience. This can lead to increased engagement and motivation to learn, as well as a greater willingness to take risks and make mistakes.

Another key benefit of using mobile gamification tools is their ability to promote language proficiency and intercultural competence. By providing learners with opportunities to practice and apply their language skills in authentic contexts, instructors can help learners develop the communicative competence necessary to effectively interact with speakers of the target language. Additionally, gamification can promote intercultural competence by encouraging learners to explore and appreciate cultural differences and similarities.

However, the use of mobile gamification tools also presents several challenges. One major challenge is ensuring that the games and activities are technically and pedagogically feasible. Instructors must ensure that the tools and games used are compatible with the learners' devices and operating systems, and that the games align with course objectives and syllabus. Additionally, instructors must ensure that the games are pedagogically sound and promote authentic language use.

Another challenge is ensuring that learners remain engaged and motivated, without being distracted or disengaged by the games. Instructors must ensure that the games provide learners with clear instructions, goals, and feedback, and that they are appropriately scaffolded to promote learners' sense of progress and achievement. In addition, instructors must ensure that the games balance the intrinsic motivation and enjoyment of gameplay with the extrinsic motivation and learning objectives of the course.

Finally, the use of mobile gamification tools also presents potential issues of unequal access and participation. Instructors must ensure that the tools and games used are accessible to all learners, regardless of their backgrounds, needs, and abilities. In addition, instructors must ensure that learners' privacy and data protection rights are respected, and that learners are provided with informed consent and choice in the use of their personal information.

Foreign language teachers and researchers must be familiar with the various gaming platforms and gamification elements available to develop digital gamified language learning activities for their students. In other words, English teachers should possess digital literacy skills to effectively utilize these resources. Embracing digital technology to create gamified language learning activities is an integral aspect of English teachers' digital literacy.

In conclusion, the use of mobile gamification tools in higher education foreign language teaching offers a range of benefits and challenges. By carefully considering and balancing the affordances and constraints of these tools, instructors can enhance learners' motivation and engagement, promote language proficiency and intercultural competence, and enable flexible and convenient access to learning resources and tools.

REFERENCES

Abramovich, S., Schunn, C., & Higashi, R. M. (2013). Are badges useful in education? It depends upon the type of badge and expertise of learner. *Educational Technology Research and Development*, *61*(2), 217–232. doi:10.100711423-013-9289-2

Arnold, B. J. (2014). Gamification in education. *Proceedings of the American Society of Business and Behavioral Sciences*, *21*(1), 32–39.

Azzouz, N., & Gutierrez-Colon Plana, M. (2020). Effect of Gamification on students' motivation and learning achievement in Second Language Acquisition within higher education: a literature review 2011-2019. *The EuroCALL Review*, *28*, 40. . doi:10.4995/eurocall.2020.12974

Beck, A., Lakkaraju, K., & Rai, V. (2017). Small is big: Interactive trumps passive information in breaking information barriers and impacting behavioral antecedents. *PLoS One*, *12*(1), e0169326. doi:10.1371/journal.pone.0169326 PMID:28099478

Chen, C. H., Wang, K. C., & Lin, Y. H. (2015). The Comparison of solitary and collaborative modes of game-based learning on Students' science learning and motivation. *Journal of Educational Technology & Society*, *18*(2), 237–248.

Chen, C.-M., Li, M.-C., & Chen, T.-C. (2020). A web-based collaborative reading annotation system with gamification mechanisms to improve reading performance. *Computers & Education*, *144*, 103697. doi:10.1016/j.compedu.2019.103697

Chou, Y. (2015). *Gamification & behavioral design. Learn how to use gamification to make a positive impact on your work and life.* Retrieved on September 17, 2018 from https://yukaichou.com/gamification-examples/octalysis-complete-gamification-framework/

Dehganzadeh, H., & Dehganzadeh, H. (2020). Investigating effects of digital gamification-based language learning: A systematic review. *J. Eng. Lang. Teach. Learn.*, *12*, 53–93.

Dehghanzadeh, H., Fardanesh, H., Hatami, J., Talaee, E., & Noroozi, O. (2019). Using gamification to support learning English as a second language: A systematic review. *Computer Assisted Language Learning*, *34*(7), 1–24. doi:10.1080/09588221.2019.1648298

Deterding, S., Dixon, D., Khaled, R., & Nacke, L. (2011). From game design elements to gamefulness: defining "gamification". *Proceedings of the 15th International Academic MindTrek Conference: Envisioning Future Media Environments*, 9-15. 10.1145/2181037.2181040

Dicheva, D., Dichev, C., Agre, G., & Angelova, G. (2015). Gamification in education: A systematic mapping study. *Journal of Educational Technology & Society*, *18*(3), 75–88. doi:10.1109/EDUCON.2014.6826129

Dindar, M., Ren, L., & Järvenoja, H. (2020). An experimental study on the effects of gamified cooperation and competition on English vocabulary learning. *Britiish J. Educ. Technol.*, *2020*(52), 142–159.

Edwards, E., Lumsden, J., Rivas, C., Steed, L., Edwards, L., Thiyagarajan, A., Sohanpal, R., Caton, H., Griffiths, C.J., Munafo, M.R., Taylor, S., & Walton, R. (2016). Gamification for health promotion: Systematic review of behaviour change techniques in smartphone apps. *BMJ Open*, 6. doi:10.1136/bmjopen-2016-012447

Ferrara, J. (2013). Games for persuasion argumentation, procedurality, and the lie of gamification. *Games and Culture*, *8*(4), 289–304. doi:10.1177/1555412013496891

Figueroa Flores, J. F. (2015). Using gamification to enhance second language learning. *Digital Education Review*, *27*, 32–54.

Hamari, J., Koivisto, J., & Sarsa, H. (2014). Does gamification work? A literature review of empirical studies on gamification. *Proceedings of the 47th Hawaii International Conference on System Sciences*. 10.1109/HICSS.2014.377

Hew, K., Huang, B., Wah Samuel Chu, K., & Chiu, D. (2016). Engaging Asian students through game mechanics: Findings from two experiment studies. *Computers & Education*, *92-93*, 221–236. doi:10.1016/j.compedu.2015.10.010

Kapp, K. (2012). *The gamification of learning and ınstruction: Game-based methods and strategies for training and education*. Pfeiffer.

Klopfer, E. (2008). *Augmented learning: research and design of mobile educational games*. MITPress. doi:10.7551/mitpress/9780262113151.001.0001

Koster. (2004). *A theory of fun*. Paraglyph Press.

Landers, R. N. (2014). Developing a theory of gamified learning: Linking serious games and gamification of learning. *Simulation & Gaming*, *45*(6), 752–768. doi:10.1177/1046878114563660

Landers, R. N., & Armstrong, M. B. (2017). Enhancing instructional outcomes with gamification: An empirical test of the technology-enhanced training effectiveness model. *Computers in Human Behavior*, *71*, 499–507. doi:10.1016/j.chb.2015.07.031

Landers, R. N., Bauer, K. N., Callan, R. C., & Armstrong, M. B. (2015). Psychological theory and the gamification of learning. In T. Reiners & L. Wood (Eds.), *Gamification in education and business* (pp. 165–186). Springer. doi:10.1007/978-3-319-10208-5_9

Lee, J., & Hammer, J. (2011). Gamification in education: What, how, why bother? *Academic Exchange Quarterly*, *15*, 1–5.

Lister, C., West, J. H., Cannon, B., Sax, T., & Brodegard, D. (2014). Just a fad? Gamification in health and fitness apps. *JMIR Serious Games*, *2*(2), e9. doi:10.2196/games.3413 PMID:25654660

Nair, S., & Mathew, J. (2019). A theoretical framework for gamified learning. *International Journal of Innovative Technology and Exploring Engineering*, *8*(12).

O'Donovan, S., Gain, J., & Marais, P. (2013). A case study in the gamification of a university-level games development course. In *Proceedings of SAICSIT 2013* (pp. 242-251). ACM Press.

Pachler, N. (2010). The socio-cultural ecological approach to mobile learning: an overview. In B. Bachmair (Ed.), *Medienbildung in neuen Kulturräumen. Die deutschsprachige und britische Diskussion* (pp. 153–167). Wiesbaden VS Verlag für Sozialwissenschaften. doi:10.1007/978-3-531-92133-4_11

Pappas, C. (2013). *Gamify the classroom.* Retrieved on February 12, 2015 from https://elearningindustry.com/gamify-the-classroom

Pegrum, M., Oakley, G., & Faulkner, R. (2013). Schools going mobile: A study of the adoption of mobile handheld technologies in western Australian ındependent schools. *Australasian Journal of Educational Technology*, *29*(1). Advance online publication. doi:10.14742/ajet.64

Purgina, M., Mozgovoy, M., & Blake, J. (2020). WordBricks: Mobile technology and visual grammar formalism for gamification of natural language grammar acquisition. *Journal of Educational Computing Research*, *58*(1), 126–159. doi:10.1177/0735633119833010

Reynolds, E. D., & Taylor, B. (2020). Kahoot!: EFL instructors' implementation experiences and impacts on students' vocabulary knowledge. *Comput. Assist. Lang. Learn. Electron. J.*, *2020*(21), 70–92.

Rixon, S. (1981). *How to use games in language teaching?* The Macmillan Press Ltd.

Severengiz, M., Roeder, I., Schindler, K., & Seliger, G. (2018). Influence of gaming elements on summative assessment in engineering education for sustainable manufacturing. In Procedia Manufacturing (pp. 429–437). doi:10.1016/j.promfg.2018.02.141

Sharples, M., Inmaculada, A. S., Marcelo, M., & Giasemi, V. (2009). Mobile learning: small devices, big issues. In N. Balacheff, S. Ludvigsen, T. de Jong, & S. Barnes (Eds.), *Technology enhanced learning: principles and products* (pp. 233–249). Springer. doi:10.1007/978-1-4020-9827-7_14

Subhash, S., & Cudney, E. (2018). Gamified learning in higher education: A systematic review of the literature. *Computers in Human Behavior*, *87*, 192–206. Advance online publication. doi:10.1016/j.chb.2018.05.028

Sun, J. C.-Y., & Hsieh, P.-H. (2018). Application of a gamified interactive response system to enhance the intrinsic and extrinsic motivation, student engagement, and attention of English learners. *Journal of Educational Technology & Society*, *21*(3), 104–116.

Werbach, K., & Hunter, D. (2012). *For the win: how game thinking can revolutionize your business.* Wharton Digital Press.

Werbach, K., & Hunter, D. (2015). *The gamification toolkit: Dynamics, mechanics, and components for the win*. Wharton Digital Press.

Zarzycka-Piskorz, E. (2016). Kahoot it or not? Can games be motivating in learning grammar? *Teaching English with Technology*, *16*(3), 17–36.

KEY TERMS AND DEFINITIONS

Foreign Language: A foreign language is a language that is not the native language of a particular speaker or community, but rather is learned as a second or additional language.

Game: A game is a structured activity or form of play that is undertaken for enjoyment, entertainment, or competition.

Gamification: Gamification is the use of game design elements, mechanics, and principles in non-game contexts to motivate and engage people to achieve desired behaviors, tasks, or goals.

Mobile Tool: A mobile tool, also known as a mobile application or app, is a software program designed to run on mobile devices such as smartphones, tablets, or smartwatches.

Chapter 2
The Essence of Learning Management Systems Acceptance:
A Phenomenological Study of Pre-Service Teachers' Experiences

Tugra Karademir Coşkun
Sinop University, Turkey

ABSTRACT

In this study, the answer to the question "Why should teachers know learning management systems?" is investigated from the perspective of pre-service teachers, aiming to understand the acceptance and underlying implicit meanings of LMS usage. Seventy pre-service teachers studying in different fields participated voluntarily in the study. The research data was collected through a semi-structured interview form after ten weeks of LMS installation, management, LMS usage, and digital instructional material development training. The research findings show that pre-service teachers consider the knowledge and skills of teachers regarding LMS important and necessary. The reasons for this necessity are addressed and detailed under the themes of effective learning processes, flexibility in access, content management, assessment and evaluation/feedback, time management and organization, and professional development.

INTRODUCTION

Education and technology are becoming increasingly interdependent in modern societies. This shift aims to enhance student success by making the learning process more interactive, efficient, and customizable (Selim, 2007). Changes in technology and education necessitate the adoption of new approaches to managing learning processes. Education holds a significant position in the advancement of a society, and its uninterrupted continuation is essential (Karatepe, Küçükgençay & Peker, 2020). The role of technology in ensuring the seamless continuation of education in both ordinary and extraordinary circumstances is

DOI: 10.4018/978-1-6684-8861-4.ch002

undeniable. In this context, learning management systems (LMS) serve as a critical building block that brings together students and teachers and facilitates and organizes learning processes (Al-Adwan et al., 2018). Learning management systems offer many advantages, such as reaching multiple people simultaneously, enabling students to self-assess, providing synchronous and asynchronous learning options, saving time, and offering easy accessibility (Seven, 2012; Tuncer & Taşpınar, 2008). It can be said that they are the most important alternative for preventing interruptions in education during both ordinary and extraordinary situations. For instance, Chan, So, and Wong (2007) reported in a study conducted during the SARS pandemic that remote education platforms enabled information sharing and replication, thereby fulfilling educational needs. Similarly, Kaden (2020) summarizes the most critical aspect of remote education as supporting learning during the Covid-19 pandemic.

Learning Management Systems (LMS) are software applications that assist in managing, monitoring, and evaluating educational processes (Coates, James, & Baldwin, 2005). LMSs are defined as online learning technologies for creating, managing, and delivering course materials (Turnbull, Chugh, & Luck, 2019). An LMS provides a consolidated environment that enables a range of tasks, such as distributing educational materials, monitoring, publishing designed information, tracking grades, updating, making course announcements, and control (Watson & Watson, 2012).

There are several reasons, supported by literature, why Learning Management Systems (LMS) are important. For example, one such reason is flexibility, where LMSs provide access to learning processes independent of time and place (Bates, 2015). As a result, students and teachers can access educational materials and continue their classes whenever and wherever they want (Dalsgaard & Paulsen, 2009). Another feature is trackability and assessment. Through LMSs, teachers can closely monitor and evaluate student performance. LMSs are essential tools for providing transparency regarding students' learning experiences (Dalsgaard & Paulsen, 2009). By continuously monitoring students' progress, they can provide data that enables timely interventions (Black & Wiliam, 2009). This is important not only for teachers but also for informing parents about their children's academic progress (Emelyanova & Voronina, 2014).

Another contribution of LMSs to the educational process is their ability to provide learning experiences that adapt to individual learning rates and styles (Blau & Hameiri, 2010). For example, various measurements and assessments can be conducted through LMSs to gauge student success, and this data can be used for improving instructional strategies (Siemens, Gasévić & Dawson, 2015). This is an important step towards individualizing instruction. Another feature is the inclusion of tools that support collaboration and communication. A typical LMS combines interactive learning environments that support interaction, collaboration, communication, and knowledge sharing (Dias & Diniz, 2014). LMSs not only allow for the presentation of instructions and electronic resources to improve and enhance student learning in a collaborative environment but also provide teachers the opportunity to focus on designing meaningful pedagogical activities (Kattoua, Al-Lozi & Alrowwad, 2016).

Learning Management Systems offer teachers and students an online classroom that strengthens learning processes (Bradley, 2021) by making access to materials and information published in synchronous or asynchronous environments more accessible (Jung & Huh, 2019).

Although there are concerns about the negative aspects of LMSs in terms of social presence and socialization, the synchronous and asynchronous interaction tools available on newly developed LMS systems enable interaction in every situation by eliminating time and space constraints. Consequently, they facilitate collaboration in group work and projects for students while also assisting teachers in establishing better communication with students and responding more quickly to their needs (Slavin, 2014). In addition to these, Dias and Diniz (2014) reveal in their study that synchronous (chat) and asynchronous (web-based mail) environments used on LMS provide mutual interaction benefits (student-teacher information sharing), accessible education opportunities (self-regulated learning), and ease of use (accessibility and efficiency in learning). Emelyanova and Voronina (2014) claim that LMS usage allows for information exchange between home and school and positively impacts student performance. LMS systems are known to increase student satisfaction and positively influence student performance by enabling students to track their learning process, access learning materials, and communicate with instructors (Strakos, Douglas, McCormick & Wright, 2023).

Although the advantages of LMS systems are emphasized, it is certain that these benefits are largely associated with the effective design and use of LMS systems by educators. No matter how well-equipped LMS systems are with infrastructure and add-ons, if they are not designed appropriately for the target audience and course, various issues can be encountered. For instance, potential problems may include a lack of assessment and evaluation, loss of motivation, inequality of opportunities, lack of socialization (Özdoğan & Berkant, 2020), and insufficient interaction (Muhammed & Kainat, 2020). Additionally, infrastructure problems can cause frequent disruptions to education while attempting to provide a seamless learning environment. This is because studies show that numerous technical issues can arise during the remote education process (Tadesse & Muluye, 2020; DePaepe et al., 2018; Arat & Bakan, 2014).

The relationship between LMS systems and teachers directly affects the effectiveness and efficiency of the education provided. One of the critical points here is for teachers to use LMS systems correctly to enhance the learning experience, adapt them to meet students' needs, and believe in the usefulness of LMS systems. Teachers should balance the curriculum with LMS usage (Bradley, 2020) and pay attention to their professional development to effectively integrate LMSs into their courses (Dias & Diniz, 2014). This is because when instructors resist using new technologies in their courses, encountering challenges and disadvantages associated with LMS usage is more likely (Rubin, Fernandes, Avgerinou, & Moore, 2010). Studies show that the system content, interaction, and technology quality have a significant and positive impact on students' perceived usability, which in turn leads to student satisfaction with the LMS (Torin et al., 2022). This finding substantiates the risks that could arise from the inadequate design of LMS systems by educators.

The stronger the positive attitudes of those who will use LMS, the more positive their impressions about the ease of use of LMS become (Al-Mamary, 2022). Therefore, teachers and teacher candidates need to develop positive attitudes towards LMS usage. Despite significant research efforts in understanding students' LMS acceptance at universities, very limited studies have been conducted on LMS acceptance (Cheng & Yuen, 2018). This reveals the need for more research to understand teacher candidates' LMS acceptance, who will become future teachers (Cheng et al., 2022). More research is needed to determine LMS experiences at the primary and secondary education levels, particularly qualitative studies focusing on teachers' perspectives (Gamage, Ayres & Behrend, 2022). Developing positive attitudes towards LMS systems among teachers and teacher candidates will contribute to more effective use of these systems in education.

From this focus, the present study aims to understand the attitudes towards LMS usage and the underlying implicit meanings by investigating the answer to the question 'why should teachers know about LMS systems' from the perspective of teacher candidates. In the context of this question, the responses have been addressed and justified under the themes of effective learning processes, flexibility in access, content management, assessment and feedback, time management and organization, and professional development.

METHOD

Method Research

In this study, the phenomenological design, one of the qualitative research methods, was utilized with the aim of reaching the essence of the experiences of teacher candidates in LMS setup, management, and usage through the phenomena of 'LMS,' 'teacher,' and 'acceptance.' Qualitative research is a research approach that focuses on the in-depth examination of human experiences, thoughts, and behaviors (Denzin & Lincoln, 2005). Phenomenology, as a sub-method of qualitative research, focuses on examining individuals' understanding and thoughts about a particular phenomenon, experience, or event (Moustakas, 1994). Phenomenological research aims to elicit detailed descriptions of participants' experiences and reveal common themes and understandings (van Manen, 1990). In this study, teacher candidates were first given a series of training sessions covering LMS setup, management, and usage skills, and after these trainings, they were asked to answer the question, 'Why should teachers know about LMS systems?' The phenomenological design was deemed the most appropriate method for this study, as the aim was to reach the essence of the teacher candidates' experiences, conduct research on specific phenomena, and obtain qualitative data through interviews.

Participants

Seventy-three teacher candidates studying at the faculty of education participated in the research. During this process, the study was conducted with teacher candidates who voluntarily took the elective course 'Open and Distance Learning.' Among the participating teacher candidates, 71.2% were female, while 28.8% were male. 56.2% were in their third year of study, 26% in their final year, and 17.8% in their second year. When examining the distribution of teacher candidates by their fields of study, 35.6% were studying Classroom Teaching, 20.5% Social Studies Teaching, 13.7% Preschool Teaching, 11% Turkish Language Teaching, 11% English Language Teaching, 5.5% Elementary Mathematics Teaching, 1.4% Guidance and Psychological Counseling, and 1.4% Science Teaching. The distribution of the students' course achievements in the research is as follows: 46.6% received "BA" grade, 28.8% "BB," 8.2% "AA," 5.5% "CB," 4.1% "CC," 2.7% "FD," 2.7% "FF," and 1.4% "DC."

Procedure and Data Collection

The research steps carried out during the research process are shown in Figure 1.

As can be understood from Figure 1, the research began with the preparation of the content of the training to be given to the students. The trainings were structured to gain experience in four themes: LMS system usage, LMS system installation, LMS management, and LMS content development. Moodle was chosen for the installation and XAMPP for local operation. The reasons for choosing Moodle were its well-established and advanced tools as a multidisciplinary learning platform, strong evidence of increased student participation, performance, and satisfaction (Gamage, Ayres & Behrend, 2022), easy course management, low cost, and time management (Devi, Lakshmi, & Aparna, 2020). The details of the ten-week training content plan are provided in Figure 2 below.

Figure 1. Research process

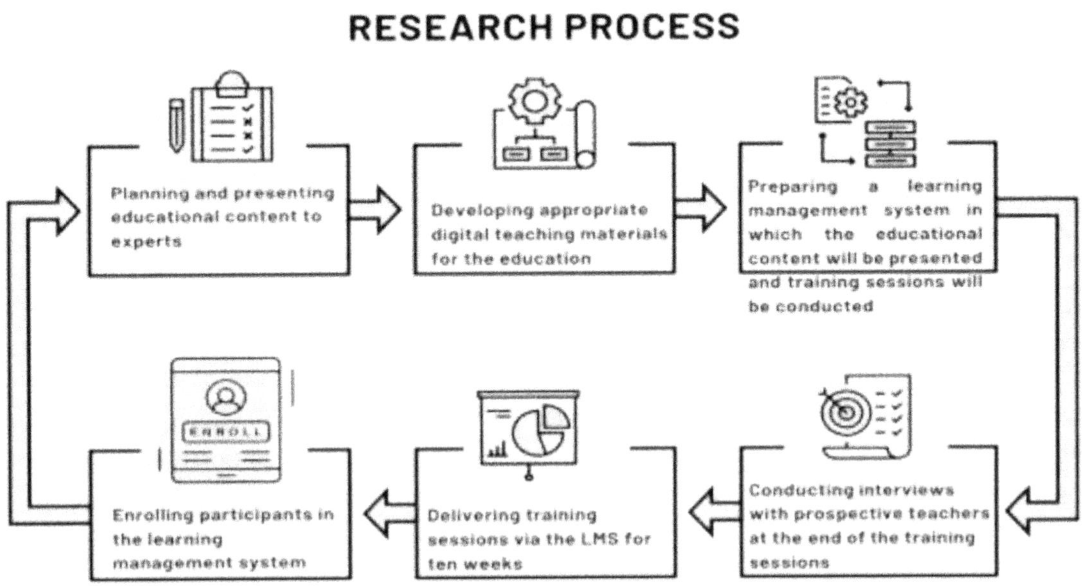

Figure 2. Ten-week training content plan

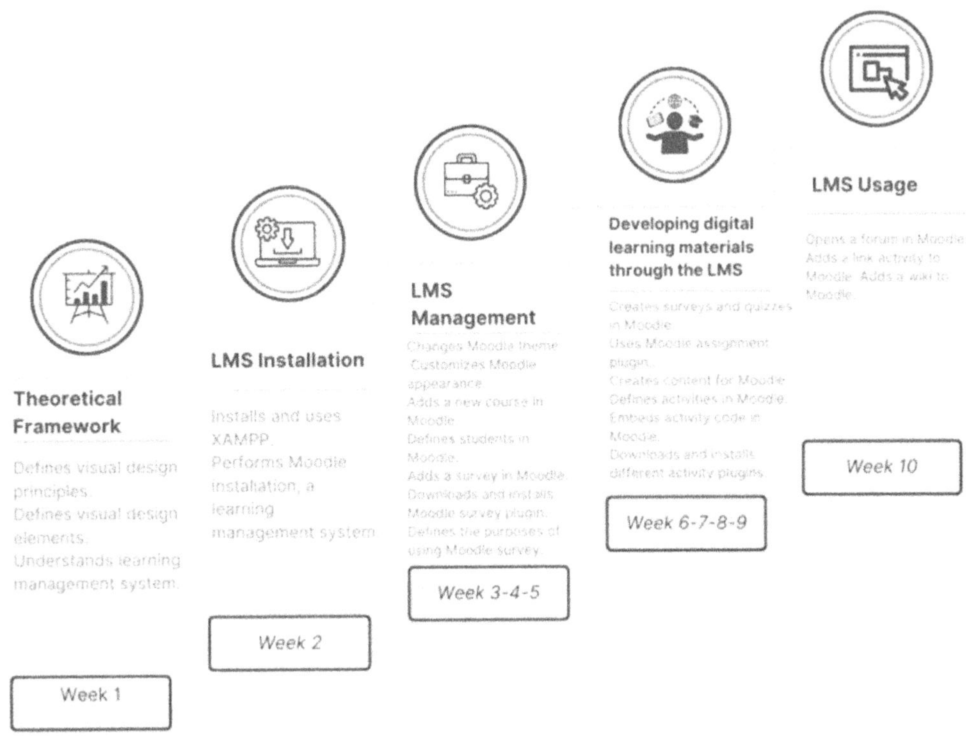

As seen in Figure 2, a one-hour learning content per week has been prepared, totaling 10 weeks and covering 22 objectives. All trainings were given in accordance with this content. In the second step, teaching materials (videos, written documents, assessment materials) suitable for the prepared learning content were developed. A few example video images are provided in Table 1. After receiving expert opinions on the prepared materials, a learning management system was created for students to experience LMS systems and access learning materials on a weekly basis. Students were officially enrolled in this LMS system. Relevant materials, assignments, and forums for the respective weeks were added by educator. The reason for providing the trainings through the LMS system is to enable students to learn how to set up and manage the LMS system with admin and instructor roles and to gain a more detailed experience in using the LMS system as end-users. This is because the use of online environments allows students to contribute to their learning experiences while reaching intended and unintended learning outcomes (Devi, Lakshmi, & Aparna, 2020).

After outlining the training sessions, we initiated the process of conducting interviews with participating students. We used a semi-structured interview form, which was specifically developed by the researcher for this purpose. This form was designed to foster a dialogue that was both guided by the researcher's inquiries and open to unique insights from the students. The interview data were collected in an innovative way; students were asked to write their responses on paper. This decision was taken to create an environment in which students felt completely at ease to express their thoughts. We believed this method could stimulate candid responses, as it freed students from the potential pressure of an immediate spoken response, giving them the freedom to reflect and articulate their thoughts at their own pace. Understanding the importance of student comfort and thoroughness, we did not impose any time restrictions during the interview form completion process. This unhurried approach further ensured that students had ample opportunity to introspectively engage with each question and provide comprehensive responses. Upon completion, the data gathered from these written interviews were transformed for analysis in a digital environment. We utilized a specialized 'text to speech' software to transcribe the hand-written responses. This methodical conversion ensured the data were readily accessible and suitable for in-depth analysis, while maintaining the integrity of the students' original written expressions.

Data Collection Tools

A semi-structured interview form was used for data collection. The semi-structured interview method allows participants to express their thoughts and feelings in their own words and helps researchers gather in-depth information (Galletta, 2013). Preparing a semi-structured interview form provides researchers with flexibility while enabling them to understand participants' experiences and thoughts (Cohen & Crabtree, 2006). In this study, the semi-structured interview form was considered the most appropriate data collection tool to reach the essence of the teacher candidates' experiences.

The interview form was structured based on phenomena. Attention was paid to structuring questions within the framework of LMS, teacher, and acceptance phenomena. The main question, "Why should teachers know about LMS systems?", was accompanied by several auxiliary questions. The preparation of questions was initiated by writing interview questions in accordance with the general objective and literature (Whiting, 2008), and 4 open-ended questions along with 4 follow-up questions were prepared for the preliminary interview guide. The aim of the follow-up questions is to facilitate the participant's understanding of the main themes (Turner, 2010) and to guide the conversation towards the topic of study (Baumbusch, 2010).

In preparing the interview form, the nature of the target audience and research topic was considered (Roulston, 2010), questions were written using clear and simple language that participants could understand (Longhurst, 2003), and ethical issues were addressed (Kvale & Brinkmann, 2009). The initial set of questions were reviewed by subject matter experts to evaluate their relevance and comprehensiveness in terms of the research goals and themes. Modifications were implemented in the phrasing of the questions, and one follow-up question was amended in response to the received feedback. To ensure participants felt comfortable (Dunn, 2005) and to encourage rich descriptions (Legard, Keegan, & Ward, 2003), a limited number of questions were provided, with participants given as much time as needed to respond.

Data Analysis

Following the interviews, the transcribed data obtained through a specialized software was subjected to content analysis. Content analysis, as one of the qualitative research methods, aims to systematically examine and analyze written, visual, or verbal communication materials (Krippendorff, 2013). Qualitative content analysis was chosen among content analysis methods. Qualitative content analysis was preferred because it aimed to reach the essence of the teacher candidates' experiences and present these experiences through an in-depth and interpretative analysis (Schreier, 2012). During the content analysis of the data obtained from the interviews, two coding cycles were utilized. The first cycle employed in vivo coding, while the second cycle utilized axial coding. In vivo coding was used to obtain information fragments by directly coding the interview data obtained from the teacher candidates using their own words and expressions (Saldaña, 2015). In this method, phenomena were elaborated by remaining particularly faithful to the participants' perspectives and experiences (Charmaz, 2014). As a result of the first cycle of coding, a total of 12 codes were reached. In the second cycle, these 12 codes were subjected to axial coding, where related codes were grouped together.

Axial coding is an analytical process that researchers use to classify, categorize, and associate collected data (Charmaz, 2006). Researchers, when analyzing data, try to uncover complex structures and processes by establishing connections between categories and subcategories in axial coding (Bryant & Charmaz, 2007). In this study, as a result of axial coding, nine codes that were thought to be related to each other were combined under a higher code, while it was decided to keep three codes unchanged. All codes obtained and quotes from the interviews used during the coding process were elaborated under the "Findings" heading, along with their reasons.

Validity and Reliability Studies

In qualitative research, ensuring objectivity is difficult because researchers use their own insights to interpret participants' experiences (Bryman, 2012). Therefore, the use of different validity and reliability methods is recommended. In this study, in order to eliminate subjectivity, students were observed during the process, and findings were provided with literature support, attempting to achieve triangulation (Denzin, 1978). Detailed and comprehensive descriptions were given under the "Findings" heading, allowing readers to better understand the research context and processes (Geertz, 1973). In this way, transferability was sought to be ensured. To provide reflexivity, in vivo coding was chosen in the first cycle, and the researchers' direct expressions were used as themes. In this way, the researcher aimed to increase the reliability of the process and outcomes by setting aside their own biases, expectations, and values (Morrow, 2005). Direct participant statements and observations were used when interpreting the data (Eisner, 1991), and interpretations and inferences were supported by expressions from students' real-world experiences.

FINDINGS

The general pattern of themes obtained from the answers given by prospective teachers to the question of why teachers should know the learning management system during the research process is presented in the concept map below.

Figure 3. Theme for the rationale of using learning management system concept map

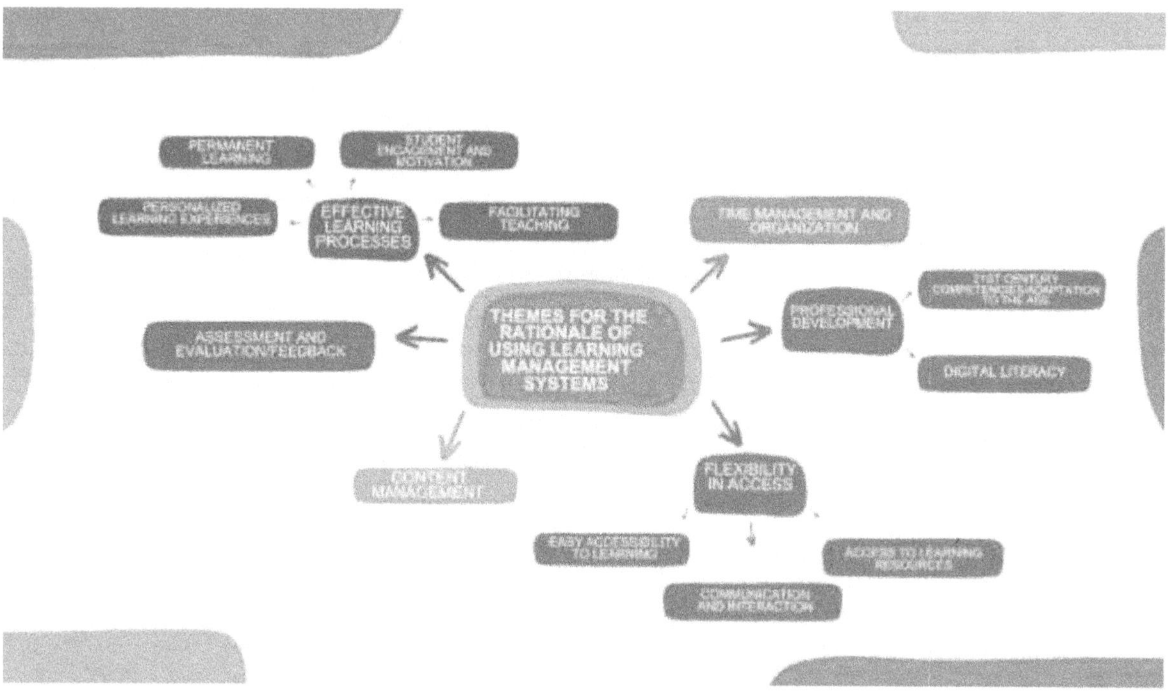

The reasons for teachers to know about LMS systems are organized under six main themes and sixteen sub-themes. These themes include Effective Learning Processes, Professional Development, Assessment and Evaluation/Feedback, Flexibility in Access, Time Management and Organization, and Content Management as the main themes. Among these themes, the Effective Learning Processes theme is divided into four sub-themes: Permanent Education, Student Engagement and Motivation, Customized Learning Experiences, and Facilitating Instruction. The Professional Development theme encompasses the sub-themes of 21st Century Competencies/Adaptation to the Age and Digital Literacy, while the Flexibility in Access theme is divided into three sub-themes: Easy Accessibility to Learning, Access to Learning Resources, and Communication and Interaction. Details related to each theme are provided below.

Table 1. Effective learning processes theme

Main Themes	Sub-Themes	Participant	Direct Quotes
Effective Learning Processes	Permanent Learning	SS1	Enriching education and teaching increases the persistence of the teacher's impact on the student; remote education systems enable the teacher to achieve this.
		ES1	The teacher must be familiar with remote education systems as an alternative environment to engage students in the subject matter and promote more lasting learning.
	Student Engagement and Motivation	T1	We must know remote education systems to provide more effective lessons to our students in the future.
		P1	Teachers should be familiar with remote education systems because they provide a good alternative for interacting with students and conducting lessons.
		E1	Remote education systems can give students the opportunity to take responsibility for their own learning.
		P2	Knowing and effectively using remote education systems can attract students to a more active and freer environment and provide information there.
		ES2	Teachers must be knowledgeable about remote education systems to make lessons more engaging, increase student participation, and provide the most effective instruction.
		P3	Knowing remote education systems is crucial for increasing students' attention, motivating them, and ensuring their participation in the course.
	Customized Learning Experiences	T2	Teachers should be proficient in remote education systems to contribute to students' education and engage their visual and auditory intelligences through different applications and techniques.
		P4	Teachers should know remote education systems to address individual differences among students.
		E2	Providing different learning environments for students is important; therefore, teachers must be knowledgeable about remote education systems.
		T3	Teachers can use remote education systems to enable students to learn how to learn and personalize their learning experiences. As a result, teachers should be prepared at all times and under all conditions.
		E3	In today's technology, it is important for teachers to know remote education systems to provide resources to students with different learning styles without restricting their learning to physical classroom environments.
		E4	Teachers should know remote education systems to choose the best system for their specific needs.
	Facilitating Instruction	SS2	Moving from traditional teaching methods to contemporary ones, teachers can utilize technology to facilitate student learning and create engaging activities that make learning enjoyable.
		T4	If teachers know remote education systems, instruction becomes easier, allowing them to provide additional lessons and make-up classes.
		E5	Remote education systems enable different teaching methods, developing students' higher-order thinking skills, facilitating the creation of diverse activities, and easing the process of relating to everyday life.
		SS3	Remote education systems are a resource that facilitates instruction and caters to everyone; therefore, teachers should know and use these systems to easily reach all students and provide education and instruction.

1. Effective Learning Processes Theme

Table 1 presents sub-themes and direct quotes related to the first theme, "Effective Learning Processes."

It has been determined that the "effective learning processes" theme is divided into four sub-themes: "sustainable learning," "student participation and motivation," "personalized learning experiences," and "facilitating teaching." Some prospective teachers detail the necessity of teachers knowing the learning management system under the theme of providing sustainable learning. SS1 explains the emphasis on sustainability with the reason "to enrich education and teaching...", while ES1 states that teachers need to know LMS systems as an alternative environment for sustainable learning.

On the other hand, according to prospective teachers, LMS systems also have an impact on student participation and motivation, and this effect should be utilized by teachers. For example, P3 states that it is essential to know distance education systems for increasing students' attention to the course, motivating them, and ensuring their participation in the course. According to the opinions of prospective teachers, the reasons for influencing participation and motivation include "P1-effective course processing," "P3-providing an active and free environment," and "ES2-making it appealing."

Another interesting theme obtained is the "Personalized Learning Experience" theme. According to prospective teachers, LMS systems offer personalized learning experiences in terms of "P4-addressing individual differences," "E2-providing different learning environments," and "T3-personalizing learning." According to T2, offering "...different applications and techniques that can activate visual-auditory intelligence..." personalizes teaching. E3 explains this situation by relating it to resource sharing as follows: "...it is important not to limit it to the physical classroom environment in schools and to provide resources for students with different learning styles."

Lastly, it is observed that they state LMSs "facilitate teaching." According to SS2, LMSs facilitate teaching by "...offering contemporary teaching methods..." and "...allowing the creation of enjoyable learning activities." Teachers can conduct "T4-makeup lessons..." and "E5- various activities" through LMS systems to facilitate teaching. According to SS2, LMSs are essential for breaking the monotony of the course. Furthermore, E5 indicates that LMS systems will "facilitate the implementation of various activities and make connections to daily life by developing students' higher-order thinking skills." Due to all these reasons provided in learning, they express that teachers need to know LMS systems.

2. Flexibility in Access Theme

A total of three sub-themes have been obtained under the "Flexibility in Access" theme. These themes are "Easy Accessibility to Learning," "Access to Learning Resources," and "Communication and Interaction." Quotes related to these themes can be found in Table 2.

The first sub-theme to be detailed under the flexibility in access theme is the "Easy Accessibility to Learning" theme. This theme can be understood from the direct expressions of many prospective teachers (ES3, ES4, ES5, E6, SS5). The most important point here is that, since this study was conducted after the pandemic, almost all participants (97%) stated that distance education is the best teaching method when extraordinary conditions like a pandemic are in question, and therefore, teachers need to know distance education systems. Based on this, it can be said that the most frequently expressed reason under this theme is a pandemic or a similar extraordinary situation. Another reason is explained by SS4 as being "helpful in solving teaching-related problems even in an off-campus location." P5 and SS5 state that it is essential for teachers to know LMS systems to "provide lessons to students who cannot attend the class due to problematic situations such as transportation distance" and "ensure the continuity of education without interruption."

Table 2. Flexibility in access theme

Main Themes	Sub-Themes	Participant	Direct Quotes
Flexibility in Access	Easy Accessibility to Learning	ES3	Even if education continues in schools, some lessons can be conducted through remote education systems, providing flexibility. Therefore, it is essential for teachers to be knowledgeable about these systems.
		SS4	Teachers should be familiar with remote education systems because they can help students solve education-related problems even outside of school.
		ES4	In the event of any pandemic, the most logical way to continue education in the technology age is through remote education. Therefore, teachers must be equipped and prepared for such situations.
		P5	Teachers should be familiar with remote education systems so that they can provide lessons to students who cannot attend class due to distance or transportation issues
		ES5	It is essential for a teacher to know various remote education systems to provide education to students under any circumstances and with any available resources.
		E6	Teachers can easily reach students using remote education systems.
		SS5	An educator familiar with remote education systems can provide education at any time they wish. It is necessary for ensuring the continuity of education without interruption
	Access to Learning Resources	P6	Teachers can more easily share information and resources using remote education systems.
		ES6	When presenting a skill to students, teachers may sometimes want to use additional materials and environments.
		E7	It is essential for teachers to know remote education systems to provide resources to students
	Communication and Interaction	ES7	Teachers should be knowledgeable about remote education systems to maintain interaction between students and teachers under any circumstances.
		SS6	Teachers should be familiar with remote education systems to establish effective communication and interaction with students.
		P7	It is essential for teachers to know these systems because it impacts the interaction between teachers and students and the success of the course.

Another sub-theme under this overarching theme is the "Access to Learning Resources" theme. We clearly understand this theme from P6's statement, "A teacher can provide easier information and resource sharing with distance education systems," and E7's statement, "It is essential to offer resources to students." On the other hand, it is understood that they see LMS systems as supportive of face-to-face education from ES6's statement, "Sometimes, when presenting the skills to be acquired by the student, they may want to get help from additional materials and environments."

The last sub-theme under this theme is the "Communication and Interaction" theme. In the reasons given by the students who expressed their opinions under this theme, references to extraordinary situations like the pandemic are observed again. For example, ES7 indirectly explains this situation as follows: "Teachers need to know this in order not to abandon student-teacher interaction under any circumstances." On the other hand, according to SS6, teachers need to know LMS usage to "...establish efficient communication and interaction." P7 relates this issue to student success by stating that "the interaction between the teacher and the student is important in terms of the success of the course."

3. Content Management Theme

Direct quotes related to the content management theme are provided in Table 3.

One of the themes obtained as a result of content analysis is "content management." According to the pre-service teachers, LMSs attract students' attention and provide environments (SS7) that prevent their interests from scattering by offering materials such as videos, animations, educational games, and similar materials (P8). On the other hand, according to ES8, "LMSs contribute to the diversification of the course by offering interactive applications like assignments, surveys, and activities to students." Teachers should be aware of these opportunities and know how to use them (ES8). P9 emphasizes the importance of knowing LMSs and supporting appropriate material selection from the perspective of pre-service teachers. Additionally, according to pre-service teachers, LMS systems support the development of quality materials (ES9) and give students the freedom to rewind and listen as much as they want. Contents supported by images or videos enhance retention (SS8). Due to all these opportunities, teachers need to know LMSs.

Table 3. Content management theme

Main Themes	Participant	Direct Quotes
Content Management	P8	It is essential to be able to share materials such as videos, animations, educational games, and similar resources with students.
	ES8	We can give our students assignments, surveys, and activities through remote education systems.
	SS7	Remote education systems provide an environment where engaging content and resources can be prepared to actively involve students in the lesson, capture their attention, and prevent their interests from dissipating.
	ES8	Remote education systems can yield results with applications such as videos, animations, assignments, forums, and surveys, diversifying the lesson. Teachers should be aware of these opportunities and know how to use them.
	P9	As a teacher candidate, it is crucial to know remote education systems to provide more effective learning environments, methods, and techniques to students during their learning and teaching processes and to support and guide them in selecting appropriate materials.
	ES9	Remote education systems support the development of higher-quality materials
	SS8	The ability to replay a lesson as many times as a student desires, along with supporting images or videos, can increase retention in students.

4. Assessment and Evaluation/Feedback Theme

Direct quotes related to the assessment and evaluation/feedback theme are provided in Table 4.

Under the "Assessment and Evaluation/Feedback" theme, one of the themes obtained, it is emphasized that it is crucial for teachers to provide activities like assignments, surveys, and similar assessment-evaluation activities to students through LMS (P10, ES10, SS10). According to P11 and ES12, teachers who know the system can "provide much faster feedback to students and provide instant feedback." On the other hand, teachers need to know distance education systems to provide feedback using features like preparing surveys, conducting exams, and opening forums (E8). According to SS9, "Distance education has the advantage of providing easy feedback," and "recording the lessons simplifies the feedback process" (ES11).

Table 4. Assessment and evaluation/feedback theme

Main Themes	Participant	Direct Quotes
Assessment and Feedback	P10	Teachers need to be familiar with the remote education system to provide students with assessment and evaluation activities such as assignments, surveys, and similar tasks.
	ES10	Teachers can give assignments to students through the system and provide feedback; knowing the system is essential for evaluation.
	P11	Remote education allows for much quicker feedback to students, which is why teachers need to be familiar with this system.
	E8	Teachers can provide feedback by utilizing features such as creating surveys, conducting exams, and opening forums.
	SS9	Remote education systems facilitate easy feedback.
	ES11	Since lessons are recorded, providing feedback becomes easier
	SS10	We can assess and evaluate students, give assignments, and create surveys.
	ES12	With remote education, teachers can provide real-time feedback to students.

5. *Time Management and Organization Theme*

Direct quotes related to the time management and organization theme are provided in Table 5.

One of the themes obtained as a result of content analysis is "Time Management and Organization." According to SS11, distance education systems "can provide good time management to prevent disruptions in education" and are essential for "teachers to better manage learning processes (P12)." Distance education offers fast data communication and easier management opportunities (ES15). Therefore, it is imperative for teachers to master distance education systems to manage the courses well (ES13). These systems save time (T5, T7) and can lighten the teacher's workload by simplifying learning (ES14). On the other hand, T6 suggests that while mistakes can be corrected in face-to-face education, this is difficult in distance education, emphasizing the importance of teachers knowing distance education systems correctly because "everything must be done according to the system in distance education.

Table 5. Time management and organization theme

Main Themes	Participant	Direct Quotes
Time Management and Organization	SS11	Remote education systems can offer good time management to prevent potential disruptions in education. Therefore, it is important for teachers to know remote education systems.
	P12	It is important for teachers to know remote education systems in order to better manage learning processes.
	ES13	To effectively manage lessons conducted through remote education, teachers must have a good understanding of remote education systems.
	T5	Remote education systems save time and allow students to watch a lesson multiple times, which can facilitate a teacher's work.
	ES14	Remote education systems help teachers save time.
	T6	In face-to-face education, mistakes can be corrected. However, in remote education, everything must be done according to the system. Therefore, teachers need to know how to use remote education systems correctly.
	T7	Every teacher should know remote education systems to save time and cater to everyone's needs.
	ES15	Remote education systems provide fast data communication and easier management

6. *Professional Development Theme:*

Another overarching theme obtained as a result of the interviews is the "Professional Development" theme. As a result of the analyses, this theme is divided into two sub-themes: "21st Century Competencies/Adaptation to the Era" and "Digital Literacy." Direct quotes related to these themes are provided in Table 6.

The first theme under the "Professional Development" theme is the "21st Century Competencies/ Adaptation to the Era" theme. Many pre-service teachers view the necessity of teachers using LMS as a requirement of the era. For example, according to T8, "Teachers must know distance education systems to keep up with the technology age in the 21st century." Similarly, ES16 emphasizes that "teachers need to know distance education systems to adapt to the evolving era and become a well-equipped teacher." It is important for teachers to always love sharing and conveying information and to adopt 21st-century technologies to keep up with Generation Z (P13). According to pre-service teachers, "Conducting lessons more efficiently and with technology-compatible teaching methods (SS12)" is essential in the field of education and teaching (SS13).

Table 6. Professional development theme

Main Themes	Sub-Themes	Participant	Direct Quotes
Professional Development	21st Century Competencies / Adapting to the Age	T8	In the 21st century, teachers must know remote education systems in order to keep up with the age of technology.
		P13	Teachers should always love sharing and transferring information everywhere and at any time. Therefore, they should adopt remote education systems to keep up with the technology of the 21st century and adapt to Generation Z.
		ES16	To keep up with the evolving age and to be a well-equipped teacher, one must know remote education systems.
		SS12	Ensuring that lessons are more efficient and conducted using teaching methods that can adapt to technology is among the necessary competencies for teachers in the new age.
		SS13	It is crucial for teachers to keep up with developments and know remote education systems in the field of education and teaching.
		ES17	Teachers need to know remote education systems for their own professional development.
	Digital Literacy	SS14	Especially today, when technology is so widespread, it is essential for teachers to be knowledgeable about technology.
		E9	Along with the constantly changing and evolving technology, the teaching techniques and materials used by teachers should change and develop in a way that is compatible with new-generation students. That is why teachers need to know remote education systems.
		P14	Computers and the internet, which are the results of our age, are part of our lives, and knowing information systems is important for teachers.
		SS14	I believe teachers should be knowledgeable in every aspect, and we are already living in a technological age. Technology is now an inseparable part of our lives, and teachers must benefit from it.
		ES18	: If we want students to develop, we must first ensure the development of teachers. They should be open to innovations and be able to use technology. Since remote education systems can reach a larger audience more quickly, it is important for teachers to know this system.

One of the themes obtained as a result of content analysis is the concept of "Digital Literacy." According to SS14, "Due to the widespread use and integration of technology today, it is imperative for teachers to have digital literacy skills." With constantly changing and updating technology, teachers need to know distance education systems to adapt and improve their teaching techniques and materials for new-generation students (E9). According to SS14, "… teachers need to be knowledgeable in every aspect and benefit from technology." ES18 explains this situation from the student's perspective by stating that "to ensure students' development" and "reach a wider audience quickly," teachers must be "open to innovations and capable of using technology."

CONCLUSION AND DISCUSSION

The aim of this research is to determine the thoughts of pre-service teachers on their knowledge of using Learning Management Systems (LMS). The research findings indicate that pre-service teachers evaluate the knowledge and skills of teachers in the field of LMS as important and necessary. This necessity and importance were explained through six main themes and nine sub-themes. For example, within the main theme of effective learning processes, the sub-themes obtained reveal that pre-service teachers believe that teachers can achieve goals such as providing lasting learning, increasing student engagement and motivation, offering personalized learning experiences, and facilitating instruction using LMSs. Studies conducted similarly show that distance education has a positive impact on success and effectiveness (Lin & Gao, 2020; Aldossary, 2021). Educators' use of LMS supports effective learning processes by increasing student engagement and motivation (Moore & Kearsley, 2011). In this context, technology can be used to create constructivist and student-centered learning environments (Jonassen, 2003). Moreover, it was emphasized that teachers can offer personalized learning experiences that take individual differences into account by using LMSs. Brusilovsky and Millán (2007) similarly argue in their studies that web-supported systems can provide effective learning by foregrounding students' individual characteristics, thanks to their features that can prioritize data such as students' knowledge levels, learning preferences, interests, and goals. As pre-service teachers highlighted, LMSs can facilitate instruction and contribute to effective learning processes with contemporary teaching methods. Of course, as Ally (2004) states, it is crucial to know educational theories and their significance in practice to create online learning experiences that respond to individual needs and encourage interaction. In line with the literature, the findings of this study demonstrate that teachers' knowledge and skills in e-learning applications can play a significant role in enriching learning processes (Garrison & Anderson, 2003). Consequently, the importance of LMS training in teacher education programs should be emphasized, and more comprehensive and careful approaches should be adopted to address students' cognitive, social, and emotional needs beyond knowledge distribution in online learning processes (Herrington et al., 2005).

According to the research findings, pre-service teachers believe that it is important for teachers to know Learning Management Systems (LMS) in terms of flexibility in access. The results indicate that this belief is primarily shaped through three main themes: Easy Accessibility to Learning, Access to Learning Resources, and Communication and Interaction. The theme of flexibility in access emphasizes the importance of teachers providing continuous and uninterrupted education to students through LMS systems, especially when considering extraordinary situations such as pandemics. Many researchers point out that emergency remote education implemented due to the COVID-19 pandemic has contributed significantly to educational and instructional activities. This process has enabled students to access their classes from home and maintain their learning (Dhawan, 2020). Teachers found that one of the most advantageous aspects of emergency remote education is that it prevents students from falling behind in their courses and offers independence from time and place (Fidan, 2020). Easy access is also provided through time savings in commuting between home and school and high-quality communication with online platforms (Horspool & Lange, 2012). Emergency remote education practices during the pandemic have also contributed to the development of students' digital skills. Apart from extraordinary situations, distance education offers the advantages of being able to learn anytime and anywhere (Todri et al., 2021).

The theme of Easy Accessibility to Learning reflects the views of pre-service teachers who value the provision of accessible education to students in different environments and times using LMS systems. This theme is also supported by studies on learning processes proposed by Cronbach and Snow (1977), which emphasize the importance of access to learning resources and responding to individual learning needs. The theme of Communication and Interaction highlights the importance of teachers using LMS systems to establish effective and efficient communication with students. As mentioned by Garrison and Anderson (2003), ensuring student-teacher interaction is a critical factor for student success. In online instruction, students' continuous interaction with each other, teachers, and content supports better understanding and learning processes (Gaytan & McEwen, 2007). Well-designed online learning experiences offer students opportunities to construct knowledge and engage in collaborative learning through group work (Herrington, Reeves, & Oliver, 2005). Emergency remote education practices have also contributed to reducing disparities in our education system. During the pandemic period, students have had the opportunity to work with students from geographically different regions and different cultures (Barbour et al., 2020). Especially in extraordinary situations like pandemics, LMSs are considered a critical tool for ensuring the continuity of education (Bozkurt et al., 2020). In this context, pre-service teachers believe that knowing LMS systems and using them effectively can provide flexibility in access.

Another theme reached as a result of the research is content management. Pre-service teachers justify the importance of LMS systems in terms of content management with benefits such as attracting students' attention, adding variety to the learning process, and increasing the permanence of learning materials. LMSs are systems that host a range of tools for developing digital materials in the teaching-learning process, distributing materials to users, managing instructional materials and user data, and evaluating the user information obtained (Dobre, 2015). The interaction of LMSs with content management provides better learning experiences for students and enhances teachers' ability to manage educational processes more effectively (Coates et al., 2005). Pre-service teachers can provide more effective and engaging learning experiences to students by using these technologies (Park & Son, 2009). Additionally, students have the freedom to use materials (such as videos and documents) repeatedly on LMSs (Lamanauskas & Makarskaitė-Petkevičienė, 2021). It is important for pre-service teachers to increase their awareness and skills regarding content management and LMS systems and to effectively use these technologies in their own courses (Koehler & Mishra, 2009). This is because focusing solely on information distribution in online learning environments is insufficient, and more comprehensive and careful approaches need to be adopted to meet students' cognitive, social, and emotional needs (Herrington, Reeves, & Oliver, 2005). Additionally, having these skills may positively impact the career development of pre-service teachers as educators (Tondeur et al., 2017).

According to pre-service teachers, it is also important for teachers to be able to use LMSs effectively in terms of "Assessment and Evaluation/feedback". It is essential for teachers to know the tools on LMS for evaluation and assessment of learning processes to provide students with rapid and instant feedback. Providing quick feedback helps deepen students' understanding and achieve learning goals, while the feedback process increases student motivation and self-confidence (Gaytan & McEwen, 2007). Pre-service teachers believe that it is important for teachers to provide feedback by using features such as conducting exams, preparing surveys, and opening forums on LMS. A study conducted by Coşkun Karademir and Filiz (2021) also reveals that a majority of instructors use systems such as assignments, surveys, and forums on LMSs. Web-based tests conducted via computers provide educators with enriched environments with multimedia objects (such as video, images, and audio) and offer convenience in evaluating test results easily and quickly (Luecht, 2001). As pre-service teachers emphasize, LMSs contribute to the realization of learning by providing opportunities for feedback (Yağcı, 2012). However, it is essential to adapt the materials and methods used for assessment to e-learning for a fair evaluation, rather than using them in the same way as in traditional classroom environments (Giray, 2021). For this, it is necessary for teachers to know LMS systems well.

According to pre-service teachers, another contribution of LMSs to teachers is in "time management and organization." These systems allow teachers to better manage the learning process and save time. In a study by Serçemeli and Kurnaz (2020), supporting the pre-service teachers' statements, it was concluded that LMSs provide time savings. Distance education offers quick data communication and easier management for teachers, thus facilitating learning and alleviating the teacher's workload. However, pre-service teachers believe that at this point, it is essential for teachers to know distance education systems correctly to correct errors in remote education. Lastly, according to pre-service teachers, teachers need to know LMS systems for professional development. This theme is explained in two sub-themes: "21st-century competencies/adaptation to the era" and "Digital literacy." In these themes, pre-service teachers emphasize the importance of teachers' professional development and mastery of distance education systems. Teachers in the 21st century need to be individuals who can adapt to technology and keep up with Generation Z (Ertmer & Ottenbreit-Leftwich, 2010). Having digital literacy skills and continuously adapting to changing technology helps teachers provide more effective and efficient educational experiences to students (Selwyn, 2016). Therefore, to educate students in the 21st century requires adaptation to the era and digital literacy skills. It is observed that the emphasis of pre-service teachers is precisely in this direction.

In conclusion, the findings of this study emphasize the importance of teachers and pre-service teachers mastering distance education systems. It is essential for teachers to demonstrate competence in content management, assessment and evaluation, time management and organization, 21st-century competencies, and digital literacy skills to enhance the effectiveness and efficiency of educational processes. Additionally, considering technological developments and the expectations of Generation Z students, it is emphasized that teachers need to master LMS systems to provide students with more compatible, flexible, and engaging educational experiences, ensure the continuity of education in extraordinary situations, eliminate some disadvantages in online environments, and ensure time management. Therefore, it is of critical importance for teachers and pre-service teachers to continuously improve their skills and knowledge in distance education systems in their professional development processes to be successful in future educational environments.

Suggestions

The findings show that pre-service teachers attach importance to LMS knowledge and skills. Therefore, teacher education programs should prioritize providing training on LMS usage to prepare for modern educational practices. Future studies can examine the impact of teachers' abilities to effectively use

LMSs in teaching practices. LMS can improve learning experiences by providing flexible access to learning materials, promoting interaction between students and teachers, and offering personalized learning experiences based on individual needs. Teachers should be encouraged to use LMSs effectively to increase the quality of education. Therefore, future studies can investigate how LMS can be used to create and distribute learning materials effectively and how it can enhance the learning experience for students. The COVID-19 pandemic has highlighted the importance of online learning and demonstrated the potential of LMS to provide continuous and uninterrupted education. Hence, institutions should be prepared to use LMS in emergencies.

REFERENCES

Al-Adwan, A., Al-Adwan, A., & Smedley, J. (2018). Exploring students' acceptance of e-learning using Technology Acceptance Model in Jordanian universities. *International Journal of Education and Development Using Information and Communication Technology*, *9*(2), 4–18.

Al-Mamary, Y. H. S. (2022). Why do students adopt and use Learning Management Systems?: Insights from Saudi Arabia. *International Journal of Information Management Data Insights*, *2*(2), 100088. doi:10.1016/j.jjimei.2022.100088

Aldossary, K. (2021). Online distance learning for translation subjects: Tertiary level instructors' and students' perceptions in Saudi Arabia. *Turkish Online Journal of Distance Education*, *22*(6), 96–109. doi:10.17718/tojde.961821

Ally, M. (2004). Foundations of educational theory for online learning. In T. Anderson (Ed.), *The theory and practice of online learning* (pp. 3–31). Athabasca University.

Arat, T., & Bakan, Ö. (2014). Uzaktan Eğitim ve Uygulamaları. *Selçuk Üniversitesi Sosyal Bilimler Meslek Yüksekokulu Dergisi*, *14*(1-2), 363–374.

Barbour, M. K., LaBonte, R., Kelly, K., Hodges, C., Moore, S., Lockee, B., Trust, T., Bond, A., & Hill, P. (2020). *Understanding pandemic pedagogy: Differences between emergency remote, remote, and online teaching*. A special report of the State of the Nation: K-12 E-Learning in Canada Project. Canadian eLearning Network.

Bates, A. W. (2015). *Teaching in a Digital Age: Guidelines for Designing Teaching and Learning*. Tony Bates Associates Ltd.

Black, P., & Wiliam, D. (2009). Developing the theory of formative assessment. *Educational Assessment, Evaluation and Accountability*, *21*(1), 5–31. doi:10.100711092-008-9068-5

Blau, I., & Hameiri, M. (2010). Implementing technological change at schools: The impact of online communication with families on teacher interactions through learning management system. *Interdisciplinary Journal of E-Learning and Learning Objects*, *6*, 245–257. doi:10.28945/1313

Bozkurt, A. (2020). A global outlook to the interruption of education due to COVID-19 pandemic: Navigating in a time of uncertainty and crisis. *Asian Journal of Distance Education*, *15*(1), 1–126. doi:10.5281/zenodo.3878572

Bradley, V. M. (2021). Learning Management System (LMS) use with online instruction. *International Journal of Technology in Education*, *4*(1), 68–92. doi:10.46328/ijte.36

Brusilovsky, P., & Millán, E. (2007). User models for adaptive hypermedia and adaptive educational systems. In *The Adaptive Web* (pp. 3–53). Springer. doi:10.1007/978-3-540-72079-9_1

Bryant, A., & Charmaz, K. (Eds.). (2007). *The SAGE handbook of grounded theory*. Sage Publications. doi:10.4135/9781848607941

Bryman, A. (2012). *Social research methods* (4th ed.). Oxford University Press.

Chan, S. S., So, W. K., Wong, D. C., Lee, A. C. K., & Tiwari, A. (2007). Improving older adults' knowledge and practice of preventive measures through a telephone health education during the SARS epidemic in Hong Kong: A pilot study. *International Journal of Nursing Studies*, *244*(7), 1120–1127. doi:10.1016/j.ijnurstu.2006.04.019 PMID:16857203

Charmaz, K. (2006). *Constructing grounded Theory: a practical guide through qualitative analysis*. Sage Publications.

Cheng, M., Tao, D., Xie, S., Cao, X., & Yuen, A. H. (2022). Exploring students' learning management system acceptance patterns: Antecedents and consequences of profile membership. *Computers in Human Behavior*, *135*, 107374. doi:10.1016/j.chb.2022.107374

Cheng, M., & Yuen, A. H. K. (2018). Student continuance of learning management system use: A longitudinal exploration. *Computers & Education*, *120*, 241–253. doi:10.1016/j.compedu.2018.02.004

Coates, H., James, R., & Baldwin, G. (2005). A critical examination of the effects of learning management systems on university teaching and learning. *Tertiary Education and Management*, *11*(1), 19–36. doi:10.1080/13583883.2005.9967137

Cohen, D., & Crabtree, B. (2006). *Semi-structured interviews*. Retrieved from http://www.qualres.org/HomeSemi-3629.html

Coşkun Karademir, T., & Filiz, O. (2021). Öğretim Elemanlarının Uzaktan Eğitim Sürecinde Ölçme Değerlendirmeye Yönelik Yaşadığı Sorunlar ve Çözüm Önerileri. In Uzaktan Eğitimde Güncel Sorunlar ve Çözüm Önerileri, ASBÜ yayınları.

Cronbach, L. J., & Snow, R. E. (1977). Aptitudes and instructional methods: A handbook for research on interactions. Academic Press.

Dalsgaard, C., & Paulsen, M. F. (2009). Transparency in cooperative online education. *International Review of Research in Open and Distance Learning*, *10*(3). Advance online publication. doi:10.19173/irrodl.v10i3.671

De Paepe, L., Zhu, C., & DePryck, K. (2018). Drop-out, retention, satisfaction and attainment of online learners of Dutch in adult education. *International Journal on E-Learning*, *17*(3), 303–323.

Denzin, N. K. (1978). *The research act: A theoretical introduction to sociological methods*. McGraw-Hill.

Denzin, N. K., & Lincoln, Y. S. (2005). Introduction: the discipline and practice of qualitative research. In N. K. Denzin & Y. S. Lincoln (Eds.), The Sage Handbook of Qualitative Research (3rd ed.). SAGE Publications, Inc.

Devi, S., Lakshmi, V., & Aparna, M. (2020). Moodle: an effective learning management system for 21st century learners. *Alochana Chakra Journal, 9*(6).

Dhawan, S. (2020). Online learning: A panacea in the time of COVID-19 crisis. *Journal of Educational Technology Systems, 49*(1), 5–22. doi:10.1177/0047239520934018

Dias, S. B., & Diniz, J. A. (2014). Towards an enhanced learning management system for blended learning in higher education incorporating distinct learners' profiles. *Journal of Educational Technology & Society, 17*(1), 307–319.

Dunn, L. (2005). The importance of reflexivity in the changing field of qualitative research. *Forum Qualitative Sozialforschung/Forum: Qualitative Social Research, 6*(3).

Eisner, E. W. (1991). *The enlightened eye: qualitative inquiry and the enhancement of educational practice*. Macmillan.

Emelyanova, N., & Voronina, E. (2014). Introducing a learning management system at a Russian university: Students' and teachers' perceptions. *International Review of Research in Open and Distance Learning, 15*(1), 272–289. doi:10.19173/irrodl.v15i1.1701

Fidan, M. (2020). Covid-19 belirsizliğinde eğitim: İlkokulda zorunlu uzaktan eğitime ilişkin öğretmen görüşleri. *Uşak Üniversitesi Eğitim Araştırmaları Dergisi, 6*(2), 24–43. doi:10.29065/usakead.736643

Galletta, A. (2013). *Mastering the semi-structured interview and beyond: From research design to analysis and publication*. NYU Press.

Gamage, S. H. P. W., Ayres, J. R., & Behrend, M. B. (2022). A systematic review on trends in using Moodle for teaching and learning. *International Journal of STEM Education, 9*(1), 9. doi:10.118640594-021-00323-x PMID:35096513

Garrison, D. R., & Anderson, T. (2003). *E-learning in the 21st century: A framework for research and practice*. Psychology Press.

Gaytan, J., & McEwen, B. C. (2007). Effective online instructional and assessment strategies. *American Journal of Distance Education, 21*(3), 117–132. doi:10.1080/08923640701341653

Geertz, C. (1973). *The interpretation of cultures: Selected essays*. Basic Books.

Giray, G. (2021). An assessment of student satisfaction with e-learning: An empirical study with computer and software engineering undergraduate students in Turkey under pandemic conditions. *Education and Information Technologies, 26*(6), 6651–6673. doi:10.100710639-021-10454-x PMID:33686329

Herrington, J., Reeves, T. C., & Oliver, R. (2005). Online learning as information delivery: Digital myopia. *Journal of Interactive Learning Research, 16*(4), 353–367.

Horspool, A., & Lange, C. (2012). Applying the scholarship of teaching and learning: Student perceptions, behaviours and success online and face-to-face. *Assessment & Evaluation in Higher Education, 37*(1), 73–88. doi:10.1080/02602938.2010.496532

Jonassen, D. H. (2003). *Learning to solve problems with technology: A constructivist perspective.* Prentice Hall.

Jung, S., & Huh, J. H. (2019). An efficient LMS platform and its test bed. *Electronics (Basel), 8*(2), 154. doi:10.3390/electronics8020154

Kaden, U. (2020). COVID-19 school closure-related changes to the professional life of a K–12 teacher. *Education and Information Technologies, 10*(6), 1–13.

Karatepe, F., Küçükgençay, N., & Peker, B. (2020). Öğretmen Adayları Senkron Uzaktan Eğitime Nasıl Bakıyor? Bir Anket Çalışması. *Journal of Social and Humanities Sciences Research, 7*(53), 1262–1274.

Kattoua, T., Al-Lozi, M., & Alrowwad, A. (2016). A review of literature on e-learning systems in higher education. *International Journal of Business Management and Economic Research, 7*(5).

Koehler, M. J., & Mishra, P. (2009). What is technological pedagogical content knowledge? *Contemporary Issues in Technology & Teacher Education, 9*(1), 60–70.

Krippendorff, K. (2013). *Content analysis: an introduction to its methodology* (3rd ed.). Sage Publications.

Kvale, S., & Brinkmann, S. (2015). *Interviews: learning the craft of qualitative research interviewing* (3rd ed.). Sage Publications.

Lamanauskas, V., & Makarskaitė-Petkevičienė, R. (2021). Distance lectures in university studies: Advantages, disadvantages, improvement. *Contemporary Educational Technology, 13*(3), ep309. Advance online publication. doi:10.30935/cedtech/10887

Legard, R., Keegan, J., & Ward, K. (2003). In-depth interviews. In Qualitative research practice: A guide for social science students and researchers (pp. 138-169). Academic Press.

Lin, X., & Gao, L. (2020). Students' sense of community and perspectives of taking synchronous and asynchronous online courses. *Asian J. Distance Educ., 15*, 169–179. doi:10.5281/zenodo.3881614

Longhurst, R. (2003). Semi-structured interviews and focus groups. *Key Methods in Geography, 3*, 143-156.

Moore, M., & Kearsley, G. (2012). *Distance education: a systems view of online learning* (3rd ed.). Wadsworth.

Morrow, S. L. (2005). Quality and trustworthiness in qualitative research in counseling psychology. *Journal of Counseling Psychology, 52*(2), 250–260. doi:10.1037/0022-0167.52.2.250

Moustakas, C. (1994). Phenomenological research methods. *Sage (Atlanta, Ga.).*

Muhammed, A., & Kainat, A. (2020). Online learning amid the COVID-19 pandemic: Students' perspectives. *Online Submission, 2*(1), 45–51.

Özdoğan, A. Ç., & Berkant, H. G. (2020). Covid-19 pandemi dönemindeki uzaktan eğitime ilişkin paydaş görüşlerinin incelenmesi. *Milli Eğitim Dergisi, 49*(1), 13–43. doi:10.37669/milliegitim.788118

Park, C. N., & Son, J. B. (2009). Implementing computer-assisted language learning in EFL classroom: Teachers' perceptions and perspectives. *International Journal of Pedagogies and Learning*, *5*(2), 80–101. doi:10.5172/ijpl.5.2.80

Reeves, T. C., & Oliver, R. (2005). Online learning as information delivery: Digital myopia. *Journal of Interactive Learning Research*, *16*(4), 353–367.

Roulston, K. (2010). *Reflective interviewing: A guide to theory and practice*. Sage Publications. doi:10.4135/9781446288009

Rubin, B., Fernandes, R., Avgerinou, M. D., & Moore, J. (2010). The effect of learning management systems on student and faculty outcomes. *The Internet and Higher Education*, *13*(1-2), 82–83. doi:10.1016/j.iheduc.2009.10.008

Saldaña, J. (2015). *The coding manual for qualitative researchers* (3rd ed.). Sage Publications.

Schreier, M. (2012). *Qualitative content analysis in practice*. Sage Publications.

Selim, H. M. (2007). Critical success factors for e-learning acceptance: Confirmatory factor models. *Computers & Education*, *49*(2), 396–413. doi:10.1016/j.compedu.2005.09.004

Serçemeli, M., & Kurnaz, E. (2020). Covid-19 pandemi döneminde öğrencilerin uzaktan eğitim ve uzaktan muhasebe eğitimine yönelik bakış açıları üzerine bir araştırma. *International Journal of Social Sciences Academic Researches*, *4*(1), 40–53.

Seven, M. A. (2012). Uzaktan ve örgün eğitime devam eden öğrencilerin İngilizce dersindeki başarı düzeylerinin karşılaştırılması. *Ekev Akademi Dergisi*, *6*(50), 215–228.

Siemens, G., Gasévić, D., & Dawson, S. (2015). *Preparing for the digital university: a review of the history and current state of distance, blended and online learning*. Athabasca University Press.

Slavin, R. E. (2014). Cooperative learning and academic achievement: Why does groupwork work? *Anales de Psicología*, *30*(3). Advance online publication. doi:10.6018/analesps.30.3.201201

Strakos, J. K., Douglas, M. A., McCormick, B., & Wright, M. (2023). A learning management system-based approach to assess learning outcomes in operations management courses. *International Journal of Management Education*, *21*(2), 100802. doi:10.1016/j.ijme.2023.100802

Tadesse, S., & Muluye, W. (2020). The impact of COVID-19 pandemic on education system in developing countries: A review. *Open Journal of Social Sciences*, *8*(10), 159–170. doi:10.4236/jss.2020.810011

Todri, A., Papajorgji, P., Moskowitz, H., & Scalera, F. (2021). Perceptions regarding distance learning in higher education, smoothing the transition. *Contemp. Educ. Technol.*, *13*. doi:10.30935/cedtech/9274

Tondeur, J., van Braak, J., Ertmer, P. A., & Ottenbreit-Leftwich, A. (2017). Understanding the relationship between teachers' pedagogical beliefs and technology use in education: A systematic review of qualitative evidence. *Educational Technology Research and Development*, *65*(3), 555–575. doi:10.100711423-016-9481-2

Toring, H., Legaspi, G., Omolon, J., Amadeo, R., Amadeo, E., Opolentisima, Q., Barina, V., Cacho, T., Illustrimo, F., & Cortes, S. (2022). *Evaluation of students' satisfaction toward an adopted learning management system at Indiana Aerospace University: A structural equation modelling approach*. Asia Pacific Management Review., doi:10.1016/j.apmrv.2022.12.002

Tuncer, M., & Taşpınar, M. (2008). Sanal ortamda eğitim ve öğretimin geleceği ve olası sorunlar. *Sosyal Bilimler Dergisi*, *10*(1), 125–144.

Turnbull, D., Chugh, R., & Luck, J. (2020). Learning management systems, an overview. Encyclopedia of Education and Information Technologies, 1052-1058.

Van Manen, M. (1990). *Researching lived experience: human science for an action sensitive pedagogy*. Althouse Press.

Watson, R., & Watson, S. (2012). An argument for clarity: What are learning management systems, what are they not, & what should they become? *TechTrends*, *51*(2), 28–34.

Yağcı, M. (2012). Çevrimiçi Sınav Ortamlarının Öğrencilerin Akademik Başarılarına Etkisi. *E-Journal of New World Sciences Academy*, *7*(1), 331–339.

Chapter 3
Use of Online Active Methodologies in Higher Education

Patricia Barcena-Toyos
https://orcid.org/0000-0003-2942-1768
Universidad Internacional de La Rioja, Spain

ABSTRACT

This chapter presents the results of a study that examines the effects of an online gamification of the end-of-unit tests of a course in a master's degree. Learners completed a survey about the effects of the online gamified experience on their motivation and their progress and results in the course. Participants (n = 41) were in- or pre-service teachers, so the survey examined their teachers' sense of self-efficacy in terms of the application of active methodologies and ICTs in their classes after the gamified experience. Results show that the gamification had positive effects in terms of increasing motivation for the subject and, therefore, their perceived academic results. Also, participants felt more confident to use gamification in their classes. Results also showed significant differences in terms of motivation between nontraditional and traditional learners.

INTRODUCTION

In the last decades, there has been an exponential growth of online learning programs in virtual learning environments (VLE), particularly in higher education. The profile of most students enrolling in online higher education is that of nontraditional learners, that is, adult students—commonly over 25 years old—with financial and familiar responsibilities (Chen, 2017). For these learners, VLEs provide an ideal setting for time management, at the same time that adults can pursue higher education with the aim of improving their professional conditions or to find a job (Zamecnik et al., 2022). However, nontraditional learners have to make an even bigger effort to keep up with their assignments and school workload, so the drop-out rates for this group of students is higher than for other higher-education learners (Beluce & Oliveira, 2015). Finding active methodologies that motivate and engage nontraditional learners becomes

DOI: 10.4018/978-1-6684-8861-4.ch003

a requirement that higher education teachers must consider. In fact, the effectiveness of online learning relies on three pillars: technology, students and teachers, who are in charge of selecting and implementing the appropriate digital tools to facilitate learning and motivate students (Tham & Werner, 2015). Adult learners commonly show different interests from younger students, and they tend to be more pragmatic and practical, but emotions still play an important role in their motivation for learning, as well as their socioeconomic status, which can also be leading factor in the increasing dropout rates for this group of students (Padilla-Carmona et al, 2022). For this reason, the learning experience must be relevant for their personal and professional goals, being able to put into practice the learning acquired in VLE in their own work settings (Wlodkowski & Ginsberg, 2017). The use of active methodologies becomes a need to foster adult learners' interest for the content learned at higher education and to attend to principles of autonomous and meaningful learning.

Adult education, including higher education, has grown exponentially over the last decade across the world, and online higher education has definitely experienced this growth, mostly in the last few years due to the COVID-19 pandemic. However, online higher education is not a new phenomenon and, in two decades, the number of students who learn remotely has increased 900% (Varea, 2021), most of them being adults and nontraditional learners. The term 'nontraditional learners' is used to refer to those adult students who are over 25 years old and, most of the times, are working full time, are financially dependent or have dependents, such as spouses or children, or even resume studies after a long period of time (Padilla-Carmona et al, 2022). In fact, while there are not many studies on nontraditional learners in Spain (see Navarro-Mateos et al., 2021), recent reports from the Ministry of Universities of Spain found that 84.4% of the 50.9% of students enrolled in a Master's degree at a distance-education university are 30 years old or older (Ministry of Universities, 2023). The number of adult learners registered in Master's degrees is increasing as are those enrolled in distance education universities, a group that represents 17.4% of students. According to the Ministry of Universities of Spain (2023), these universities and their VLEs provide accessibility opportunities that are reflecting on the rising numbers of students from other countries (23.8%) who enroll in distance Spanish universities; most of them coming from Latin America and the Caribbean (15.8%). Most nontraditional learners usually seek higher education as a way to improve their careers and socio-economic status, struggling with the balance between a student life and the responsibilities of adulthood (Tenorio-Rodríguez et al., 2021). Thus, motivation is even more crucial for these students, in order to fulfil their necessities and avoid dropout, and even more so in VLEs, because learners need to undertake actions to monitor the regulation of their own learning (Beluce & Oliveira, 2015). It is well known that emotions play a critical role in learning, and this is also true in adults, but in their case, motivation is also based on the practicality, relevance and application in real work settings of what they learn (Zamecnik et al., 2022). The use of active methodologies offers limitless opportunities for fostering motivation and could be especially significant for nontraditional adults in VLE as a means to foster autonomous learning and attend to different learning needs and styles (Ranilla, 2018).

Despite the fact that motivation is the most important factor to determine students' success and academic performance (Beluce & Oliveira, 2015), there are scarce studies on its effects on learning in adult education. In fact, research shows that motivation has a direct relationship with learning and that such relationship grows with the age of students (Wlodkowski & Ginsberg, 2017). The purpose of this study is to test the effectiveness of active methodologies, such as gamification, to enhance or sustain students' motivation in a VLE, following the premise that if their motivation increases it will have a direct impact on their academic results. In addition to that, the study pretends to fill in the gap in adult motivation in VLE. The chapter examines the perceptions of pre- and in-service teachers on the potential use of online active methodologies and digital tools in their classes after undergoing an online gamification as a tool for continuous assessment. Through the gamified experience, students will be actively learning and being assessed on the academic contents of the course at the same time that they develop digital competence. It is expected that students who undergo the gamification will apply this type of active methodology to assess their learners' academic progress in their future or current teaching practice.

In order to achieve those aims, the present chapter examines the results of the implementation of a gamified experience used to replace the end of unit tests that are part of the continuous assessment in a Master's degree in bilingual education from an online university in Spain, aimed at future and current teachers from Spain and other countries in Latin America. So, the study will answer the following research questions:

- RQ1: What are the perceptions of student teachers from an online Master's degree after participating on a gamified experience on how it impacted their motivation, their learning progress and results, and their sense of self-efficacy to use active methodologies?
- RQ2: Are these perceptions different for nontraditional and for traditional learners?

BACKGROUND

Adult Motivation and Active Methodologies

Adults are practical learners, and nontraditional learners are even more so, as they seek usefulness and practicality on what they learn, so they will be able to apply it to their professional lives. For adults, the highest level of motivation is achieved when the following conditions are met: success, validation, value and enjoyment (Wlodkowski & Ginsberg, 2017). That is, effective learning situations for adults should help them "successfully learn what they value and want to learn in an enjoyable manner" (Wlodkowski, 2008, p. 101), increasing their motivation towards the learning situation. According to Ryan and Deci's (2000) Self-Determination Theory (SDT), there are three conditions that must be met in order for any type of motivation to take place, which are autonomy, competence and relatedness. The degree to which each of those psychological needs are fulfilled will determine the level of motivation of an individual (Ede, 2022). While intrinsic motivation has usually been considered the only motivation that leads to successful learning, SDT argues that autonomy is the most important pillar of motivation, and it can even have positive and long-lasting effects on extrinsic motivation. The more autonomous motivation becomes, the more actively committed an individual will be. Following SDT, there is a continuum of levels of motivation going from amotivation, or total lack of motivation, to intrinsic motivation. In between, there are four levels of extrinsic motivation, according to the degree of autonomy involved.

According to the above, the use of active methodologies becomes even more needed in adult education, and research shows evidence of the effect of such methodologies on increasing learners' participation and motivation (Barata et al., 2015). Active methodologies foster the active involvement of individuals in their own learning process, supporting autonomy and engagement, which is a combination of focused attention and emotion that involves the individual in learning (Woolfolk Hoy, 2013). An example of active methodology is gamification. Gamification is the application of game elements to a nongame situation, such as learning—either physical or online, that enhances learners' engagement and motivation (Willig et al., 2021), which is expected to improve learners' academic achievement as a result of increased involvement in the subject (Santos et al., 2020). Sailer et al. (2013) provide an account of the motivational mechanisms that can be triggered by gamification, according to different motivational perspectives. In terms of rewards, in gamification players receive badges when they complete a challenge and immediate feedback, which acts as positive or negative reinforcement, and generates motivation. Regarding the achievement of goals, learners will be more motivated if the gamified experience "provides … [clear and achievable goals]" (Santos et al., 2020, p. 33) and underlines the consequences for achieving those outcomes, i.e., the assessment of learning objectives and the consequences implied. Finally, Santos et al (2020) also report the role of gamification in increasing motivation when the player's interests and the feeling of flow (by delivering feedback, clear outcomes, and appropriate challenge) are met. So, it can be assumed that challenges in gamification provide a setting to foster learners' competence, social relatedness, and autonomy, because the participant must take decisions throughout the experience that will determine the learning outcome of the experience, which will have positive and long-lasting effects on their extrinsic motivation. In fact, gamification is becoming a popular methodology also at postgraduate levels and in different domains, and there are several studies that have examined the effects of implementing gamified learning experiences with adult students at university, both in VLEs and in face-to-face classes (see Calvo Roselló & López Rodríguez, 2021; Santamaría & Alcalde, 2020). These studies recorded the positive effects of the use of active methodologies on adult learners' motivation and, consequently, their increased participation and expected improved results.

Gamification

In the planning and design of a gamified experience, like for any other serious game-like experience, there are elements that must be considered, so the gamification ends up producing the desired outcomes in terms of motivation and modification of behaviors, through the reinforcement of the desired behavior (Duhigg, 2012), using resources such as repetition, point boards or badges. According to Robson et al. (2015), there are three main elements to consider in the design of any gamification: mechanics, dynamics, and emotions. Based on their framework, mechanics are the goals set beforehand, the narrative and characters, and the challenges that players need to overcome. When players follow these mechanics, different behaviors emerge—which are called dynamics—for example, competition or cooperation. While gamification and game-based learning share some elements, the former includes immediate feedback on the learners' performance, which contributes to students' self-assessment (Moreno & Lopezosa, 2020). In fact, gamification is considered an active methodology that fosters the learner's motivation and participation in the task, as well as their autonomy in the learning process (Santamaría & Alcalde, 2020). Finally, the way players follow the mechanics of the game and create dynamics while playing will generate emotions in the player, which are usually associated with the participant's performance in the gamification, such as excitement or frustration (Trigueros et al., 2019). Therefore, the interaction between these three elements will determine the player's continuation on the game or quitting the experience, so choosing an attractive narrative and appealing characters will have an effect on the player's motivation and emotions.

Throughout the different challenges of an online gamification, learners must complete tasks that resemble real-life situations. Additionally, when used as authentic assessment (Mahoney, 2017), an online gamification reinforces active, self-regulated and meaningful learning, thanks to the immediate feedback received that allows learners to modify their learning and pay attention to areas of improvement (Kolb & Kolb, 2009). In the case of pre- or in-service teachers, they can take the online gamification as alternative forms of assessment that integrate ICTs can be used as models to be used in their own lesson design. Additionally, the use of an online gamification as an assessment tool will increase motivation and, therefore, students' involvement in the course (Santos et al, 2020), which will eventually impact the academic results of adult learners.

Gamification as an active methodology has experienced an exponential growth also in higher education, because of the wide array of possibilities it offers in terms of improving motivation and academic results (Fernández-Portero & Castillo, 2022). Through engaging and motivating activities, learners acquire content and receive immediate feedback, which is also expected to positively impact their academic results and autonomy, so the use of gamification at university levels has increased exponentially in the last years (Navarro-Mateos et al., 2021). Despite the scarcity of empirical studies supporting the effectiveness of gamification in improving academic results, it is expected that the higher the motivation the more engaged learners will be. Therefore, students' learning outcomes will also improve as a result of increased participation in the subject (Santos et al, 2020). In fact, research shows that students in higher education are keen on participating in gamified learning experiences, due to their positive impact on their motivation and interest for the subject matter, as well as their academic results and the improvement of professional skills (Pegalajar, 2021). So, the use of an online gamified experience as continuous assessment in teacher higher education is expected to have positive effects on the students future teaching practices, improving their digital competence (Juan-Lázaro & Área, 2021) and management of active methods as alternative assessment tools (Barata et al., 2015). Teacher beliefs are intrinsically influenced by their own training experiences as learners themselves (Bárcena-Toyos, 2022), so if learners have a positive experience being assessed through an online gamification, it is expected that they will use the same methodologies with their own students.

METHODOLOGY

The study used quantitative methodology to examine the students' responses to a Likert-scale questionnaire about their perceptions on the gamified experience in terms of motivation and self-perceived academic progress, as well as their sense of self-efficacy on the use and implementation of online active methodologies in their classes as teachers. Data were initially analyzed through descriptive statistical analysis of the responses to the items on the questionnaire, followed by a comparison of results by groups (nontraditional and traditional learners) to understand whether there were any statistically significant differences between groups, based on the variable of age.

Procedure of the Gamified Experience

The gamification replaced the end-of-unit tests used as part of the continuous assessment in an online MA. Originally, there were 12 tests, one for each unit in the syllabus, but the gamification comprised eight challenges—some challenges covered two units. At the end of the corresponding units, learners found the gamification, which used different templates from the online tool Genially, such as escape rooms, interactive photographs or comics, among others. Each challenge provided an extensive revision of the main concepts of the unit, which they navigated by completing different steps to move along the challenge, which also included additional ICT tools and resources, like interactive videos, games, activities or fake social media. All these online tools provided learners with immediate feedback on their performance and were thought to be a source of inspiration for participants to use in their own classes as future teachers. The design of the experience followed the principles of serious games and Robson et al. (2015) MDE framework explained above, which describes the mechanics, dynamics and emotions of any gamification. As part of the mechanics of the game, the narrative and characters of the gamification were chosen following Campbell's (2008) principles to intensify learners' engagement in the experience and the content of the course. The hero was a fictional character whose name was a play on words from the main content of the course and a celebrated British author of detective novels. In the first challenge, participants were presented with the character and his mission, and the role that the participants would have in helping him during his quest to return back with the elixir (Campbell, 2008).

At the end of each challenge, learners had to send a short task answering to a question that was revealed to them after successfully completing all previous patterns. As shown in Table 1, challenges were organized around topics and learning objectives that should be achieved at the end of each challenge. So, at the end of the challenge, together with the immediate feedback they received within each mini-task embedded in the challenge (e.g. a video with true or false questions), learners also received feedback form their professor on their achievement of the learning objective, and received a badge for each challenge completed, following a reward system that is characteristic of any gamified experience. Each badge was originally designed and had engraved the knowledge they had acquired in each challenge, which also referred to the elixir that the hero was searching for and hoping to find before returning back home (Campbell, 2008).

Table 1. Challenges and learning objectives

Challenges	Units and Topics	Learning Objective
Challenge 1	Units 1 & 2. CLIL as an approach.	Understand the principles of effective CLIL teaching.
Challenge 2	Units 3 & 4. The role of motivation in language learning and CLIL.	Evaluate the suitability of methodologies to be implemented in the CLIL classroom.
Challenge 3	Units 5 & 6. Adaptation authentic language and materials.	Adapt language and materials in CLIL to make content accessible for learners.
Challenge 4	Unit 7. The CLIL Formal Format Planner.	Plan a CLIL unit attending to content and language.
Challenge 5	Unit 8. The 4 Cs Framework: Content.	Identify what qualifies as content in CLIL and introduce it lesson planning.
Challenge 6	Unit 9. The 4 Cs Framework: Cognition.	Apply cognitive processes to the design of learning objectives in lesson planning.
Challenge 7	Unit 10 & 11. The 4 Cs Framework: Communication.	Differentiate the types of language that learners need in order to acquire content learning objectives.
Challenge 8	Unit 12. The 4 Cs Framework: Culture.	Support the development of intercultural communicative competence and integrate it in CLIL lesson planning.

Source: Own elaboration.

Participants

The gamified experience was implemented in a course at a Spanish online university during the first semester of a master's degree in bilingual education, so all students were in- or pre-service teachers. At the end of the gamified experience students could fill out an online survey about their perceptions on the self-perceived impact of the gamification on their motivation and academic progress, as compared to other subjects where active methodologies were not used. Additionally, some questions inquired about the participants' sense of self-efficacy regarding their use of online active methodologies, such as gamification, and their digital competence after participating in the gamification. At the end of the semester, when the gamified experience was over, all 127 students who were enrolled in the course received the link to answer the online questionnaire, but it only received 46 responses (36.2%), but the final sample was made of 41 participants (n = 41) (32.3%) because one participant answered twice and three did not consent to participate in the study. From these 41 participants, there were 7 male and 34 female. As for age, since there is not an agreement on a specific age from which a learner can be considered nontraditional. Instead, a learner is considered nontraditional due to the fact that, commonly, they have not followed the regular path of attending university full time after graduating from high school. So, for the aim of this chapter, we considered the age of 26 as the minimum age for the group of nontraditional learners. Participants (n = 41) were divided into two groups, according to their age: NTL (n = 21) and TL (n = 20).

Data Collection Instrument

After the gamification was complete at the end of the semester, students got access to an online survey, which was also embedded on the VLE, to evaluate the gamified experience. The survey consisted of two parts. The survey included demographic questions, such as the year of birth, and 14 Likert-scale questions (see Table 2) about learners' experiences with the gamification in terms of motivation (3 items), and the self-perceived effects of the methodology on their learning progress and academic results (3 items), as compared to other subjects in the Master's degree that followed traditional assessment methods. Questions also addressed the participants sense of teacher's self-efficacy for the use of online active methodologies as assessment tools (4 items) and other digital resources (4 items) in their classes. Learners had to select the degree of agreement with the statements on a scale from 1 (totally disagree) to 6 (totally agree). This scale was chosen in order to avoid the neutral response (neither agree nor disagree) and, therefore, also avoid central tendency. Considering the fact that learners' first language was Spanish, the survey was written in Spanish, so items and scales were translated into English for the aim of this chapter.

The items in the instrument measuring the participants' sense of self-efficacy in the use of digital tools and active methodologies were designed considering the theoretical framework of self-efficacy, a term defined as "beliefs in one's capabilities to organize and execute the courses of action required to produce given attainments" (Bandura, 1997, p. 3). According to this concept, an individual's beliefs intervene on their perception of what they can do and, in turn, on the activities they choose to do and their investment in such activities. So, in the case of teachers, it is believed that their sense of self-efficacy, will impact their pedagogical choices.

The questionnaire was validated by a group of ten experts in online education and active methodologies, who measured the validity of each item to measure the constructs related to the objectives of the study. To measure the reliability and internal consistency of the items in the instrument, the Cronbach alpha was calculated, obtaining high reliability ($\alpha = 0.96$).

RESULTS

Data were first analyzed using descriptive statistics from Excel of all the items, measuring the means (M) and the standard deviation (SD) as shown on Table 2. The items were also categorized under four groups, according to the following topics: the self-perceived impact of the gamified experience on learners' 1) motivation (items 01, 02, 04) and their 2) learning progress and academic results (items 03, 05, 06), and their teachers' sense of self-efficacy in terms of 3) digital competence (items 07, 09, 12, 13) and 4) use of active methodologies (08, 10, 11, 14).

Table 2. Descriptive statistics of each item in the questionnaire (n = 41)

Item	M	SD
1. The gamified experience increased learners' motivation:		
01. As compared to other courses in this Master's, my experience with the gamification of the end-of-unit tests in this course has been positive.	5.34	0.96
02. As compared to other courses in this Master's, my experience with the gamification of the end-of-unit tests in this course has increased my motivation for this course.	5.17	1.09
04. As compared to other courses in this Master's, my experience with the gamification of the end-of-unit tests in this course has increased my interest for this course.	5.22	1.19
2. The gamified experience increased learners' self-perceived progress and results:		
03. As compared to other courses in this Master's, the feedback received in the gamification of the end-of-unit tests has helped me acquire the contents of this course.	5.10	1.09
05. As compared to other courses in this Master's, with the gamification of the end-of-unit tests in this course has increased my participation in this course.	5.02	1.23
06. As compared to other courses in this Master's, with the gamification of the end-of-unit tests in this course has contributed to improving my academic results in this course.	5.05	1.16
3. Teachers' sense of self-efficacy regarding their digital competence:		
07. After participating in the gamification of the end-of-unit tests in this course, I am able to use more digital tools in my classes.	5.44	0.95
09. After participating in the gamification of the end-of-unit tests in this course, I feel more capable of applying the knowledge I acquired about the use of ICT tools in my classes.	5.39	0.92
12. After participating in the gamification of the end-of-unit tests in this course, I feel more capable of designing digital teaching and learning resources for my students.	5.34	0.99
13. After participating in the gamification of the end-of-unit tests in this course, I feel more capable of providing detailed feedback to my students through digital resources tools.	5.32	0.72
4. Teachers' sense of self-efficacy regarding the use of active methodologies:		
08. After participating in the gamification of the end-of-unit tests in this course, I believe that gamification can be used as a useful tool for continuous assessment.	5.63	0.73
10. After participating in the gamification of the end-of-unit tests in this course, I feel more capable of designing a gamified experience in my classes.	5.29	0.96
11. After participating in the gamification of the end-of-unit tests in this course, I feel more capable of assessing the contents of my subject through active methodologies like gamification.	5.29	0.93
14. My self-concept about my ability to integrate active methodologies and digital tools in my classes has improved after my participation in the gamified experience in this course.	5.17	0.83

Source: Own elaboration

In general, results showed that participants' level of satisfaction with the gamification was high for all items (more than 5 points out of 6) and that, in general terms, the use of active methodologies integrated with digital tools had a positive impact on the participants self-perceived academic achievement but also on their sense of self-efficacy regarding the use of active methodologies and digital competence. So, the online gamified experience was a positive experience for them as learners and as teachers in training. On the one hand, participants shared their overall satisfaction with their participation in the gamification as part of their continuous assessment in comparison with other subjects in the Master's, which did not apply active methodologies (M = 5.36) and agreed on the fact that the gamified experience had a positive impact on their motivation (M = 5.17) and interest for the subject (M = 5.22). The participants' responses for the second category showed that their self-perceived academic progress and involvement in the subject were positively influenced by the gamified experience, especially because of the immediate feedback received (M = 5.10). On the other hand, considering the fact that participants were pre- or in-service teachers, it was important to test the effect of the use of active methodologies, such as gamification, not only on their results and motivation as learners, but also its impact on participants as teachers and how likely they are to use active methodologies in their own classes. For that, two categories of questions asked about the participant teachers' sense of self-efficacy. The construct of self-efficacy, originally coined by Bandura (1977), is a person's belief that their actions in a specific situation will lead to a given outcome, so the higher a person's sense of self-efficacy, the more persistent they will be in achieving the task or outcome at hand. Teacher self-efficacy makes teachers examine the teaching task while considering their own teaching competence (Tschannen-Moran et al., 1998), so teacher self-efficacy relates to teachers' beliefs of whether they are able to impact learning results due to their own competence and ability in that particular teaching-learning task. Therefore, teacher self-efficacy is considered a key element of teacher beliefs and professional competence. Participants' sense of teacher self-efficacy in this survey was measured in terms of their digital competence and use of active methodologies. Results show that, after participating in the online gamification and using different digital resources (e.g. Genially gamification templates, Educaplay interactive games, and EdPuzzle videos with questions), participants felt capable of using different online tools in their classes (M = 5.44), not only the same tools that were used in the gamification they participated in (M = 5.39) but their sense of self-efficacy was also high for their ability to create new digital resources for their students (M = 5.34) that can help teachers provide detailed feedback for their students (M = 5.32), implying that they consider online gamification to be a good tool to give students timely feedback. Finally, as Table 2 shows, participants' sense of self-efficacy was also high for the use of gamification in the classroom, particularly for its effectiveness to be used as part of the continuous assessment of their subject (M = 5.63). Results show that participants felt capable of designing a gamified experience in their classroom (M = 5.29) and assess contents of their subject through gamification or other active methodologies (M = 5.29). All in all, participants felt that their participation in the gamified experience of the subject in the Master's degree had improved their ability to integrate active methodologies and ICT tools in their own classes as teachers, and use them with their own students in their future or current practice (M = 5.29), which means that the gamification of the end of unit tests had a positive impact not only on the students' motivation and participation in the subject, but also on their future practice, something that is particularly valued by adult learners, being even more relevant in VLE, where motivation is key to success in keeping up with the workload of a Master's.

Chi-square was used to analyze the relation between two categorical variables (age and degree of agreement with statements in the survey) and test whether there is any statistically significant association between them. The test was performed comparing NTL (n = 21) and TL (n = 20) groups, according to their responses to all four categories of questions (see Table 2). In this first chi-square test, results showed no significant difference between groups ($\chi 2 = 0.12$, $\chi 2crit = 7.81$, df = 3, p = 0.05), so the null hypothesis, that there is no significant difference between the observed and expected frequencies in both groups, was accepted. Then, on a second analysis, four chi-square tests were performed comparing responses of both age groups by category (i.e. motivation, learning progress and results, and sense of self-efficacy regarding digital competence and use of active methodologies), one for each of these categories. The results showed a significant difference between NTL (n = 21) and TL (n = 20) in the motivation category ($\chi 2 = 15.72$, $\chi 2crit = 11.07$, df = 5, p = 0.05). The null hypothesis, that there is no significant difference between the observed and expected frequencies in terms of impact of gamification on learners' degree of motivation, was rejected in favor of the alternative hypothesis, that there is a significant difference between the observed and expected frequencies in terms of impact of gamification on learners' degree of motivation. The remaining three chi-square test results showed no significant difference between groups, so the null hypotheses that there are no significant differences between the observed and expected frequencies in terms of self-perceived progress and results ($\chi 2 = 5.45$, $\chi 2crit = 11.07$, df = 5, p = 0.05), sense of self-efficacy of their teacher digital competence ($\chi 2 = 2.05$, $\chi 2crit = 11.07$, df = 5, p = 0.05), and sense of self-efficacy of their use of active methodologies ($\chi 2 = 5.61$, $\chi 2crit = 11.07$, df = 5, p = 0.05) were accepted.

SOLUTIONS AND RECOMMENDATIONS

This study aims at examining teachers' perceptions on the use of active methodologies and digital tools in the classroom, based on the impact that their participation on a gamified experience as students of a Master's degree. Motivation is considered a predictor of success, based on the fact that it increases participation and engagement in the course (Botte et al, 2020), and a lack of motivation in certain scenarios (for example, in VLE with adult and nontraditional learners) might lead to students dropping out. So, active methodologies become an ideal resource for teachers to engage and motivate their students (Peña-Acuña, 2022), and even more so, if it is integrated with digital resources. Online gamification is one of such methodologies and through the use of rewards (in the case of this study, badges and immediate feedback) and fostering autonomous and meaningful learning, it promotes extrinsic motivation, one of the pillars of SDT (Ryan and Deci, 2000). In order to measure motivation, this study asked participants about their observed levels of motivation and their engagement in the course, as compared to other courses they were enrolled in, after they participated in the gamification of the end-of-unit tests. As expected, gamification increased participants' motivation and interest for the subject and its contents (Fernández-Portero & Castillo, 2022), with in turn is expected to have positive effects on their academic results (Díaz Ramírez, 2020). In the same way, results from the study corroborate the idea that gamification is an effective methodology to use in VLE and with adult learners, who can also benefit from learning in a ludic environment (Looyestyn et al., 2017). Also as expected, learners' self-perceived academic progress and participation in the course were also higher than in other courses thanks to the gamified experience (Santos et al, 2020). In fact, these results align with findings that gamification is more effective in terms of engagement and motivation in adult learners (Balci et al., 2022).

A second part of this study analyzes teachers' sense of self-efficacy, considering the fact that participants are current or future teachers, since they were enrolled in a teacher-training Master's degree. Teacher cognition is made up of not only the knowledge they acquired through training, but an important part of teacher cognition is also their own experience as students (Borg, 2005). This means that, together with learning about teaching in formal settings (e.g., university), teachers' use their own experience as learners in their lessons, too. So, teachers' own experiences as learners have an impact on their own teacher beliefs and consequently on the methodological choices we make (Bárcena-Toyos, 2022). Therefore, it is expected that in this case, as results show from teachers' sense of self-efficacy, participants are more likely to use active methodologies such as gamification, after participating in the experience themselves and understanding the mechanics of the different challenges and the possibilities that this methodology offer, even as a tool for continuous assessment (Jurgelaitis et al., 2019) and integrated with digital tools, that they also learned about thanks to their participation in the gamified experience of this subject.

Results from the comparison of results in two groups of participants, who were divided according to their age and grouped under NTL and TL seemed inconclusive. While the sample in this study was small, the difference between groups in terms of motivation can be understood from their giving high importance to autonomy and practicality of what they learn (Landrum et al., 2000), which are all indicators of integrated regulation (Ryan & Deci, 2000). Also, adult NTL are commonly more pragmatic, but emotions must still be part of the learning situation, so in this sense, results align with previous studies that report the importance of fostering extrinsic motivation for NTL (Wlodkowski & Ginsberg, 2017) and the benefits of using gamification as a motivational methodology (Fernández-Portero & Castillo, 2022; Ramos Navas-Parejo et al., 2019).

On the other hand, there were no significant differences between both groups in terms of their responses in participation and results or their teachers' sense of self-efficacy in the integration of active methodologies and digital tools. It can be understood, then, that gamification is equally effective in increasing the participation of adult learners, regardless of their background and situation, which aligns with previous studies, like that of Santos et al (2020). Similarly, the lack of any significant differences in terms of reported teachers' self-efficacy between groups implies that all learners felt confident about applying what they had learned as students. These results can have implications for future teacher education programs because teacher identity and teacher cognition are determinant of the methodologies implemented in the classroom (Bárcena-Toyos, 2022).

FUTURE RESEARCH DIRECTIONS

It is recommended that future research examines the impact of gamification in students' learning with control groups and larger samples that allow us to experimentally test the effectiveness of this active methodology. Examining students' perceptions provides an insight into the effects of gamification at different levels, such as motivation, but measuring learners' academic progress and to what extent it was influenced by the gamified experience, can inform future pedagogical decisions.

Similarly, it would be interesting to carry out longitudinal studies to understand to what extent teachers put into practice the active methodologies (e.g., online gamification) that they experienced as students themselves, in their own classes with their students.

In order to get more conclusive results, future research should include a larger sample of participants who openly share demographic information that can help researchers understand their personal and professional situation to fit under TL or NTL categories. This would allow researchers to understand the effect of active methodologies in the academic progress and engagement of nontraditional learners, and advise policy makers and course designers choices.

CONCLUSION

It becomes clear that the use of active methodologies is an important resource that university teachers and professors should consider in their classes, not only to increase their motivation and participation in their classes, which in turn are expected to have a positive impact on their academic results, but also as methodologies that increase engagement and autonomous learning, giving learners a sense of autonomy that will also impact their interest for the course. Furthermore, active methodologies are also a great resource to foster meaningful learning, as they provide an ideal setting to tackle real-life experiences that they will have to face eventually in their future professional lives. While a VLE provide is an ideal setting to integrate online tools and active methodologies, the use of these methodologies and ICT should not be limited to these learning environments. As this study showed, gamification can be integrated in the continuous assessment of the course as an ideal way of fostering engagement and facilitating immediate feedback about their own performance.

ACKNOWLEDGMENT AND FINANCING

This work is funded by the Educational Innovation Project *La gamificación como alternativa a los test de evaluación continua (GAMITEST)* (Ref. No. B0041), Universidad Internacional de La Rioja.

REFERENCES

Bandura, A. (1977). Self-efficacy: Toward a unifying theory of behavioral change. *Psychological Review*, *84*(2), 191–215. doi:10.1037/0033-295X.84.2.191 PMID:847061

Barata, G., Gama, S., Jorge, J., & Gonçalvez, D. (2015). Gamification for smarter learning: tales from the trenches. *Smart Learning Environments, 2*(10).

Bárcena-Toyos, P. (2022). Teacher identity in CLIL: A case study of in-service teachers. *Latin American Journal of Content and Language Integrated Learning*, *15*(2), e1516. doi:10.5294/laclil.2022.15.1.6

BeluceA. C.OliveiraK. L. D. (2015). Students' motivation for learning in virtual learning environments. *Paidéia (Ribeirão Preto), 25*, 105-113. doi:10.1590/1982-43272560201513

Borg, S. (2005). *Teacher cognition in language teaching*. doi:10.1057/9780230523470_10

Botte, B., Bakkes, S., & Veltkamp, R. (2020). Motivation in gamification: Constructing a correlation between gamification achievements and self-determination theory. *Games and Learning Alliance: 9th International Conference, GALA 2020, Laval, France, December 9-10, 2020 Proceedings*, *9*, 157–166.

Calvo Roselló, V., & López Rodríguez, M. I. (2021). Gamificación en el aula universitaria: Una experiencia de "escape room inversa.". *REDU Revista de Docencia Universitaria, 19*(1), 45–74. doi:10.4995/redu.2021.15666

Chen, J. C. (2017). Nontraditional adult learners: The neglected diversity in postsecondary education. *SAGE Open, 7*(1), 2158244017697161. doi:10.1177/2158244017697161

Duhigg, C. (2012). The power of habit: Why we do what we do in life and business (Vol. 34). Random House.

Ede, S. (2022). Gamification and Motivation. *Issues and Trends in Learning Technologies, 10*(1).

Fernández-Portero, I., & Castillo, C. (2022). Gamification in the English language class: Analysis of pre-service teachers' perceptions. *Computer Assisted Language Learning Electronic Journal, 23* (1), 425-444.

Juan-Lázaro, O., & Área, M. (2021). Gamificación superficial en e-learning: evidencias sobremotivación y autorregulación [Thin layer gamification in e-learning: evidence on motivation and self-regulation]. *Pixel-Bit. Revista de Medios y Educación, 62*, 146–181.

Jurgelaitis, M., Čeponienė, L., Čeponis, J., & Drungilas, V. (2019). Implementing gamification in a university-level UML modeling course: A case study. *Computer Applications in Engineering Education, 27*(2), 332–343. doi:10.1002/cae.22077

Kolb, A. Y., & Kolb, D. A. (2009). The learning way: Meta-cognitive aspects of experiential learning. *Simulation & Gaming, 40*(3), 297–327. doi:10.1177/1046878108325713

Landrum, R. E. (2000). Motivational differences among traditional and nontraditional students enrolled in metropolitan universities. *Metropolitan Universities, 11*(1), 87–92.

Looyestyn, J., Kernot, J., Boshoff, K., Ryan, J., Edney, S., & Maher, C. (2017). Does gamification increase engagement with online programs? A systematic review. *PLoS One, 12*(3), e0173403. doi:10.1371/journal.pone.0173403 PMID:28362821

Mahoney, K. (2017). *The assessment of emergent bilinguals: supporting English language learners*. Multilingual Matters.

Ministry of Universities of Spain. (2023, July). *UNIVBase*. Retrieved July 4, 2023, from https://estadisticas.universidades.gob.es/jaxiPx/Datos.htm?path=/Universitaria/Alumnado/EEU_2023/Master/Matriculados//l0/&file=1_1_Mat_Master_Sex_Edad(2)_Rama.px

Moreno, E., & Lopezosa, M.D. (2020). Gamificación a través de un proyecto de aprendizaje-servicio: diseñando un breakout educativo desde la universidad para el alumnado de primaria. *Revista científica electrónica de Educación y Comunicación en la Sociedad del Conocimiento, 20*(1). . doi:10.30827/eticanet.v20i1.15524

Navarro-Mateos, C., Pérez-López, I., & Marzo, P. (2021). La gamificación en el ámbito educativo español: revisión sistemática [Gamification in the Spanish educational field: a systematic review]. *Retos, 42*, 507–516. doi:10.47197/retos.v42i0.87384

Padilla-Carmona, T., González-Monteagudo, J., & Tenorio Rodríguez, M. (2022). Lights and shadows: The inclusion of invisible students in Spanish universities. *Autonomy and Responsibility Journal of Educational Sciences*, *7*(1), 37–44. doi:10.15170/AR.2022.7.1.3.

Pegalajar Palomino, M. C. (2021). Implicaciones de la gamificación en Educación Superior: Una revisión sistemática sobre la percepción del estudiante. *Revista de Investigación Educacional*, *39*(1), 169–188. doi:10.6018/rie.419481

Peña-Acuña, B. (2022). Indagación evaluativa de una intervención con metodologías activas para estudiantes universitarios. *Revista de estilos de aprendizaje, 15*(29), 5-18. . doi:10.55777/rea.v15i29.4054

Ramos Navas-Parejo, M., Rodríguez García, A. M., & Gómez García, G. (2019). La gamificación en el aula universitaria. Una metodología activa e inclusiva. Percepciones por parte del alumnado universitario. In R. Roig Vila (Ed.), *Investigación e Innovación en la Enseñanza Superior* (pp. 669–678). Octaedro.

Ranilla, M. (2018). Gamificación de la alfabetización digital en mayores según los estilos de aprendizaje y actividades polifásicas. *Revista de estilos de aprendizaje, 11*(22). . doi:10.55777/rea.v11i22.1085

Robson, K., Plangger, K., Kietzmann, J. H., McCarthy, I., & Pitt, L. (2015). Is it all a game? Understanding the principles of gamification. *Business Horizons*, *58*(4), 411–420. doi:10.1016/j.bushor.2015.03.006

Ryan, R. M., & Deci, E. L. (2000). Self-determination theory and the facilitation of intrinsic motivation, social development, and well-being. *The American Psychologist*, *55*(1), 68–78. doi:10.1037/0003-066X.55.1.68 PMID:11392867

Sailer, M., Hense, J., Mandl, H., & Klevers, M. (2013). Psychological perspectives on motivation through gamification. *Ixd&a, 19*(1), 28-37.

Santamaría, A., & Alcalde, E. (2020). Una experiencia universitaria de gamificación en línea o en el aula presencial: ¿es este recurso de aprendizaje posible en ambos entornos? *Revista Brasileira de Lingüística Aplicada*, *20*(4), 761–786. doi:10.1590/1984-6398202016390

Santos, M. J., Leiva, J. J., Ramos, M., & Benítez-Márquez, M. D. (2020). Higher education students' assessments towards gamification and sustainability: A case study. *Sustainability (Basel)*, *12*(20), 8513. doi:10.3390u12208513

Tenorio-Rodríguez, M. A., González-Monteagudo, J., & Padilla-Carmona, T. (2021). Employability and inclusion of non-traditional university students: Limitations and challenges. *IAFOR Journal of Education: Inclusive Education*, *9*(1), 133–151. doi:10.22492/ije.9.1.08

Tham, C. M., & Werner, J. M. (2005). Designing and evaluating e-learning in higher education: A review and recommendations. *Journal of Leadership & Organizational Studies*, *11*(2), 15–25. doi:10.1177/107179190501100203

Trigueros, R., Aguilar-Parra, J. M., Cangas, A. J., Bermejo, R., Ferrandiz, C., & López-Liria, R. (2019). Influence of emotional intelligence, motivation and resilience on academic performance and the adoption of healthy lifestyle habits among adolescents. *International Journal of Environmental Research and Public Health*, *16*(16), 2810. doi:10.3390/ijerph16162810 PMID:31394722

Tschannen-Moran, M., Hoy, A. W., & Hoy, W. K. (1998). Teacher efficacy: Its meaning and measure. *Review of Educational Research*, *68*(2), 202–248. doi:10.3102/00346543068002202

Varea, R. (2021, May 17). Aulas virtuales y contenidos personalizados: Así será la Universidad en unos años. *El País*. Retrieved April 8, 2023, from https://elpais.com/sociedad/educacion-online/2021-09-21/aulas-virtuales-y-contenidos-personalizados-asi-sera-la-universidad-en-unos-anos.html

Vergara, D., Mezquita, J. M., & Gómez Vallecillo, A. I. (2019). Metodología innovadora basada en la gamificación educativa: Evaluación tipo test con la herramienta Quizizz. *Profesorado. Revista de Currículum y Formación de Profesorado*, *23*(3), 363–387. doi:10.30827/profesorado.v23i3.11232

Willig, J. H., Croker, J., McCormick, L., Nabavi, M., Walker, J., Wingo, N. P., Roche, C. C., Jones, C., Hartmann, K. E., & Redden, D. (2021). Gamification and education: A pragmatic approach with two examples of implementation. *Journal of Clinical and Translational Science*, *5*(1), e181. doi:10.1017/cts.2021.806 PMID:34849256

Wlodkowski, R., & Ginsberg, M. (2017). *Enhancing adult motivation to learn: A comprehensive guide for teaching all adults* (4th ed.). Jossey-Bass.

Woolfolk Hoy, A. (2013). A reflection on the place of emotion in teaching and teacher education. In Emotion and School: Understanding how the Hidden Curriculum Influences Relationships, Leadership, Teaching, and Learning (pp. 255-270). Emerald Group Publishing Limited. doi:10.1108/S1479-3687(2013)0000018017

Zamecnik, A., Kovanović, V., Joksimović, S., & Liu, L. (2022). Exploring non-traditional learner motivations and characteristics in online learning: A learner profile study. *Computers and Education: Artificial Intelligence*, *3*, 100051. doi:10.1016/j.caeai.2022.100051

KEY TERMS AND DEFINITIONS

Active Methodologies: They are educational approaches in which the learner is at the center of the teaching-learning situation. Some examples of active methodologies are project based learning, case-study or gamification, among others.

Digital Competence: It is the ability to confidently and critically use information and communication technologies for work, teaching and learning, or social use.

Gamification: It is the use of game-like elements in situations which are not ludic.

Motivation: It is a force that determines the reason why an individual starts and continuous a particular behavior, like, for example, learning.

Nontraditional Learners: Adult learners who enroll in higher education programs a few years after graduating and who usually have personal and professional obligations.

Teacher Self-Efficacy: It is a teacher's beliefs of their own ability to effectively handle any teaching tasks and obligations related to their profession.

Virtual Learning Environment (VLE): An online teaching and learning platform used by educational institutions to deliver academic content.

Chapter 4
Educational Innovation in Teacher Classrooms:
A Systematic Review

Maria del Pilar Díaz-López
https://orcid.org/0000-0002-0888-925X
Universidad de Almería, Spain

Ana M. Martínez-Martínez
Universidad de Almería, Spain

Nuria del Mar Torres López
Universidad de Almería, Spain

Noelia Navarro Gómez
Universidad de Almería, Spain

ABSTRACT

The teaching paradigm is undergoing a real revolution in addition the pandemic situation suffered from COVID-19. Therefore, both teachers and faculty must adapt and train so as to give an adequate response that improves the results and favors the motivation for learning. They are putting their focus on the permanent training of teachers and on educational innovation. One of the most current and innovative teaching innovation tools is gamification. Therefore, this study aims to analyze the existing scientific literature, carrying out a systematic review of the empirical studies that show experiences carried out in university classrooms, especially in careers related to teaching in different specialties. Discarding the texts that did not meet the eligibility criteria, six articles were selected that are committed to teaching innovation and the practice of educational strategies such as gamification in university classrooms to promote meaningful learning as well as the involvement and motivation of students towards learning.

DOI: 10.4018/978-1-6684-8861-4.ch004

INTRODUCTION

The educational system is facing an unprecedented challenge, primarily because society is evolving faster than the system itself and more rapidly than teachers can manage. The students who currently attend lessons do not resemble the students of twenty years ago. One of the most important explanations is the impact of the COVID-19 pandemic, which greatly accelerated the process of change at all educational levels. During the pandemic, all educational centers, from nursery schools to universities, students, families, teachers, and teaching staff had to adapt to a situation that they had not experienced before, leading to the adoption of alternative teaching methods. For instance, telematic teaching was established at all educational levels, with new modalities being implemented, such as e-learning, m-learning, and b-learning (Davies, Cotton & Korte, 2016; Torres-López, Díaz-López and Méndez, 2020; Keskin & Yurdugül, 2020).

Both the impossibility of attending class in person and the new system of teaching remotely allowed educators to reflect on the contents and methods that were most suitable to telematic teaching, leading them to reconsider aspects such as the role of teachers themselves. This change opened the possibility to modify the methods, materials, and formats that, until not long ago, had only been used in more flexible traditional lessons, as new forms of evaluation, student participation, online group work, etc. had become necessary given the circumstances.

All the repercussions involved in the pandemic aside, society and students who attend universities today need tools, strategies and skills that will make them resilient, creative, entrepreneurial, and versatile in a constantly changing world in which professional opportunities are more diversified and innovative. In short, they must be able to evolve in a world that is very different from what we were used to in the past.

The students who attend schools nowadays are still seriously affected by the consequences of COVID-19 in all areas (social, linguistic, motor, psychological). The pandemic has affected everyone, and future professionals dedicated to teaching must receive adequate training that makes it easier for them to implement the tools and strategies that they will need to face their own students in the classroom. That is why it is necessary for all teachers, from early years to higher education, to adapt and train for the challenges of the future.

In this regard, organizations, governments, teaching centers, etc. are focusing on both the permanent training of teachers and educational innovation. One of the latest and most innovative teaching innovation tools is gamification, which is understood as "the application of game resources (design, dynamics, elements, etc.) in non-recreational contexts to modify the behavior of individuals through actions on their motivation" (Teixes, 2014, p. 23).

Hence, the aim of this study is to identify and analyze empirical studies on the use of innovative methodologies, specifically the use of gamification as a teaching and learning tool in university classrooms in master's degree courses. This subject is of particular relevance since the students in these programs will someday be teaching future generations of pupils.

To conduct this systematic review, four databases were consulted: Web of Science; Scopus; ERIC and Dialnet, following the strictest standards of rigor in the search for information. The search period was limited to works dating between the years 2015 and 2023. The keywords chosen to carry out the review were the following: "higher education", "gamification", "innovation" and "primary school", using the Boolean operators, AND/OR, in order to delimit the information necessary for the object of study.

After excluding the texts that did not meet the eligibility criteria, six articles were selected. The articles focus on teaching innovation and the practice of educational strategies such as gamification in university classrooms to promote meaningful learning, as well as the involvement and motivation of students towards learning.

BACKGROUND

Teaching Innovation, and Transformation of Teaching and Learning

In the last decade, the teaching paradigm has undergone a veritable revolution, and the students who currently attend classes do not resemble those of twenty years ago. In addition, the COVID-19 pandemic greatly accelerated the process of change at all educational levels (Alvarez-Herrero & Fernández-Herrero, 2020). During the pandemic, all educational centers, from nursery schools to universities, as well as students, families, and teachers, had to adapt to a situation that they had not experienced before, leading to the adoption of alternative teaching methods. Telematic teaching was established at all educational levels, implementing new modalities such as e-learning, m-learning and b-learning (Davies, Cotton & Korte, 2016; Torres-López, Díaz-López & Méndez, 2020; Keskin & Yurdugül, 2020). In many cases, these new methodologies have remained as teaching tools, such as in most teacher training courses. The delivery of distance learning led teachers to reflect on what content and methods were most appropriate to carry out this type of telematic teaching, reconsidering aspects such as the role of teachers themselves, the prioritization and selection of content, tools and instruments, evaluation, etc. This change opened the possibility of making all aspects of the teaching process more flexible (i.e., methods, materials, formats, evaluation). Until that point, all aspects of teaching had only been used in traditional classes. In the words of García-Alons et al. (2023):

It is a fact that nowadays technology has invaded our lives and is gradually modifying the requirements of university students, which is why university educational innovation is a necessity to respond to these new social requirements, implying a deep change in the mentality of teachers and students. In contrast to teaching approaches centered on the lecturer and the master class, the European Higher Education Area proposes a teaching model centered on the student autonomous activity and participation in learning. (p. 1)

Considering this perspective, and in light of the circumstances of the past several years, society and students who attend universities today need tools, strategies and skills that make them resilient, creative, entrepreneurial and versatile in a constantly changing world, in which professional profiles are more diversified and innovative ("drone pilot, cybersecurity specialist or being an architect expert in renderings and 3D designs, E-commerce specialists, digital marketing experts or copywriting profiles and content creators"). In short, they must be able to evolve in a world that is very different from that of previous generations (Garagui, 2021; René, 2022).

As mentioned before, both teachers and teaching staff must adapt and train constantly. In doing so, they transform the paradigm of current education by working to improve results and promote motivation for learning, given that the performance of a teacher in the classroom directly influences the academic performance of their students (Fierro-Suero, et al., 2019; Ocampo, et al., 2018; Trigueros et al., 2019).

In response to this need for change, governments, teaching centers and associations are committed to training in educational innovation. Such actions are supported by a multitude of scientific articles that continue to insist that teaching innovation must exist in the classroom (Manzano, Sánchez, Trigueros, Álvarez and Aguilar-Parra, 2020; Moreno 2019, Páramo, 2020). However, it must be noted that educational innovation does not only imply the inclusion of information and communication technologies or learning technologies in the classroom, but instead it is much more complex. More specifically, educational innovation produces a change in teaching that is based on four essential elements: people, knowledge, processes, and technology. These four elements orchestrated together will influence the success or failure of the innovation. In the words of Tello & Aguaded "the incorporation of ICTs means much more than providing the centers with equipment and infrastructures: it is also necessary to encourage and develop the ability to reflect on the information received" (2009, p. 41).

There are many innovative tools and strategies that are regularly implemented daily in classrooms of all educational levels around the world. Some of these methods include escape room (consists of students having to solve a series of tests and enigmas to be able to escape from a closed space, for example, a classroom). As described in Martínez et al. (2023, p. 715), "students considered the escape rooms a motivating way to acquire knowledge and consolidate concepts, appreciated the time dedicated by professors to design the games and encouraged them to create escape rooms for the following topics"). In this sense, Sierra & Fernández-Sánchez, in their study "Gamifying the University classroom. Analysis of an Escape Room experience in higher education" show similar results: "the results show an improvement of the motivation of the students with the application of escape room in the classroom" (2019, p. 1).

Unlike escape games, in Breakout Edu (Cornellà et al., 2020) the purpose is to open or unlock a lock, a box or chest after also solving a series of challenges or puzzles. Breakout Edu could be a test within an escape game. Some of the tests that are usually carried out are: ordering words or letters; translating secret phrases; crosswords; encrypted messages; mazes; solving equations or problems; or searching for clues with ultraviolet light. Another one of the latest teaching innovation strategies is the flipped classroom (Hodgson et al., 2017; Hung et al., 2019), where students become the protagonists of learning, requiring them to prepare the subject agenda outside of class to later carry out other types of tasks or activities in the classroom, such as participating, raising questions, collaborating as a team, organizing and planning to carry out projects or solve problems, presenting what has been learned, etc. In addition, project-based learning (Miller et al., 2021; Toledo & Sánchez, 2018; Rugh et al., 2021; Wan et al., 2020; Warin et al., 2016) is a strategy that implies a task-oriented teaching and learning model. It involves a collaborative process of consultation between the participants with the main objective of achieving a final product. Apart from these innovations, an active commitment to service-learning can be observed (Chiva-Bartoll & Fernández-Río, 2021; Courville et al., 2021; Diaz, A., 2021; Francisco & Moliner 2010). It consists of an educational proposal that combines learning processes and community service within a well-structured framework in which participants learn by actively working to address the real needs of the environment in order to improve it. Finally, we wish to highlight two very innovative strategies, namely game-based learning and gamification, the latter being the focus of the present study.

The use of these strategies in the classroom offers a series of benefits:

- Any type of curricular content is susceptible to adaptation to these methodological strategies. They can be used to work on any type of content as well as to develop procedures and attitudinal learning (Negre i Walczak, 2017; Pérez Vázquez et al., 2019).

- They serve to improve linguistic competence, since students must decipher messages, structure texts, engage in conversations, speeches, debate, or orally present content (Vesga et al., 2021).
- They allow improvement within STEAM competence as well as information processing and digital competence: it is necessary to work on reasoning, discern between what is important and what is secondary, solve challenges and calculations or operations (Spinales, 2018).
- They promote values such as fellowship, solidarity, collaboration, respect, perseverance, patience, and teamwork (Ausín et al. 2016; Saz & Ramo, 2015).
- They facilitate the principles of activity (Sánchez-Rodríguez, Ruíz-Palmero & Sánchez-Vega, 2017) and globalization, as well as allowing meaningful learning.
- They develop deductive thinking as they allow learning to be generalized and applied to daily life (López, 2018).
- They promote autonomy and allow students to be the main protagonists of their own learning (Vergara et al., 2019).
- Lasting learning over time (UNIR, 2020).
- Deep learning (Toledo & Sánchez, 2018).
- Greater student motivation (Galeana, sf; Martínez et al. (2023); Sierra & Fernández-Sánchez (2019)
- Improved socializing capacity (Caparrós, 2017; Negrey Carrión, 2020).

Gamification

"Gamification" consists of "applying game strategies (thoughts and mechanics) in non-playable contexts, unrelated to games, in order for people to adopt certain behaviors" (Ramírez, 2014).

On many occasions, gamification is confused with game-based learning, but they are different concepts and methodologies. Game-based learning (ABJ) is based on the continuous use of games in teaching as a resource to develop a certain competence or teach specific content (Cornellà et al., 2020). There are a multitude of commercial games that can be used for all areas of learning. These commercial games can be adapted, or new ones invented following the guidelines of any given game. Several applications have even been created to this end. Various types of games can be used, such as video games, virtual reality games, augmented reality games and board games. In contrast, gamification differs from game-based learning in that gamification aesthetics and narrative are designed, making use of game components such as cards, prizes, points, medals, badges, levels, avatars, characters, goals, and strategies.

We can consider that gamification is a more complex activity than applying a game. A deep reflection on the objectives to be achieved is necessary: once determined, the rules that will govern the process will be established. Therefore, carrying out a gamification project requires thorough planning and can sometimes be met with resistance to its implementation. (Ortíz- Cólon, Jordán & Agredal, 2018, p. 13)

While an important aspect of gamification is having fun during the development of the game itself, its main purpose is to influence people's behavior (Hamari and Koivisto, 2013, Ramírez, 2014).

There is a growing tendency to include recreational elements in different fields, such as education, business, and human resources. Gamification is being used as a motivating element for employees to achieve proposed objectives (Molina, 2023), and there are numerous examples of companies and brands that implement this approach, for example, Nike with its Nike Fuel application, BBVA, Correos (Spain's postal service) or Samsung.

The use of gamification as an innovative teaching practice has been analyzed by the scientific community (Minovié. Et al., 2016; Sanchez, 2015) for different reasons. Research has been conducted in order to clarify its benefits and to determine whether its application is advisable (or not) in university classrooms, as well as to evaluate the suitability of various applications and online platforms created in this line, such as Kahoot!, ClassDojo and Classcraft (Barahona Mora, 2020; da Rocha, L. et al., 2016; Dillon et al., 2019). Landers (2014) highlights that its use does indeed improve results in education. Moreover, gamification incorporates attractive and motivating environments designed to appeal to students and gain their involvement, thereby obtaining better results (Borras, 2015; Beltrán & Romero-Riaño, 2020; Ortíz-Colón, Jordán & Agredal, 2018). Indeed, it stimulates reflective critical learning and promotes student motivation, as well as meaningful learning (Vega, M. & Guerrero, G. 2023). The introduction of these gamification strategies into university classrooms is especially important in teaching degrees. For this reason, the present study focuses on this particular area, as future teachers need to be familiar with this type of strategy, and what better way to achieve this than with first-hand experience. Manzano and Arrifano (2022) reaffirm this point of view in their study:

There is a need for initial teacher training to show different teaching-learning methodologies so that students are capable in their professional future of generating transformative and effective educational proposals to meet the needs of their classroom, thus contributing to an improvement in the quality and equity of the educational system.

However, it must be taken into consideration that the use of a single methodology, however effective it may be, can lead to the opposite effect and fall into reductionism, maintaining the structure of the traditional qualifications/awards/rewards system (Fasce, 2018).

Aim

To know and analyze empirical studies on the use of innovative methodologies, specifically, the use of gamification as a teaching and learning tool in university classrooms in master's degrees.

METHODOLOGY

This study presents a brief overview of the main research that has been carried out in recent years on educational innovation strategies, specifically on the use of gamification as a teaching and learning tool in different subjects of the teaching degrees taught at universities, preferably in Spain. For this purpose, several tasks of a diverse nature were performed, the first of which was a thorough search in the main education databases, followed by a descriptive systematic review on the empirical studies found. Systematic reviews aim to provide, in a single document, a complete and unbiased synthesis of various relevant studies published in databases of greater recognition and scientific impact using rigorous and transparent methods. Moreover, they apply explicit methods to minimize bias, thus providing more reliable results from which conclusions can be drawn and decisions made, obtaining a reliable perspective of the current state of affairs of a given subject.

To carry out this systematic review, following the strictest standards of rigor in the search for information, four main databases were consulted which are references of excellence in scientific literature: Web of Science; Scopus; ERIC and Dialnet. The search period was limited to works dating between 2015 and 2023. The keywords chosen to carry out the review were the following: "higher education", "gamification", "innovation" and "primary school", using the Boolean operators. AND/OR to delimit the information necessary for the object of study. The selected eligibility criteria were the following:

- Studies published between 2015-2023.
- Empirical works.
- Articles published in English and Spanish.
- Open access and full text.

Direct relationship with the search elements, excluding those that had only one of the requested criteria. These eligibility criteria allowed for an exhaustive selection to be made from the results obtained, which was 19 works. Of the 19 articles found, 3 were rejected with the first reading of the abstract and an additional 10 were rejected after reading the full text and applying the eligibility criteria.

RESULTS

Of the four databases consulted, a total of 19 results were extracted, of which 6 were ultimately selected after applying the inclusion and exclusion criteria established.

From the selected studies, positive conclusions were obtained that support teaching innovation and the practice of educational strategies such as gamification in university classrooms to promote meaningful learning, as well as the involvement and motivation of students towards learning.

Next, a summary table is presented in which the most important aspects (Authors; Year; Qualification, Sample; Research characteristics; Goals and Results) are indicated from each of the studies selected following the search. All the works are in descending order by date of publication.

DISCUSSION

Many studies have been carried out on educational innovation and gamification, however, the existing scientific literature to date regarding gamification experiences in teaching courses is still scarce. However, after the bibliographic review of the databases consulted, it is observed that existing studies show positive results and conclusions for innovation and the practice of gamification in university classrooms.

Table 1. Results, most important aspects of the studies selected after the search

Authors (Year)	Qualification	Sample	Research Characteristics	Goals	Results
Gómez, IM, Ruíz, M., Ortega, D., García-Valero, B., Penalva, J., Sagrelles, JA & Formoshino, MD (2017)	Transmedia proposals and gamification applied to the EHEA: New active methodologies to implement competences in Social Sciences and Literature	Collaboration was maintained with various centers and departments of the areas involved in the Network, such as Social Sciences, Geography Degree); Language and Literature (Didactics of Language and Literature and the department of Spanish Philology, General Linguistics and Theory of Literature) from the University of Alicante and from other universities (University of Burgos; Universidade Portucalense Infante D. Henrique from Porto).	Implementation of didactic proposals based on interdisciplinary work, where the focus was on the importance of ICTs and transmedia, as well as the use of gamification, for the development, acquisition and attainment of the contents and competences of the different higher education programs.	Analyze, verify, and demonstrate the need for methodological transformation used in undergraduate studies, in initial teacher training, as well as the essential incorporation of ICTs, transmedia and gamification in EA processes.	The didactic proposals and the educational action of said Research Network have been aimed at Social Science, Language and Literature; the curricular contents have been worked on indicated in the respective study plans, as well as the key competences and ICT. This type of classroom intervention provides new keys for the acquisition of interdisciplinary content, as well as the digital competence of students, in addition to allowing future teachers to increase the motivation of students
Ortíz-Colón, A., Jordán, J. & Agredal, M. (2018)	Gamification in education: an overview about the state of the art.		Qualitative methodology based on the analysis of content, selecting relevant content from scientific publications on the subject, found in databases and repositories, such as EBSCOHost, Proquest, Web of Science, Scopus, ScienceDirect, Google Scholar, ACM Digital library.	The aim of this paper is to review the benefits of the use of gamification and its application in education.	37 gamification experiences carried out in different contexts in the last six years were analyzed, each of them with different characteristics, whose common axis is the benefit of its application in education, the motivation of students in learning and the incorporation of game elements in the educational activity. To this end, a catalog of the five selected studies is presented, on the one hand, and an analysis of them, on the other, following the three categories of game elements defined by Werbach and Hunter (2012). The principles and benefits that gamification presents in different areas are established, as well as some of the risks and factors to take into account when gamifying.
Moreno, E. & Lopezosa, MD (2020)	Gamification through a service-learning project: designing an educational Breakout from the university for primary school students.	Students studying degrees in early childhood education and primary education at the Sagrada Familia University Center, as well as primary school students.	Descriptive study of the implementation of the service-learning methodology and gamification in groups of students in early years and primary grades.	Educational experience that was intended to motivate and provide a firm base that would promote the involvement of university students in performing tasks similar to those they will carry out in the performance of their profession as primary and early years education teachers.	The experience allowed the student body to participate in their own learning process and investigate the implementation of gamification in primary education. They developed digital skills and knowledge, responsibility, didactic programming skills and the ability to correct their mistakes. Clear benefits of the service-learning experience and gamification are observed. Participatory, interactive, and gamified activities favor an improvement in learning and are highly valued by the participating students.

continues on following page

Table 1. Continued

Authors (Year)	Qualification	Sample	Research Characteristics	Goals	Results
Moreno, E., Perales, R., Hidalgo, J. (2019)	Qualitative study on the use of gamification in higher education to promote motivation in students.	21 students in the third year of the early childhood education teaching degree at the Sagrada Familia University Center.	It is an action research study. An ad hoc survey related to motivation and validated by experts was used.	Plan motivating activities that they assumed an advance in the knowledge of the students; design and put into practice a Breakout Edu for children in early childhood education, by students of a teaching degree in early childhood education and, finally, promote and increase motivation for learning the contents of the university subjects taught and involved in the investigation project.	The students maintain high motivation throughout the entire project, from the very beginning to the end. Their motivation was mainly due to the methodology used and the design of gamification practices. In addition, the students point out that the fact of being able to carry out the practice and not just simulate it increased their motivation and involvement throughout the project.
Alsina, M. & Farrés, I. (2021)	Playing or learning? Playful learning in the musical training of teachers.	Group of second-year students (n = 90) from a Spanish university.	Following a mixed, qualitative and quantitative case study methodology, different aspects of the designed activities are coded and evaluated in order to be analyzed. The sample comprises a regular group of second year students (N=90), divided into three sub-groups of a similar size (A1, A2 & A3), enrolled in the second year of the degree program at a Spanish university. Each of the three sub-groups was then divided into random working groups of five members. This meant that the teacher-researcher was able to maintain regular practice in the classroom while also acting as an investigator, without changing the usual dynamics of the course or the distribution of the pre-established groups, and presenting a report at the end of the process answering the research questions set.	Overcome discouragement; provide opportunities for deep thought and reflection; change behavior to positive attitudes; promote real and contextualized practices.	Games as a resource are highly valued and used in schools considering the benefits highlighted up to now. This kind of practice, produced in collaboration with schools, and with a final application with the target students, provides an excellent opportunity to develop specific professional competencies. The study shows that gamification in teacher training degrees is a suitable and effective way of working on the practical application of the basic musical content that should be acquired by future generalist teachers. At the same time, students' initiative and creativity is also promoted, fostering connections with the social reality, critical thinking, as well as links with their future professional context. Finally, the results highlight the fact that game-based music learning which follows the strategies developed in the practice analyzed encourages the school children who carry out these activities to self-regulate their own learning.
Cangalaya, L. & Casazola-Cruz, OD & Farfán, JA (2022)	Gamification in the teaching and learning process of university students.	30 students of the robotics course at the National University of Callao.	Confirmatory research that examined the correlational analyse to determine the relationship between the variables of gamification and the teaching-learning process. The design used was quasi-experimental.	Implement gamification to improve the teaching-learning process in a distance course based on the perception of the participating students.	It was shown that the implementation of gamification improves the teaching-learning process in the studied sample. Knowing what students think and how they assimilate this new way of learning allows for more precise results and clearer conclusions about the importance of gamification in improving teaching-learning, especially in the case of university students. The results also indicated that these students showed interest in gamification in higher education.

The study by Gómez Trigueros, Isabel María, et al. (2017) consists of the presentation of a project of participatory collaboration in teaching in the Degrees of Early Childhood Education and Primary Education as well as in the Degrees of Language and Literature, Geography and in the master's degree course of Secondary Education. This proposal was approached from different learning areas and education departments of several universities (University of Alicante; University of Burgos; Universidade Portucalense Infante D. Henrique de Porto). The objective pursued by said research was to study, verify and demonstrate the need for a methodological transformation of teaching in university classrooms, especially in undergraduate studies, which is the primary focus of the present review. In addition, the aforementioned project analyzed the inclusion of ICTs, transmedia and/or gamification as fundamental tools to achieve this transformation and with it improve the teaching and learning processes of university students.

For their part, Moreno, Perales and Hidalgo (2019) carried out an action-research project with third-year students of an Early Childhood Education Teaching degree. A total of 21 students participated in this research. The project had the objective of including teaching innovation practices and gamification in the teaching of the degree, with the proposal that university students create a Breakout Edu for young students. After the implementation, a qualitative analysis of the results was carried out, noting the motivation of the students before, during and after the educational practice. No major differences were found in the levels of motivation between the three moments, but it is important to note that the level of motivation of the participating students was high throughout the entire process. They conclude that the use of gamification in teaching degree classrooms not only allows improving the teaching and learning process, as Cangalaya-Sevillano, Casazola-Cruz, & Aguilar (2022) highlight, but also does so with high motivation of the students.

In line with the previous study, Moreno and Lopezosa (2020) carry out a new investigation at the same university, Centro Universitario Sagrada Familia, with the students studying degrees specializing in early childhood education and others specializing in primary education. The project, which included a service-learning element, intended that university students could collaboratively design and implement escape games created on a website with Google Sites that included a variety of tests, puzzles, and riddles related to the curricular content of primary education. Later, the games could be sent to an educational center and put into practice using learning technologies. After the implementation phase, and the analysis of results, the authors concluded that the experience was a success for both the students of the two grades and for the primary school students of the educational center. They highlight in their discussion that not only motivation is important, but also the search for solutions to a problem, cooperative work and following the rules of the game. Moreover, the activity facilitated more effective learning.

In their study carried out at the University of Jaén, Agreda-Montoro, Ortiz-Colon & Rodriguez (2020) show the benefits of implementing an escape game project in the Degrees of Social Education and Early Childhood Education. Their conclusions are related to the belief that playful experiences are a good strategy for acquiring skills in the educational field, as well as an opportunity to teach and learn using ICTs. As they themselves explain, "Escape rooms promote communication, collaborative and cooperative work, the development of critical thinking and lateral thinking, as well as the need to pay attention to detail. In addition, any individual can access them, regardless of age".

Another of the investigations conducted in the university field is the study by Alsina & Farrés (2021), in which they present mixed research (qualitative/quantitative) of a descriptive/interpretative type, aiming to study and examine three aspects, which are: the relationships that occur between the research participants; the curricular contents of the subject of music, finally, the game strategies used in the study. The sample consisted of 90 students in the second year of the Primary Education Degree, in the subject of music education. The research consisted of several phases, from the training of undergraduate students in game experiences and methodology to the creation by the students themselves of a gamified experience which they would initially put it into practice among themselves. Thus, it could be tested in order to correct and add modifications and improvements to later be applied with a group of primary school students from a collaborating center with the assistance of faculty. This study shows positive results for the use of gamification for teaching different content, in this case the elementary contents that all primary education teachers must master. The findings demonstrate the effectiveness in the motivation and learning of future teachers in the acquisition of this subject, which is generally characterized by a low self-perception of mastery and efficacy. The experience also allowed the development of creativity and initiative in the participants since the application was carried out in a real context, similar to future teacher practice in the classroom. Finally, it should be noted that self-regulation of learning by students is promoted, which is a useful strategy to apply the contents of this teaching.

Finally, Cangalaya-Sevillano, Casazola-Cruz, & Aguilar (2022) conducted confirmatory research with which they intended to analyze the correlation between gamification and the teaching-learning process. It was a quasi-experimental design with a sample of 30 students from the robotics course at the National University of Callao. The results reveal a positive correlation between both variables, that is, the use of gamification as a teaching and learning strategy is a success, in addition, the students showed an attitude of involvement and participation, which had a positive impact on the results obtained.

CONCLUSION

The natural evolution of society, advances in technology and new professional perspectives, which were all affected by the crisis caused by the COVID-19 pandemic, indicate that the educational paradigm must adapt to changes in modern-day life in order to provide effective methods to future generations. The use of innovative strategies, such as gamification in the classroom of all educational levels but especially in university classrooms, reports benefits in learning and motivation. Moreover, it improves creativity and active participation, as demonstrated by the analysis of the articles found. The first step for the education system to change is to implement these methodologies in the initial training of teachers, that is, in the education degrees of all specialties. In this way, future teachers could learn first-hand what they must later execute during their day-to-day routine in the classroom. In addition to this, it is necessary to invest in ongoing training and encourage active teachers to include these methodologies in their syllabuses. However, there is still a great deal of work to be done in this regard as all changes are received with some degree of resistance that must eventually be overcome. The articles found show several experiences related to teaching innovation and the use of game experiences in university classrooms, however, the experiences of gamification in its broadest concept are very scarce, finding a greater number of Breakout Edu projects or escape room games, not only in teaching degrees but in many others.

Therefore, more research in this field would be essential, as well as the implementation of education/research projects, that is, action research that is carried out in parallel at universities and educational centers. It is essential that connections and relationships be strengthened between universities and schools where work groups are organized for current and future teachers in a way that they can be trained and, all those involved—students and teachers from both institutions—can benefit each other.

The limitations of this study are those of a systematic review, as it will always be limited by the quantity and quality of the studies included, as well as the databases consulted or even regarding the range of years considered for the study. From our point of view, it is quite necessary to continue researching in this line by carrying out research-action studies.

REFERENCES

Agreda-Montoro, M., Ortíz-Colón, A. M., & Rodríguez, J. (2020). Gamificación y competencia digital Escape Rooms para la formación de los profesionales educativos. La tecnología como eje del cambio metodológico, 693-696.

Alsina M. & Farrés, I. (2021). ¿Jugar o aprender? El aprendizaje lúdico en la formación musical del maestro. *Revista electrónica complutense de investigación en educación musical, 18*, 83-96. . doi:10.5209/reciem.67853

Alvarez-Herrero, J. F., & Fernández-Herrero, J. (2020). *Aprendizaje online versus presencial. Percepciones del alumnado del grado de Magisterio en Educación Infantil.* Octaedro.

Ausín, V., Abella, V., Delgado, V., & Hortigüela, D. (2016). Project-based learning through ICT: An experience of teaching innovation from university classrooms. *Formación Universitaria, 9*(3), 31–38. doi:10.4067/S0718-50062016000300005

Barahona, A. (2020). Gamification for classroom management: An implementation, using ClassDojo. *Sustainability (Basel), 12*(22), 9371. Advance online publication. doi:10.3390u12229371

Beltrán, A., & Romero-Riaño, E. (2020). El papel de la gamificación en la conciencia ambiental: Una revisión bibliométrica. Evolución de la estructura intelectual. *Prisma Social, 30*, 161–185. https://revistaprismasocial.es/article/ view/3764

Borras, O. (2015). *Fundamentos de la gamificación.* Universidad Politécnica de Madrid.

Cangalaya-Sevillano, L. M., Casazola-Cruz, O. D., & Aguilar, J. A. F. (2022). Gamificación en el proceso de enseñanza y aprendizaje de estudiantes universitarios. *Horizontes revista de investigación en ciencias de la educación, 6*(23), 637-647. . doi:10.33996/revistahorizontes.v6i23.364

Caparrós, M. (2017). *Gamificación en educación: Una guía práctica.* Retrieved from https://www.trespuntoelearning.com/gamificacion-en-educacion-guia-practica/

Chiva-Bartoll, O., & Fernández-Rio, J. (2021). Advocating for service-learning as a pedagogical model in physical education: Towards an activist and transformative approach. *Physical Education and Sport Pedagogy*. Advance online publication. doi:10.1080/17408989.2021.1911981

Cornellà, P., & Meritxell, B. (2020). Gamificación y aprendizaje basado en juegos. *Enseñanza de las ciencias de la tierra, 28*(1), 5-19.

Courville, K. A., Lowe, A., Dannelley, B., Sellers, J., & Deal, B. (2021). Learning-by teaching and service learning to promote bleeding control education: An academic-community partnership. *Nurse Educator*. Advance online publication. doi:10.1097/NNE.0000000000001038 PMID:33958559

Da Rocha Seixas, L., Gomes, A. S., & de Melo Filho, I. J. (2016). Effectiveness of gamification in the engagement of students. *Computers in Human Behavior*, *58*, 48–63. doi:10.1016/j.chb.2015.11.021

Davis, L. (2020). Don't hide the hurt: Using storytelling for trauma-informed teaching. In A. Ellsworth (Ed.), *Cases on emotionally responsive teaching and mentoring* (pp. 13–26). IGI Global., doi:10.4018/978-1-7998-2971-3.ch002

Daza, Caridad, & Fernández-Sánchez. (2019). Gamificando el aula universitaria. Análisis de una experiencia de Escape Room en educación superior. *Revista de estudios y experiencias en educación, 18*(36), 105-115. doi:10.21703/rexe.20191836sierra15

Diaz, A. (2021). Pandemic times: Knowledge management and technology in service learning. *Revista Virtual Universidad Católica Del Norte*, *63*, 1–4. doi:10.35575/rvucn.n63a1

Dillon, M. B., Radley, K. C., Tingstrom, D. H., Dart, E. H., & Barry, C. T. (2019). The effects of tootling via ClassDojo on student behavior in elementary classrooms. *School Psychology Review*, *48*(1), 18–30. doi:10.17105/SPR-2017-0090.V48-1

Espinales, A. (2018). Gamificación en el desarrollo de la competencia matemática Plantear y Resolver Problemas. *La revista científica del ITSUP, 12*(1), 1390-9770.

Fasce, P. (2018). Portare il gioco a scuola, non la gamification. *Educazione Aperta,* 4.

Fierro-Suero, S., Almagro, B.J., & Sáenz-López, P. (2019). Necesidades psicológicas, motivación e inteligencia emocional en Educación Física. *Revista electrónica interuniversitaria de formación profesorado, 22*, 167-186.

Francisco Amat, A., & Moliner Miravet, L. (2010). El aprendizaje servicio en la Universidad: una estrategia en la formación de ciudadanía crítica. *Revista electrónica interuniversitaria de formación del profesorado, 13*(4), 69-77.

Galeana de la O. L. (n.d.). *Aprendizaje basado en proyectos*. Retrieved from http://ceupromed.ucol.mx/revista/PdfArt/1/27.pdf

Garagui, E. (2021). *Las profesiones del futuro, la nueva generación de empleos.* https://www.linkedin.com/pulse/las-profesiones-del-futuro-la-nueva-generaci%C3%B3n-de-empleos-everylgbt?trk=public_post-content_share-article

Garcia-Alonso, A., Mateos-Aparicio, I., Pérez-Rodríguez, M. L., Sánchez-Paniagua, M., Rodriguez-Rodriguez, E., Moreno-Guzmán, M., Sanz Landaluce, J., Rueda, C., Gallo, A., & De la Peña-Armada, R. (2023). Gamificación y desarrollo de contenidos virtuales para la dinamización del aula y mejora del aprendizaje universitario. *Actas INTED*, 788–791. Advance online publication. doi:10.21125/inted.2023.0252

Hamari, J., & Koivisto, J. (2013). Social motivations to use gamification: An empirical study of gamifying exercise. *Proceedings of the 21st European Conference on Information Systems*.

Hodgson, T. R., Cunningham, A., McGee, D., Kinne, L., & Murphy, T. J. (2017). Assessing behavioral engagement in flipped and non-flipped mathematics classrooms: Teacher abilities and other potential factors. *International Journal of Education in Mathematics Science and Technology*, *5*(4), 248–261. doi:10.18404/ijemst.296538

Hung, C. Y., Sun, J. C. Y., & Liu, J. Y. (2019). Effects of flipped classrooms integrated with MOOCs and game-based learning on the learning motivation and outcomes of students from different backgrounds. *Interactive Learning Environments*, *27*(8), 1028–1046. doi:10.1080/10494820.2018.1481103

Keskin, S., & Yurdugül, H. (2020). Factors affecting students' preferences for online and blended learning: Motivational vs. cognitive. *European Journal of Open Distance and E-Learning*, *22*(2), 72–86. doi:10.2478/eurodl-2019-0011

Landers, R. (2014). Developing a theory of gamified learning: Linking serious games and gamification of learning. *Simulation & Gaming*, *45*(6), 752–768. doi:10.1177/1046878114563660

López, P. (2018). E*l Aprendizaje Basado en Proyectos en la enseñanza de la historia. Propuesta de dos unidades didácticas para la Educación Secundaria Obligatoria* [Master's thesis]. Retrieved from http://file:///C:/Users/danij/Downloads/tfm_2017-18_MFPR_plm970_1987.pdf

Manzano León, A., Arrifano Tadeu, P. J. (2022). Escape Rooms educativos: una experiencia en una universidad portuguesa. *International journal of developmental and educational psychology INFAD revista de psicología, 1*(1), 281-288. . doi:10.17060/ijodaep.2022.n1.v1.2374

Manzano León, A., Sánchez, M., Trigueros Ramos, R., Álvarez Hernández, J., & Aguilar-Parra, J. M. (2020). Gamificación y Breakout Edu en formación profesional. El programa "Grey Place" en integración social. *EDMETIC*, *9*(1), 3–20. doi:10.21071/edmetic.v9i1.12067

Martínez, E., Carbonell, V., Álvarez, J., Flórez, M., & Montoya, E. (2023). Let's escape! the use of a digital Escape Room to increase student's motivation and learning on kinematics concepts. *INTED2023 Proceedings.* 10.21125/inted.2023.0236

Miller, E. C., Reigh, E., Berland, L., & Krajcik, J. (2021). Supporting equity in virtual science instruction through Project-Based Learning: Opportunities and challenges in the era of COVID-19. *Journal of Science Teacher Education*, *32*(6), 642–663. Advance online publication. doi:10.1080/1046560X.2021.1873549

Minović, M., García-Peñalvo, F. J., & Kearney, N. A. (2016). Gamification ecosystems in engineering education. *International Journal of Engineering Education*, *32*(1B), 308–309.

Molina, P., & Ruiz, M. C. (2023). *La gamificación como herramienta motivadora en la empresa*. https://hdl.handle.net/10953.1/19582

Moreno, E. y Lopezosa, M.D. (2020). Gamificación a través de un proyecto de aprendizaje-servicio: diseñando un Breakout educativo desde la universidad para el alumnado de primaria. *Revista científica electrónica de Educación y Comunicación en la Sociedad del Conocimiento, 20*(1).

Moreno Fuentes, E., Perales Molada, R., & Hidalgo Navarrete, J. (2019). *Aula de encuentro: revista de investigación y comunicación de experiencias educativas*. Academic Press.

Negre, C., & Carrión, S. (2020). *Desafío en el aula. Manual práctico para llevar los juegos de escape educativo a clase*. PAIDÓS Educación.

Negre i Walczak, C. (2017). *'Breakout Edu', microgamificación y aprendizaje significativo.* https://www.educaweb.com/noticia/2017/07/26/breakoutedu-microgamificacion-aprendizaje-significativo-15068/

Ocampo, F., De Luna, R., & Pérez, M.G. (2018). Factores de relación entre la inteligencia emocional y el rendimiento académico en estudiantes de la ESCOM del IPN. *Revista electrónica cuerpos académica grupos investigación,* 5.

Ortíz-ColónA.-M.JordánJ.AgredalM. (2018). Gamification in education: an overview on the state of the art. *Educ. Pesqui.* 44. https://doi.org/ doi:10.1590/S1678-4634201844173773

Páramo Rengifo, C. A. (2020). Gamificación e innovación educativa: apuntes para la práctica. *Revista Conrado, 16*(S1), 241-248. Retrieved from https://conrado.ucf.edu.cu/index.php/conrado/article/view/1547

Pérez Vázquez, E., Gilabert Cerdá, A., & Lledó Carreres, A. (2019). Gamificación en la educación universitaria: El uso del escape room como estrategia de aprendizaje. In R. Roig-Vila (Ed.), *Investigación e innovación en la Enseñanza Superior. Nuevos contextos, nuevas ideas.* Octaedro.

Ramírez, J. L. (2014). *Gamificación. Mecánicas de juegos en tu vida personal y profesional.* SC Libro.

René, A. (2022). *Profesiones para nuevas generaciones postpandemia.* https://reneaguirrephd.com/2022/06/18/profesiones-para-nuevas-generaciones-postpandemia/

Rugh, M. S., Beyette, D. J., Capraro, M. M., & Capraro, R. M. (2021). Using DIME maps and STEM project-based learning to teach physics. *Interactive Technology and Smart Education, 18*(4), 553–573. Advance online publication. doi:10.1108/ITSE-07-2020-0109

Sánchez i Peris, F. J. (2015). Gamification. *Education in the Knowledge Society, 16*(2), 13–15.

Sánchez-Rodríguez, J., Ruiz-Palmero, J., & Sánchez-Vega, E. (2017). Flipped Classroom. Claves para su puesta en práctica. *EDMETIC, Revista de educación mediática y TIC, 6*(2), 336-358. . doi:10.21071/edmetic.v6i2.5832

Saz, I., & Ramo, R.M. (2015). Aproximación a los impactos y beneficios del aprendizaje servicio en la Universidad de Zaragoza. *RIDAS, Revista iberoamericana de aprendizaje y servicio, 1,* 9-27. . doi:10.1344/RIDAS2015.1.2

Teixes, F. (2014). *Gamificación: fundamentos y aplicaciones.* Editorial UOC.

Tello, J. & Aguaded, J. I. (2009). Desarrollo profesional docente ante los nuevos retos de las tecnologías de la información y la comunicación en los centros educativos. *Pixel Bit. Revista de medios y educación, 34*, 31-47.

Toledo, P., & Sánchez, J.M. (2018). Aprendizaje basado en proyectos: una experiencia universitaria. *Revista de currículum y formación de profesorado, 22*(2), 471-491. . doi:10.30827/profesorado.v22i2.7733

Torres-López. Díaz-López y Méndez, (2020). Educación a distancia: e-learning, m-learning y b-learning, conceptos, características y necesidades actuales ante el COVID.19. *XIV International Conference on Education and Innovation.*

Trigueros, G., & María, I. (2017). Propuestas transmedia y gamificación aplicadas al EEES: Nuevas metodologías activas para implementar las competencias en Ciencias Sociales y en Literatura. In Memorias del Programa de Redes-I3CE de calidad, innovación e investigación en docencia universitaria. Convocatoria 2016-17 = Memòries del Programa de Xarxes-I3CE de qualitat, innovació i investigació en docència universitària. Convocatòria 2016-2017. Alicante: Universidad de Alicante, Instituto de Ciencias de la Educación (ICE).

Trigueros, R., Aguilar-Parra, J. M., Cangas, A. J., Bermejo, R., Ferrandiz, C., & López-Liria, R. (2019). Influence of emotional intelligence, motivation and resilience on academic performance and the adoption of healthy lifestyle habits among adolescents. *International Journal of Environmental Research and Public Health, 16*(16), 2810. doi:10.3390/ijerph16162810 PMID:31394722

Uni. (2020). *Flipped Classroom, las claves de una metodología rompedora.* https://www.unir.net/educacion/revista/flipped-classroom/

Vega Pardo, M. L., & Guerrero Castillo, G. S. (2023). La gamificación y su incidencia en la calidad educativa, en la escuela Manuel Reinaldo Benalcázar Castillo. *Revista social fronteriza, 3*(2), 38–62. . doi:10.5281/zenodo.7686135

Vergara, D., Mezquita, J.M. & Gómez Vallecillo, A.I. (2019). Metodología innovadora basada en la gamificación educativa: evaluación tipo test con la herramienta Quizizz. *Revista de currículum y formación de profesorado, 23*(3), 363-387. doi: .v23i3.11232 doi:10.30827/profesorado

Vesga Aldana, I., Herrera Montero, M., & Horta Guette, M. (2021). *Estrategia pedagógica mediada por un recurso de gamificación para el fortalecimiento del componente gramatical del inglés en estudiantes de quinto grado*. Universidad de Cartagena.

Wan, Z. H., So, W. M., & Zhan, Y. (2020). Developing and validating a scale of STEM project-based learning experience. *Research in Science Education, 17*. Advance online publication. doi:10.100711165-020-09965-3

Warin, B., Talbi, O., Kolski, C., & Hoogstoel, F. (2016). Multi-role project (MRP): A new Project-Based Learning method for STEM. *IEEE Transactions on Education, 59*(2), 137–146. doi:10.1109/TE.2015.2462809

Chapter 5
A MOOC on Promoting Vaccination for Healthcare Professionals in Higher Education:
Project IENE11 PROVAC

Remedios López-Liria
https://orcid.org/0000-0002-3416-2791
University of Almería, Spain

Patricia Rocamora-Pérez
https://orcid.org/0000-0003-0662-2487
University of Almería, Spain

María de los Ángeles Valverde-Martínez
University of Almería, Spain

María Jesús Benzo-Iglesias
https://orcid.org/0000-0002-6114-6981
University of Almería, Spain

ABSTRACT

The European project entitled "Empowering Nurses and Healthcare Professionals to Promote Vaccination and Tackle Vaccine Hesitancy. IENE 11 PROVAC" is an initiative that has been developed by three European Universities (from Spain, Cyprus, and Greece) and an educational organization from Romania. Its goal is to train healthcare professionals to become reliable sources of information as well as address the doubts of the general population. Furthermore, this program aims to create a curriculum, information tools, and a massive online open course (MOOC) targeting healthcare professionals, individuals from disadvantaged groups, and those who are more vaccine hesitant to help them become better informed, overcome their fears, and ultimately strengthen confidence in vaccines. This MOOC aspires to be an example for innovative online courses for healthcare professionals around the world.

DOI: 10.4018/978-1-6684-8861-4.ch005

INTRODUCTION

Social Context

The year 2020 unexpectedly turned the world upside down, affecting every aspect of our day-to-day lives. The arrival of the pandemic caused by COVID-19 has meant worldwide destabilization of public health, an economic crisis and a rethinking of the way we live and even the methods traditionally used to carry out our work.

At present, more than 600 million confirmed cases of COVID-19 have been reported, with Omicron being the dominant variant, although it is estimated that the actual number of positive cases is much higher (Datosmacro.com., n.d.). Every demographic of the population has been affected by the fear of falling ill. What is more, they were deprived of their liberty and paralyzed in their activities for many months while possible solutions were sought to alleviate the contagion, along with the only possible cure/solution to fight the virus—a vaccine. An arduous year had to pass before this long-awaited event arrived, during which time many people were forced to adapt their jobs and their day-to-day tasks. The general population was subjected to severe restrictions on social contact and had to make use of the possibilities offered by information and communication technologies (ICTs), which had to be applied in almost all facets of life (e.g., work, shopping, family, and social relationships) (Valverde-Martínez et al., 2023).

In this context, educational centers, including universities, were among the first institutions that were obliged to adapt to a distance teaching-learning modality to avoid the spread of the virus among children, adolescents and young people. In this way, a large part of society was protected without having to diminish the education of students over an extended period of uncertainty and isolation, until normality could be reestablished.

Today, the situation has greatly improved thanks to the massive vaccinations throughout the world. However, the percentages of immunized people vary depending on factors such as continent, country and age group, thus still keeping society as a whole far from the expected and desired figures (Greyling & Rossouw, 2022).

By the end of 2022, the percentage of the world population that had received at least one dose varied: North America (82%), Asia-Pacific (81%), Latin America (81%), Europe (70%), Middle East (58%), Africa (33%). Within the European Union, countries such as Portugal, Spain and Italy show higher complete vaccination rates; however, Bosnia-Herzegovina and Ukraine stand out for lower complete vaccination rates (Statista, 2023).

Although immunization is one of the most effective public health interventions, from 2019 to 2020 global vaccination coverage fell from 86% to 83%, according to the World Health Organization (WHO) (Datosmacro.com., n.d.). Therefore, it has been shown that there is still a great deal of reluctance in society to receive not only the COVID vaccine, but any vaccine. The WHO ranks vaccine reluctance or refusal as one of the top 10 threats to global health. Many detractors have also appeared among healthcare professionals, who have further promoted uncertainty among those who do not have sufficient knowledge or information about the benefits provided by immunizing children and the rest of the population against the new strains that emerge as well as diseases that have almost been eradicated. This skepticism seems surprising considering vaccines, along with the treatment of drinking water and the widespread and rational use of antibiotics, are the health intervention that have prevented the most illnesses and deaths in history.

Addressing Misinformation on Social Media

The propagation of such misinformation and the proliferation of false information which was rampant in the press, on the Internet and social networks during the pandemic—generated greater concerns and increased doubts about vaccines among the general population and even among healthcare workers themselves in response to a disease that was unknown. This information contradicted vaccine information from the WHO and health authorities such as the National Health Service in England (NHS). Some examples were the false claims that "the vaccine kills people, causes infertility, or contains a secret microchip that [would] be implanted into recipients." This false information has likely made the decision to be vaccinated more difficult, endangering the lives and health of the general population, and affecting the ability of countries to stop the existing crisis (Ochieng et al., 2021).

Psychological, political, cultural, and even societal factors are entangled in the reasoning and decision-making process towards vaccination, rendering vaccine hesitancy a complex issue. In the study of Kalimeri et al. (2019), administering a series of surveys via a Facebook-hosted application, they studied the worldviews of people that "liked" supportive or vaccine resilient Facebook ages. In particular, they assessed differences in political viewpoints, moral values, personality traits, and general interests, finding that those sceptical about vaccination, appear to trust less the government, are less agreeable, and tend to emphasize anti-authoritarian values. Exploring the differences in moral narratives as expressed in the linguistic descriptions of the Facebook Pages, they saw that pages that defend vaccines prioritize the value of the family while the vaccine hesitancy pages are focusing on the value of freedom. Finally, creating embeddings based on the health-related likes on Facebook Pages, they explored common, latent interests of vaccine-hesitant people, showing a strong preference for natural cures (Kalimeri et al., 2019).

Currently, a large part of the communications related to people's health occurs in digital media, mainly in social networks. Some communication strategies on social networks have been shown to be very successful in improving attitudes towards vaccines, addressing population concerns, and even increasing vaccination coverage. Among the general recommendations to debunking vaccine myths and providing trustworthy information to help individuals navigate the infodemic and make informed decisions about vaccination, we find (Pan American Health Organization, 2020):

- Face fears and doubts through dialogue and the dissemination of scientific evidence.
- Share data and evidence together with personal experiences and playful messages that facilitate the understanding of the information.
- Share information from professional accounts so that it comes from a reliable source.
- Some users may post negative comments. In these cases, it is not a matter of trying to convince, but of engaging in a constructive dialogue that can be useful for other people.
- Each social network has its own communication codes and different audience/users. In order to identify the most appropriate channels for the dissemination of information, it is necessary to take into account at least the generational differences in the use of media and social networks (Facebook, Twitter, Instagram, Pinterest, etc.). Also, it is necessary to develop a communication plan with clear objectives and tools to optimize communication and dialogue.
- In addition to adapting content, formats and channels to different needs, messages and materials must be available in the relevant languages and guarantee their accessibility to people with visual, hearing and/or cognitive disabilities.

To counteract this trend based on misinformation and false information, a European Erasmus+ project was undertaken entitled "Empowering nurses and healthcare professionals to promote vaccination and tackle vaccine hesitancy. IENE 11 PROVAC".

Uncertainty about Vaccines: Misinformation

One of the greatest advances in medicine has been the prevention of infectious diseases through vaccination. A vaccination schedule is a chronological sequence of vaccines that are systematically administered to the entire population in a given geographic area or country. In Spain, this schedule includes vaccines that protect against 14 infectious diseases: tetanus, diphtheria, whooping cough, poliomyelitis, hepatitis B, invasive disease by Haemophilus influenzae type b, pneumococcus, serogroup B and ACWY meningococcus, measles, rubella, mumps, chickenpox, human papillomavirus and flu. To decide on the inclusion of a new vaccine in these programs, an evaluation of five criteria is performed: disease burden; vaccine effectiveness and safety; repercussions of the modification in the vaccination program; ethical aspects and economic considerations. The establishment, implementation and evaluation of the vaccination schedule is the responsibility of the Spanish autonomous communities. The results for 2021 show a decrease in the vaccines administered during the first year of life and early childhood (hexavalent, pneumococcal, meningococcal, triple viral, varicella first dose and dTpa at 6 years), except for the second dose of chickenpox, for which an increase is observed due to the consolidation of vaccination against this disease. At present, there is an uptick, or rather growing skepticism against vaccines, further complicating the already complex pandemic situation in which the world has been immersed (AlShurman et al., 2021).

The global spread of COVID-19 created an urgent need for safe and effective vaccines against the disease. However, achieving immunity has become a global objective and challenge, considering vaccination is a prosocial action. Although various studies show wide variability in the rates of acceptance, rejection and hesitation towards the COVID-19 vaccine, there has been a trend towards acceptance in European countries, North America and China. On the contrary, we find high rejection rates in Middle Eastern countries (Alimohamadi et al., 2022; AlQerem & Jarab, 2021; AlShurman et al., 2021; Salomoni et al., 2021). On the other hand, it has been described in the scientific literature that there is a greater predisposition to vaccination in older adults, due to a greater risk of infection and death from COVID associated with age, and in healthcare professionals, as they represent the first line of defence in the fight against COVID-19 (AlQerem & Jarab, 2021; Nachega et al., 2021).

Several causes of rejection by the population have been highlighted. They include a lack of transparent health information and effective tools to promote a better understanding of the benefits and risks of vaccination and the use of misconceptions that flood social networks, ultimately confusing the population. The politicization of COVID-19, the manufacture of antibody tests to collect people's DNA, and the harmful effects of vaccines in children are some of the beliefs that have been disseminated among society, causing distrust of vaccination and its effects (Nachega et al., 2021). In addition, demographic and racial factors also seem to influence a lower rate of acceptance of vaccines (AlQerem & Jarab, 2021; AlShurman et al., 2021). These data indicate the need for and importance of disclosing reliable sources of information and support the idea that it is healthcare professionals who should serve as a vehicle for the dissemination of truthful information.

The path towards ending this disease is still unknown, but it is undeniable that vaccines have brought much hope and greater peace of mind to alleviate the pandemic. Nevertheless, as described, there is great global disparity in vaccination, which has favored the development of new variants and is prolonging the virus's presence among us. On the other hand, scientific evidence shows a decrease in the contagiousness of the virus and an increase in protection against severe symptoms of the disease among vaccinated people (Mohamed et al., 2022).

Coping by Universities: More Training

Today's society is immersed in the era of knowledge and globalization and is constantly undergoing technological changes. In the 21st century, individual learning is conceived through interactive and participatory technologies that allow students to learn and share their knowledge experiences with other people who are often on the other side of a screen. ICTs have become an essential element for them to interact in all spheres (e.g., family, work, social). The symbiosis between people and ICTs constantly favours development, knowledge, learning and ways of communicating. In fact, social interactions are increasingly conditioned by this medium (e.g., asynchronous communications, social networks) (Cueva-Gaibor, 2020; Hernández et al., 2022).

On the other hand, ICTs lead to a change in the role of teachers and their performance in the teaching-learning system, making prior training and advice on ICTs essential to ensure proper use in the classroom. What is more, since these technologies are continuously evolving, new educational standards are always being set, providing an innovative and renewed approach to knowledge and subjects. Consequently, higher education centers are obliged to constantly reinvent themselves and adapt to the reality of a digital world (Sharpe & Benfield, 2016). Education and the educational system constitute an essential element for the development of human beings and future society, which, if combined with ICTs, will improve the quality of teaching and provide increased access to a great number of young people from different backgrounds, but who were all born in the digital era and who are direct or indirect consumers of technology. However, it should be noted that there are still teachers who are reluctant towards using ICTs in the teaching process due to phobias of the digital world, skepticism about digital learning and lack of knowledge. This aversion is the result of inadequate and vague teacher training plans and concern for the possible risks and difficulties which their use may cause on certain occasions (Fernández-Márquez et al., 2020; Poveda-Pineda & Cifuentes-Medina, 2020).

The key to the success of using ICTs in the educational system lies in teacher involvement, taking into account the digital resources at their disposal, knowing the needs, advantages and their applications. With a proactive attitude, teachers will be able to effectively implement these tools in the teaching-learning process of the students (Granda-Asencio et al., 2019).

It is important to highlight that incorporating ICTs should prove easy for both teachers and students. To ensure the user-friendliness and effectiveness of these tools, the ICTs implemented will use accessible devices compatible with different operating systems that facilitate collaborative work and real-time interaction as well as favor working safely and efficiently. It has been verified in different investigations that the application of ICTs promotes self-learning, collaborative work and skills such as creativity, autonomy and motivation of students in their subjects (Amutha, 2020; Das, 2019; Kapur, 2019; Laro-González, 2020; Venegas-Ramos et al., 2020; Zonneveld et al., 2020).

Continuous Training on Vaccines

Healthcare professionals have an ethical responsibility before society in relation to vaccines. In addition, when they are vaccinated, they act as an example for the benefit of public health and generate confidence. Furthermore, they provide truthful information supported by scientific evidence, avoid the spread of false information, resolve doubts, uncertainties, fears and continually review the scientific information available (Jin et al., 2022).

For this reason, first it is necessary to raise health professionals' awareness about the importance of vaccination, thereby promoting their knowledge and a positive attitude for the dissemination of vaccination campaigns. This can be achieved through various methods (WHO, 2022; Pan American Health Organization, 2020; Ross & Magris, 2021):

- Preparation of protocols or practical guides.
- Distribution of posters, audios, videos, etc. through different media, such as social networks.
- Organization of awareness meetings.
- Inclusion of new strategies in the service portfolio of healthcare professionals.
- Active recruitment of professionals willing to participate in this vaccine promotion process, carried out virtually or in writing, even by offering incentives.

The Role of Healthcare Professionals

In this way, the healthcare professional becomes a closer source of truthful and reliable information related to vaccines that is in contact with the citizen (Paternotte et al., 2016). The scientific literature shows that intercultural communication is a valuable tool for the promotion of vaccination campaigns. The doctor-patient relationship establishes trust and allows evidence-based intervention strategies to be effective in the office. While doctors' attempts in the clinic to increase coverage remain crucial, vaccine hesitancy remains a huge public health problem that requires attention to be addressed successfully. Public health interventions would effectively complement strategies in the clinic to increase overall vaccination coverage (Paterson et al., 2016).

Several studies affirm that multi-component and dialogue-based interventions are the most effective in increasing vaccination rates. Still, the various strategies need to be carefully tailored according to the target population, their reasons for refusing vaccination, and the specific context (Jarrett et al., 2015; Paternotte et al., 2016). The responsibility of healthcare professionals lies in responding to vaccine-hesitant individuals with great compassion, patience and respect. Understanding and discussing reasonable and not-so-reasonable arguments is essential to build trust between the public and healthcare professionals (Dainton & Wong, 2022).

Educational interventions to improve knowledge and communication appear to be effective for healthcare professional to increase the vaccination rates (Leung et al., 2019; Lin et al., 2021). Therefore, it is necessary to design and implement effective communication skills strategies at universities so that students may learn how to approach and talk to different types of patients with different beliefs (counselling abilities), fostering increased confidence towards population vaccination. Teacher plays an essential role in promoting this learning and skills in Health Sciences degrees such as nursing, medicine and other related areas. Another method could be to allocate 30 minutes per day for nurses or doctors to speak to parents about vaccination in courses or presentations. Sharing personal examples is another strategy suggested for convincing patients (González-Juarez & Noreña-Peña, 2011). The key elements in a vaccine-counselling visit are time to listen, empathy and dissemination of adequate information (Hernández-Angulo et al., 2021).

Health promotion and disease prevention are tasks that directly and indirectly fall on healthcare professionals in their daily practice. Society's search for health information has increased considerably in recent years but, above all since the pandemic. However, there are many sources of unreliable information disseminated on different platforms, generating doubts and uncertainty among citizens (Kroger et al., 2022). Many healthcare professionals (e.g., doctors, nurses, physiotherapists, pharmacists) have generated truthful knowledge on health issues that they have disseminated on social networks (e.g., Instagram, Facebook, TikTok), taking advantage of their widespread popularity, or other technological resources (e.g., blogs, videos, web pages). However, these individuals highlight that there is a training gap in healthcare and call for the need to incorporate teaching strategies that promote community participation in healthcare adapted to ICTs through virtual means of communication. Such a task should begin in the undergraduate stages, providing students with tools for the development of communication skills, the use of new technologies and adequate teamwork (Ree & Lozano, 2019).

IENE11 PROVAC PROJECT: OBJECTIVES

The main objective of the IENE11 PROVAC project is to support and train healthcare professionals to promote vaccination and address the hesitation of vulnerable populations. This goal can be achieved by improving their knowledge about vaccination and developing their communication, cultural and social relationship skills. This will empower them to combat misinformation—aimed at people who are more indecisive or who are from disadvantaged groups—and also improve confidence in the vaccine.

METHOD: DESIGN, CREATION, PILOTING, AND IMPLEMENTATION OF THE MOOC

Four partners (Edunet Organization of Romania; Technological University of Cyprus; Kapodistrian National University of Athens and the University of Almería) will be working together until 2024 on the design of a training curriculum focused on promoting vaccination and addressing hesitancy to vaccination. These efforts will materialize in the creation of an innovative course based on digital learning and an online methodological approach (Massive Open Online Course (MOOC)) aimed at university students and healthcare professionals in four languages (English, Spanish, Greek and Romanian).

In the first phase of the project, to determine the contents to be included in the MOOC course, an exhaustive review was conducted of the scientific literature available in different health sciences databases (Pubmed, Medline, Web of Science, Cochrane, among others). This review facilitated the compilation of the most valuable and reliable sources of information on different topics of vaccination and immunization, false myths and widespread beliefs among the most reticent population to vaccination. In addition, for the development of the teaching/learning resources and the MOOC course, the national and international policies and guidelines of each of the participating countries regarding vaccines were studied. Using the Delphi method, a panel of experts from all partner countries identified the key constructs/statements used for the description of the competencies according to the partners' own national literature reviews, as well as international reviews. These statements include information on vaccine and vaccination; behaviour and attitudes towards vaccination; interventions to promote vaccination and tackling of vaccine hesitancy. The Delphi study consisted of two rounds during the month of June and July 2022. The participant experts consisted of various stakeholders involved in vaccine communication and uptake, mainly healthcare professionals, but also policymakers and community leaders. Their diverse insights, experiences, recommendations and viewpoints were incorporated to provide a more comprehensive approach and understanding when identifying the key constructs/statements for the description of the competencies.

IENE11-PROVAC MODEL AND TRAINING CURRICULUM

The IENE11, like previous IENE projects (IENE 1 to IENE 10), is based on the Papadopoulos, Tilki, and Taylor (PTT) Model (Papadopoulos, 2006). The four constructs of the model include: Cultural Awareness, Cultural Knowledge, Cultural Sensitivity and Cultural Competence (Papadopoulos, 2011; Rodrigo, 2022).

- Cultural Awareness. The idea that we must know our cultural identity and that of other cultures in order to understand its importance and avoid ethnocentrism.
- Cultural Knowledge. It encompasses various disciplines such as anthropology, sociology and psychology. Contact with people from different ethnic groups can improve knowledge about their health beliefs and behaviors and increase understanding of the problems they face.
- Cultural Sensitivity. An aspect achieved through the development of appropriate interpersonal relationships.
- Cultural Competence. The ability to provide effective medical care while considering people's cultural beliefs, behaviors, and needs.

Each module of the curriculum comprises four sub-constructs which represent the content map (see Figure 1).

The thematic units that comprise the contents of the course are described below for each of the modules (https://iene11.eu/es/training-curriculum/):

Module 1. Cultural Awareness, with Topic 1: Self Awareness; Topic 2: Myths and fears; Topic 3: Values, attitudes, and behavior; Topic 4: Acknowledgment of healthcare and social service needs for users and staff.

Module 2. Cultural Knowledge, with Topic 1: Policies and strategies; Topic 2: Advantages and disadvantages; Topic 3: Cultural aspects of immunization and hesitancy; Topic 4: Understanding vaccine and vaccination.

Module 3. Cultural Sensitivity, with Topic 1: Intercultural Communication; Topic 2: Culturally sensitive compassion and respect; Topic 3: Ethical issues and human rights; Topic 4: Working with vulnerable population and special groups.

Module 4. Cultural Competence, with Topic 1: Advocacy; Topic 2: Empowerment; Topic 3: Practice and techniques of vaccination; Topic 4: Culturally competent needs assessment.

Figure 1. IENE 11: Promoting vaccination and tackling vaccination hesitancy—PROVAC—culturally competent training model

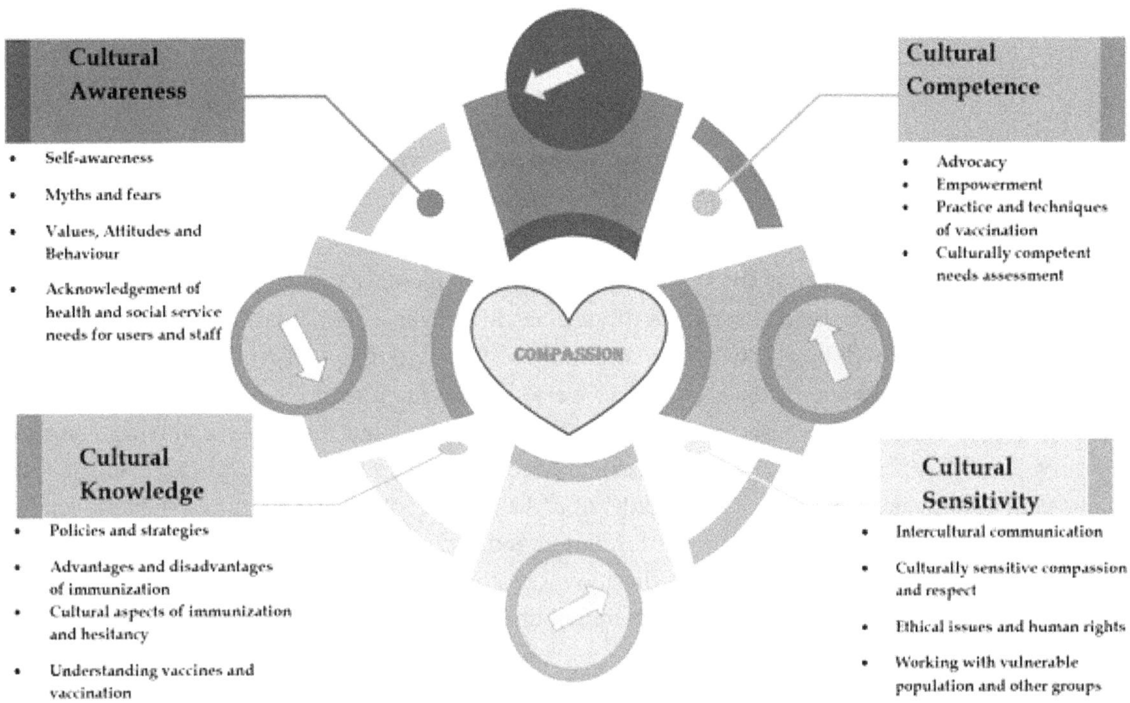

The training curriculum has guided the creation of the Bite-sized Learning Units and the training methodology of the training course. For each thematic unit, easy-to-use learning/teaching tools have been created and selected with links to interactive resources, videos, podcasts, exercises, etc.

THE MOOC SCHEDULE

The learning tools and modules produced based on the PTT/PROVAC model will be used for the development of the MOOC to be delivered by the IENE 11 project over a period of six-weeks.

During the introductory week, the participants will watch orientation videos and learn how to navigate the course and its various components. They will also complete a pre-MOOC questionnaire, produce a brief profile to share with their peers, and have a brief introduction to assessments, evaluations and how to use the chatrooms.

Each week will consist of 5 days during which participants will engage with a different learning tool, focusing on a specific aspect of the PROVAC module of the week. On the last day of the sixth week, participants will complete the post MOOC questionnaire, evaluation, and summative assessment. Activities may include short videos, podcasts, articles, reflective accounts, quizzes, etc. The MOOC aims to be an innovative online example for healthcare professionals around the world.

Since the project began in February 2022, the partners are regularly sending information to stakeholders (teachers/trainers, learners/ senior students and healthcare and social care workers) via letters, e-mails, and newsletters about the project activities and results. These activities are carried out during face-to-face meetings or at national seminars, at which time stakeholders receive information on the project's outputs and how to use the online course.

RESULTS—MOOC: METHODOLOGY, DISSEMINATION, COMPETENCES, AND LEARNING OUTCOMES, SCOPE

The training curriculum created in the IENE11 project aims to increase knowledge about immunization, promoting an understanding and acceptance of vaccines in the general population.

This MOOC course will be piloted with a minimum of 120 participants recruited from partner countries. It will be developed and evaluated over six weeks, in the months of October and November. The proposed target groups will be healthcare professionals (Family Medicine, Paediatrics, Nursing, Midwives, Preventive Medicine, Epidemiology, Occupational Medicine, Hospital Pharmacy and Primary Care), teachers and training professionals dealing with public health, and other managers and stakeholders who could subsequently disseminate adequate information and carry out information campaigns, related to the issue of vaccines, which would target the most skeptical individuals, people from disadvantaged groups and those at higher risk (immunodeficiencies, chronic diseases, institutionalized people, morbid obesity) (Madewell, et al., 2021).

The methodology of this course will be participatory, involving students in their own learning through individual and group reflection that they will share through communication tools. These group activities will allow feedback among the different participants, thereby creating a safe learning environment. The acquisition of knowledge and practical skills will be carried out through strategies that simulate a clinical teaching-learning environment for the different healthcare professionals related to primary care (Voudoukis & Pagiatakis, 2022).

On the other hand, it is also intended to modify the attitude of the professional. This process will utilize different resources, for example explanatory videos, and dynamic group activities, such as reading and commenting on press articles or other media (e.g., Twitter), and discussing the resolution of practical cases through ICTs, which will introduce the student or professional to the reality exhibited by society, helping to address situations of denial/acceptance of vaccines. Nevertheless, it is necessary to recognize that the process of rebuilding trust depends on the specific social, cultural, political and economic context of the country or region in question. It is therefore important to first take steps to obtain an accurate understanding of these factors before engaging in trust-building activities such as dialogues and exchanges of information and opinion with healthcare workers or other individuals (European Centre for Disease Prevention and Control [ECDC], 2015).

In addition, to attract potential users interested in this course and promote the dissemination of the information created from this project, 20 informative brochures have been produced that address important topics such as: vaccination and flu; vaccination during lactation or pregnancy; vaccination in the elderly, babies and children; how culture and religion influence; contraindications of vaccines; how vaccination is carried out; reliable sources of information; the European vaccination strategy; vaccine myths; vaccines to prevent human papilloma virus; compulsory vaccinations in the European

Union; natural infection and immunity; military and vulnerable groups such as migrants; vaccine performance; adverse reactions and allergies; importance of vaccination and travel. These brochures have been distributed at various events, such as the "European Researchers Night" which is held every year in Andalusia and national and international congresses, in Health Districts and Health Centers, and in online and face-to-face informative talks aimed at students in the branches of Health Sciences. In this way, a significant number of people from disadvantaged groups and those with the most difficult access to vaccination have been reached either directly or indirectly. The dissemination of the project is also being promoted through publications (newsletters) on social networks, such as Twitter (@11Iene) and Facebook, about the meetings held and the objectives to be achieved during each phase of the project. On the other hand, it is important to highlight that in Andalusia, with the implementation of the figure of the school nurse in the field of Primary Care, health promotion and prevention activities are being carried out in schools. One of these activities involves the administration of vaccines according to the immunization calendar and raising awareness of their importance among parents (Kalimeri et al., 2019; Li et al., 2021). The University of Almería is carrying out activities to support these school nurses, and thus increase the number of vaccinated children. At times, parents question the administration of several vaccines simultaneously at very early ages in a systematic way and without individualizing each case; they fear adverse effects and do not understand the variability in the vaccination schedule. The main reasons for not vaccinating, which they raise at these informative sessions, are the presence of mercury, aluminium, stabilizers or preservatives and the risk of developing autism or anaphylaxis.

Training for teachers and facilitators is scheduled for June 2023, when they will engage in a preliminary test of the MOOC platform. These same individuals will later accompany the students during their journey on the MOOC course—advising them, offering help when they request it, and motivating them during the activities. In addition, several national training seminars are planned to be held in the partner countries. At these events, the participants will learn about the objectives of the training curriculum, the methodology, the content of the MOOC and how to use the different resources independently. Furthermore, any possible doubts that may arise will be addressed by specialists in the field of vaccines. In this way, it is expected that this course will have a multiplier effect, achieving wider dissemination and reaching the largest possible population (Cataldi et al., 2019; Goss et al., 2020).

Upon completion of the course, students/healthcare professionals should be able to identify existing myths and false beliefs about vaccines in relation to issues such as adverse effects, culture, religion, or sex. In addition, they should be aware of and able to recognize their own cultural stigmas or thoughts that carry negative connotations. They should also be capable of analyzing and abandoning the former through efficient strategies fostered within collaborative online learning. Participants will also have learned about other issues, such as the indications for frequent vaccinations in risk groups (e.g., pneumococcus, influenza and COVID-19), the indications and contraindications for vaccination in immunosuppressed people, and how to manage accelerated vaccination schedules in incorrectly vaccinated populations and identify the peculiarities of vaccinations in premature infants (Lin et al, 2021).

DISCUSSION

On the one hand, it is necessary to guarantee that future healthcare professionals trained at universities will be able to clearly communicate the benefits, risks, and importance of vaccines, thereby fostering confidence among the general population. When a large portion of the population has acquired immunity to an infectious disease, then it is rather unlikely that the disease will spread from person to person. This is known as "collective immunity" (also referred to as "herd immunity"), which also allows for the indirect protection of unvaccinated individuals.

Vaccine Hesitancy in Visible Minorities

The results of this project contribute to understanding the emotional challenges associated with vaccine uptake and informing policymakers, healthcare workers and stakeholders who communicate with the public during infectious disease outbreaks. Although evidence of vaccine hesitancy exists, there is minimal focus on visible minorities, and the reasons for hesitancy in this group are unclear. Identifying these populations and their reasons for vaccine hesitancy is crucial to improving vaccine uptake and curbing the spread of COVID-19 (Ochieng et al, 2021).

In minority and immigrant communities that make up a significant portion of long-term care workers, misinformation is arriving in rampantly, fuelling scepticism, which may be related to more recent or deeper historical origins (such as the example of black people) (Dainton & Wong, 2022; Eissa, 2021; Lasco & Yu, 2021). In fact, the WHO has stated that we are facing an "infodemic" due to the amount of easily accessible information available, which makes it difficult for people to reach reliable resources when it is necessary, and information is often false and spreads quickly among individuals and through the media (Aleixandre-Benavent, 2020; *WHO*, 2020b; Pfattheicher, 2022).

With the existing limitation in medical literature, many healthcare professionals have scarce ability to defend themselves against the great amount of poor-quality information. In order to solve this problem, dialogue must be carried out, since questioning the quality of the individual's information source or classifying them as lazy or stupid will only make them show more resistance and division. Vaccine-hesitant individuals who are sceptical pride themselves on a complex belief system determined in turn by also complex environmental factors, all of which must be compassionately understood and acknowledged before any progress can be made (Dainton & Wong, 2022).

In addition, personal experience is an important driver of attitudes toward medicine. The secondary effect that a family member may have suffered generates a high emotional impact on opinions, fears and attitudes that ends up being considered more objective than the scientific data with the best evidence. These pragmatic concerns must be treated with deep empathy (Dainton & Wong, 2022).

Summarizing, strategies to improve vaccination rates in vulnerable groups should focus on the facilitation that implies ensuring an effective vaccination program, of a comprehensive nature and reducing the many barriers, in terms of cost, distance and time, to achieve high acceptance levels, especially for marginalized or vulnerable populations. Information is crucial, whether in the form of public information campaigns or interactions between health workers and target populations, they must be designed very carefully to avoid the risk of counterproductive results. There is no universal solution to achieve high levels of vaccine acceptance, but rather a combination of options. Along these same lines, accessibility can be improved by expanding the range of health professionals who administer the vaccine, provided they receive appropriate training. This diversity in provision is important for reaching remote or underserved areas and disadvantaged groups. In this regard, special attention should be paid to groups such as immigrants, especially the undocumented, and ethnic minority populations, who may face multiple barriers to obtaining care (Siciliani et al., 2020).

Vaccine Safety Communication

Vaccines have stopped the spread of multiple infectious diseases throughout history (e.g., measles, flu, smallpox). However, several studies have shown that even vaccinated individuals occasionally have doubts about the safety of vaccines. Parents, pregnant women and individuals with immunosuppressive disorders sometimes express fears about the risk of side effects and the safety of ingredients and adjuvants. They can also display a lack of understanding about the risk of the diseases they prevent (Greyling & Rossouw, 2022).

In the study realized by Writers (2022) in United States, the author admits that fears over the safety of vaccines are understandable. The CDC vaccination schedule calls for children to receive up to 14 inoculations by the age of six—many of them vaccines developed within the last twenty years. Many parents distrust these vaccines, worried about the potential for risks and long-term side effects. Research, however, shows that most of our biggest fears about vaccinations are unfounded. These eight major vaccine myths that research has shown to be baseless: Myth #1: Vaccines cause autism, Myth #2: Infant immune systems can't handle so many vaccines, Myth #3: Natural immunity is better than vaccine-acquired immunity, Myth #4: Vaccines contain unsafe toxins, Myth #5: Better hygiene and sanitation are actually responsible for decreased infections, not vaccines, Myth #6: Vaccines aren't worth the risk, Myth #7: Vaccines can infect my child with the disease it's trying to prevent, Myth #8: We don't need to vaccinate because infection rates are already so low in the United States (Writers, 2022).

In the study of Kreps et al. (2021), through a nationally representative survey of 1027 adult Americans conducted in February 2021, they examined individual misconceptions about COVID-19 vaccine safety; the demographic factors associated with these misconceptions; and the relationship between misconceptions and willingness to vaccinate. Misconceptions about vaccine safety were widespread. A sizeable minority (40%) believed that vaccine side effects are commonly severe or somewhat severe; 85% significantly underestimated the size and scale of the clinical trials; and a sizeable share believed either that the vaccines contain live coronavirus (10%) or were unsure (38%), a proxy for fears that vaccination itself may cause infection. These misconceptions were particularly acute among Republicans, Blacks, individuals with lower levels of educational attainment, and unvaccinated individuals. Perceived side effect severity and underestimating the size of the clinical trials were both significantly associated with vaccine hesitancy (Kreps et al., 2021). In addition, the study of Geoghegan et al. (2020) was shown that parents and patients have a number of concerns about vaccine safety, among the most common are fears that adjuvants like aluminium, preservatives like mercury, inactivating agents like formaldehyde, manufacturing residuals like human or animal DNA fragments, and simply the sheer number of vaccines might be overwhelming, weakening or perturbing the immune system. As a consequence, some fear that vaccines are causing autism, diabetes, developmental delays, hyperactivity, and attention-deficit disorders, amongst others.

Although immunization is based on a system of quality assurance, safety, efficiency, efficacy and supply, studies have shown that providing information on these guarantees is not enough to change people's behavior regarding the decision on whether to vaccinate, and does not generally increase vaccination coverage (Tversky and Kahneman, 1981). In this scenario, it is necessary to understand the reasons that lead people to decide not to get vaccinated or not to vaccinate their children, in order to tailor messages and initiate a two-way dialogue based on respect. Such a dialogue must incorporate emotional elements, enhance personal narratives, and use evidence-based communication techniques to achieve the goal of strengthening, maintaining, or regaining confidence in vaccines and immunization programs (Tversky and Kahneman, 1981; UNICEF, 2019).

You have to focus on interpersonal communication between health workers and individuals (patients, parents, caregivers, or children). Many studies have shown that health personnel, as they interact with individuals regarding the decision to vaccinate, constitute the most reliable source of information (UNICEF, 2019).

Vaccines are very safe. Before a vaccine is approved, there must be clinical trials showing that it does not produce serious side effects in people, and that it is effective in protecting against the specific disease or diseases for which it is indicated. In addition to very strict safety standards in the production of vaccines, there are regulations on their transport, storage, and administration that function as a quality assurance system to make vaccines as safe as possible; moreover, they are monitored by pharmacovigilance systems in each country (CDC, 2022; ECDC, 2016; Hamborsky et al., 2015; UNICEF, 2019; UNICEF, n.d.; Washington State Department of Health, n.d.).

Access to Vaccines and Vaccine Equity

As described in previous studies, more information about vaccine safety and expected side effects is needed to increase positive attitudes towards vaccines. Furthermore, it has been highlighted that government procurement of vaccines and vaccine rollout should improve, accessibility to the vaccine should be a priority and a collective effort should be made to increase positive messaging about the vaccine, especially on social media (Greyling & Rossouw, 2022). The global fight against COVID-19 might be lost if attitudes towards vaccines are not improved.

In the study of Bedford et al. (2018), they proposed more precision in the term 'vaccine hesitancy' is needed particularly since much under-vaccination arises from factors related to access or pragmatics. Only with clear terminology can we begin to understand where the problem lies, measure it accurately and develop appropriate interventions. This will ensure that our interventions have the best chance of success to make vaccines available to those who want them and in helping those who are uncertain about their vaccination decision (Bedford et al., 2018).

In industrialized countries, high vaccine uptake rates are generally observed. However, two major parental factors contribute to under-vaccination. Firstly, socioeconomic disadvantage hinders parents or caregivers from accessing adequate resources and support to overcome logistical barriers like lack of transportation or childcare (Falagas, 2008; Leask et al., 2012; Samad, 2006). Secondly, parental concerns about vaccine safety or necessity also influence under-vaccination (Gust, 2008; Leask et al., 2012; Smith, 2004).

Inequalities in access to basic vaccines persist globally, making their elimination a priority goal (Arsenault et al., 2017). Vulnerable groups, due to lack of awareness, misconceptions about vaccination, and health system limitations as well as the cost and inability of health systems, remain under-vaccinated (Doherty et al., 2016). Health workers play a crucial role in achieving high coverage for immunization, and their efforts are vital in educating the public about the importance of vaccination (Paterson et al., 2016).

GAVI, the Vaccine Alliance, focuses on increasing equitable vaccine usage in low-income countries and monitors vaccination coverage inequities within the countries it supports (GAVI Alliance, 2013). Socioeconomic indicators, such as maternal education, impact vaccination rates in households in low- and middle-income countries, necessitating monitoring and improvement efforts, especially for children of uneducated women, aligning with Sustainable Development Goals and GAVI's focus on gender equality (Brearley, 2013; WHO, 2020a).

Regarding immigrant populations, barriers to vaccination include inadequate education, financial constraints, access difficulties, and family problems (Fairhead et al., 2006). Providing vaccination through primary care, rural areas, hospital outpatient clinics, and medical-social centers can enhance vaccination rates for immigrants (Pinaka et al., 2021).

For adults, lack of access to immunization records significantly affects vaccination rates, particularly among mobile populations with multiple healthcare providers. Patient recall, the primary method for obtaining vaccination history, is known to be inaccurate, impacting coverage (MacIntyre et al., 2013; Ridda et al., 2008).

Barriers to vaccinating the elderly include access to care, mobility issues, lack of confidence in health providers, and the absence of clinical trial data for frail elderly individuals. Furthermore, the scarcity of vaccination information campaigns and resources for adults contributes to this challenge (MacIntyre, 2013; MacIntyre et al., 2016).

Through the social networks, to improve vaccination rates, the Pan American Health Organization (2020) recommended adapting content, formats, and channels to different needs, ensuring accessibility for people with disabilities. They also urged healthcare providers to simplify settings and systems for obtaining vaccinations, considering factors like costs, location, staffing, transport, scheduling, and timing as key determinants of vaccination program success. Making 'vaccination journeys' easier can enhance overall vaccine uptake.

Tailoring Vaccine Promotion Messages

We must address vaccine acceptance as a current problem that requires solutions from various perspectives which faces both structural and behavioral barriers. Examples of the former include lack of access, misinformation, and diversity of knowledge about vaccines, while the latter include denial attitude and mistrust that foster less acceptance. Fishbein and Capella (2006) emphasize the importance of vaccine promotion messages considering the action, objective, context and time frame in which they are applied, taking into account the aforementioned structural and behavioral barriers to promoting effective messages directed and adapted to given circumstances. It is essential to know the characteristics of the audience to whom the message will be delivered. This requires interdisciplinary work that allows health messages to be modified and conveyed to each target group, taking into consideration cultural and linguistic factors, addressing specific concerns or misconceptions that are prevalent within different communities, and using appropriate communication channels and formats. Understanding the diversity within the target population and employing targeted messages can improve the effectiveness of vaccine promotion efforts.

The use of ICTs offers highly valuable educational and training opportunities that foster creative and innovative educational environments. However, it also increases access to a vast number of information sources which are sometimes unreliable. The proliferation of social networks and the infodemic (abundance of information on a subject) favors, in part, finding false information about vaccines that are used to discourage people from participating in the immunization process, causing negative repercussions for public health. This phenomenon has forced different experts and healthcare institutions to publish web pages that refute this information (Gómez-Contreras et al., 2022).

On the other hand, this project addresses to the common training needs of the associated countries, filling certain gaps in the training curricula of healthcare professionals and training them with specific knowledge and communication and social skills. These objectives are achieved by means of a fully online distance learning course, with complementary resources and practical materials. It offers students the flexibility of e-learning as well as a setting where they can interact with other users and benefit from their experiences, share and exchange questions, opinions, etc. in forums where both experts, professionals and university students participate.

Healthcare and social health professionals must act as outlets for truthful information to achieve greater acceptance of vaccines. For this reason, it is essential to promote therapeutic intercultural communication and a relationship of trust between the professional and the patient and the general population, especially with those who are most vulnerable. The modules that will be presented in the MOOC course include subjects such as maintaining an effective and compassionate relationship, educating on ethnic concerns and human rights related to immunization, the reasons that cause indecision and the needs of vulnerable groups (Ledford & Nan, 2023).

The partners of this project have previous experience in the creation of other MOOC courses carried out in European projects, namely the IENE7 project: "Improving the communication and social skills of foreign and migrant caregivers who work with the elderly" and the IENE9 project "Training for culturally competent and compassionate healthcare professionals in LGBT+". These projects have addressed relevant social issues with methodologies based on co-learning and co-creation, having great social impact and obtaining very enriching results which have shown great satisfaction (Baiocco et al., 2022; Rocamora-Pérez et al., 2018).

It is necessary to focus on research that deals with vaccine health issues to identify those social groups with less acceptance or tendency not to receive the vaccine. In general, the literature shows that the elderly, with other associated pathologies, and health professionals are more susceptible to vaccination, considering that they are prone to contagion. However, there are many people who display reluctance to be vaccinated for various reasons, such as distrust, misinformation, beliefs in conspiracy theories, etc (Maatuk et al., 2022).

Monitoring and Evaluating Interventions

Greyling & Rossouw (2022), in a study of ten countries, reflects the need for interventions to change attitudes towards vaccines and increase their acceptance, as the countries analyzed tended towards low acceptance of vaccines, except for Belgium and the Netherlands. As indicated, the variables that seem to influence acceptance relate to information on safety and side effects. In this sense, it is necessary to improve confidence in vaccines, review the regulations implemented, increase confidence in governments and use social media platforms to educate people of all ages. In addition, this same study analyzed tweets related to negative feelings towards vaccines and identified that people actually have positive attitudes towards vaccines, whereas they express negativity about government incompetence regarding vaccine implementation, acquisition and access.

It is important the ongoing monitoring and evaluation of vaccine uptake interventions to assess their impact, identify successful strategies, and make evidence-based improvements. This could involve measuring vaccine coverage rates, analyzing data on vaccine acceptance, and evaluating the outcomes of communication campaigns and educational initiatives.

During the existence of the IENE11 PROVAC project, periodic publications have been made on different social networks (e.g., Twitter or Facebook), targeting a wide range of ages and social groups, with the aim of achieving greater population reach. Face-to-face and online seminars in different countries have also been conducted. All the information generated from the project and in the MOOC course as well will remain on a free and open access platform and website in the future (Aljawarneh, 2022; Zaremohzzabieh et al., 2022).

FUTURE RESEARCH DIRECTIONS

For future research and interventions on the topic, it would be interesting and necessary to delve into:

- Vaccine education in schools: Introducing comprehensive vaccine education in schools may ultimately have a major impact on shaping attitudes towards vaccination. Research can assess the impact of adding vaccine education to school curricula, assess knowledge gains, and measure long-term effects on vaccine uptake.
- Long-Term Vaccine Confidence: Understanding the long-term factors that contribute to maintaining vaccine confidence is paramount. Future research can examine the factors that influence vaccine confidence over time, explore the impact of repeated exposure to accurate information, and investigate the role of social media and the influence of the social environment in maintaining positive attitudes. towards vaccines.
- Also, future research has to keep going investigating about cultural and societal factors, vaccine hesitancy among specific populations, vaccine communication strategies, vaccines adverse effects and risk perceptions and role of healthcare providers can further advance the understanding and promotion of vaccine acceptance.

By delving into these areas of research and conducting evidence-based interventions, we can continue to advance understanding of vaccine uptake and develop effective strategies to promote vaccination and protect public health.

CONCLUSION

In this chapter several key points related to vaccine acceptance have been discussed:

1. Importance of Communication: Healthcare professionals need to effectively communicate the benefits, risks, and importance of vaccines to foster confidence in the general population.
2. Collective Immunity: Achieving a high level of population immunity, also known as herd immunity, can prevent the spread of infectious diseases and indirectly protect unvaccinated individuals.
3. Vaccine Hesitancy in Visible Minorities: Understanding the reasons for vaccine hesitancy among visible minority groups is crucial to improving vaccine uptake and controlling the spread of diseases like COVID-19.

4. Impact of Misinformation: Misinformation, especially in minority and immigrant communities, fuels skepticism and hampers vaccine acceptance. Addressing the "infodemic" and providing reliable information sources is essential.
5. Personal Experience and Emotional Impact: witnessing a family member's suffering, can significantly influence attitudes toward vaccines. Empathy and understanding are necessary to address pragmatic concerns.
6. Vaccine Safety Communication: Different safety concerns, such as fears about side effects and ingredients, contribute to vaccine hesitancy. Dispelling common vaccine myths and providing accurate information is crucial.
7. Access to Vaccines and Vaccine Equity: Ensuring equitable access to vaccines, addressing logistical barriers, and improving positive messaging about vaccines are important for increasing vaccine acceptance.
8. Role of Health Workers: Health workers play a critical role in vaccine acceptance by providing reliable information and education. They are trusted sources for individuals making vaccination decisions.
9. Tailored Messages: Vaccine promotion messages should be tailored to address the specific concerns and misconceptions prevalent within different communities. Cultural and linguistic factors should be considered, and appropriate communication channels and formats should be used.
10. Use of Information and Communication Technologies: ICTs offer opportunities for innovative educational environments and can be utilized to improve vaccine acceptance through creative and targeted approaches.

Overall, a multifaceted approach involving effective communication, addressing concerns, ensuring access, and tailoring messages is necessary to advance vaccine acceptance.

In conclusion, it is imperative for universities to prioritize the development and promotion educational resources that address essential and culturally sensitive aspects as important as vaccination. The adverse effects of vaccines are clearly inferior to their individual and collective benefits. This MOOC course challenges the denialist attitudes and behaviors towards vaccines in healthcare, social health fields, and in the general population. In order to promote trust in vaccines, this project proposes a demand for greater rigor in digital media, restricting access and dissemination of disinformation content, and the use of information campaigns carried out through different media, such as radio, television, and social networks, to disseminate messages based on verifiable scientific evidence that is clear, precise and understandable for the population. Finally, the project supports the idea that the active involvement of healthcare professionals is key to achieving good adherence and high vaccination coverage and fostering good adherence.

REFERENCES

Al-Qerem, W. A., & Jarab, A. S. (2021). COVID-19 vaccination acceptance and its associated factors among a middle eastern population. *Frontiers in Public Health*, *9*, 632914. Advance online publication. doi:10.3389/fpubh.2021.632914 PMID:33643995

Aleixandre-Benavent, R., Castelló-Cogollos, L., & Valderrama-Zurián, J.C. (2020). Information and communication during the first months of Covid-19. Timeline, infodemic and misinformation, fake news, ongoing investigations, and the role of information specialist. *The Information Professional*, e290408. doi:10.3145/epi.2020.jul.08

Alimohamadi, Y., Hosamirudsari, H., Hesari, E., & Sepandi, M. (2022). Global COVID-19 vaccine acceptance rate: A systematic review and meta-analysis. *Zeitschrift für Gesundheitswissenschaften*, •••, 1–13. doi:10.100710389-022-01757-5 PMID:36188446

Aljawarneh, S. A. (2020). Reviewing and exploring innovative ubiquitous learning tools in higher education. *Journal of Computing in Higher Education*, *32*(1), 57–73. doi:10.100712528-019-09207-0

AlShurman, B. A., Khan, A. F., Mac, C., Majeed, M., & Butt, Z. A. (2021). What demographic, social, and contextual factors influence the intention to use COVID-19 vaccines: A scoping review. *International Journal of Environmental Research and Public Health*, *18*(17), 17. Advance online publication. doi:10.3390/ijerph18179342 PMID:34501932

Amutha, D. (2020). The Role and impact of ICT in improving the quality of education. SSRN *Electronic Journal*. doi:10.2139/ssrn.3585228

Arsenault, C., Harper, S., Nandi, A., Rodríguez, J. M. M., Hansen, P. M., & Johri, M. (2017). Monitoring equity in vaccination coverage: A systematic analysis of demographic and health surveys from 45 Gavi-supported countries. *Vaccine*, *35*(6), 951–959. doi:10.1016/j.vaccine.2016.12.041 PMID:28069359

Baiocco, R., Pezzella, A., Pistella, J., Kouta, C., Rousou, E., Rocamora-Perez, P., López-Liria, R., Dudau, V., Doru, A. M., Kuckert-Wöstheinrich, A., Ziegler, S., Nielsen, D., Bay, L. T., & Papadopoulos, I. (2022). LGBT+ training needs for health and social care professionals: A cross-cultural comparison among seven european countries. *Sexuality Research & Social Policy*, *19*(1), 22–36. doi:10.100713178-020-00521-2

Bedford, H., Attwell, K., Danchin, M., Marshall, H., Corben, P., & Leask, J. (2018). Vaccine hesitancy, refusal and access barriers: The need for clarity in terminology. *Vaccine*, *36*(44), 6556–6558. doi:10.1016/j.vaccine.2017.08.004 PMID:28830694

Brearley, L., Eggers, R., Steinglass, R., & Vandelaer, J. (2013). Applying an equity lens in the Decade of Vaccines. *Vaccine*, *31*, 103–107. doi:10.1016/j.vaccine.2012.11.088 PMID:23598470

Cataldi, J. R., Sevick, C., Pyrzanowski, J., Wagner, N., Brewer, S. E., Narwaney, K. J., Shoup, J. A., Resnicow, K., Glanz, J., Dempsey, A., & Kwan, B. M. (2019). Addressing personal parental values in decisions about childhood vaccination: Measure development. *Vaccine*, *37*(38), 5688–5697. doi:10.1016/j.vaccine.2019.08.009 PMID:31421930

Centres for Disease Control and Prevention (CDC). (2022). *Answers to your most common questions about childhood vaccines*. https://www.cdc.gov/vaccines/parents/FAQs.html

Cueva Gaibor, D. A. (2020). Transformación digital en la universidad actual. *Conrado*, *16*(77), 483–489.

Dainton, C., & Wong, J. (2022). Repairing our broken relationship with the vaccine hesitant: Empathy, compassion, and humility are needed. *Canadian Family Physician Medecin de Famille Canadien*, *211-213*(3), 211–213. Advance online publication. doi:10.46747/cfp.6803211 PMID:35292461

Das, K. (2019). *The role and impact of ICT in improving the quality of education: an overview.* https://www.kdsepathsala.com/2021/09/the-role-and-impact-of-ict-in-improving.html

Datosmacro.com. (n.d.). *COVID-19—Administred Vaccines 2023.* https://datosmacro.expansion.com/otros/coronavirus-vacuna

Doherty, M., Schmidt-Ott, R., Santos, J. I., Stanberry, L. R., Hofstetter, A. M., Rosenthal, S. L., & Cunningham, A. L. (2016). Vaccination of special populations: Protecting the vulnerable. *Vaccine*, *34*(52), 6681–6690. doi:10.1016/j.vaccine.2016.11.015 PMID:27876197

Eissa, A., Lofters, A., Akor, N., Prescod, C. & Nnorom, O. (2021). Increasing SARS-CoV-2 vaccination rates among Black people in Canada. *CMAJ: Canadian Medical Association Journal*, E1220-E1221. doi:10.1503/cmaj.210949

European Centre for Disease, Prevention, and Control (ECDC). (2015). *Vaccine hesitancy among healthcare workers and their patients in Europe. A qualitative study.* https://www.ecdc.europa.eu/en/publications-data/vaccine-hesitancy-among-healthcare-workers-and-their-patients-europe

European Centre for Disease Control and Prevention (ECDC). (2016). *Let's talk about protection: Enhancing childhood vaccination uptake.* https://www.ecdc.europa.eu/en/publications-data/lets-talk-about-protection-enhancing-childhood-vaccination-uptake

Fairhead, J., Leach, M., & Small, M. (2006). Public engagement with science? Local understandings of a vaccine trial in the Gambia. *Journal of Biosocial Science*, *38*(1), 103–116. doi:10.1017/S0021932005000945 PMID:16266443

Falagas, M. E., & Zarkadoulia, E. (2008). Factors associated with suboptimal compliance to vaccinations in children in developed countries: A systematic review. *Current Medical Research and Opinion*, *24*(6), 1719–1741. doi:10.1185/03007990802085692 PMID:18474148

Fernandez-Marquez, E., Vazquez-Cano, E., Lopez-Meneses, E., & Sirignano, F. M. (2020). The digital competence of student of different European universities. *Revista ESPACIOS, 41*(13). https://www.revistaespacios.com/a20v41n13/20411315.html

Fishbein, M., & Cappella, J. (2006). The role of theory in developing effective health communications. *Journal of Communication*, *56*(suppl_1), S1–S17. doi:10.1111/j.1460-2466.2006.00280.x

GAVI Alliance. (2013). *Summary of definitions of mission and strategic goal level indicators in GAVI Alliance Strategy 2011-2015.* https://www.gavi.org/programmes-impact/our-impact/measuring-our-performance/2011-2015-indicators

Geoghegan, S., O'Callaghan, K. P., & Offit, P. A. (2020). Vaccine safety: Myths and misinformation. *Frontiers in Microbiology*, *11*, 372. doi:10.3389/fmicb.2020.00372 PMID:32256465

Gómez Contreras, J. L., Bonilla Torres, C. A., & Esteban Ojeda, Y. C. (2022). Uso de TIC y TAC en la educación superior: Un análisis bibliométrico. *Revista complutense de educación, 33*(3), 601-613.

González Juárez, L., & Noreña Peña, A. L. (2011). Comunicación intercultural como medio para favorecer el cuidado culturalmente aceptable. *Enfermería Universitaria*, *8*(1), 55–60.

Goss, M. D., Temte, J. L., Barlow, S., Temte, E., Bell, C., Birstler, J., & Chen, G. (2020). An assessment of parental knowledge, attitudes, and beliefs regarding influenza vaccination. *Vaccine*, *38*(6), 1565–1571. doi:10.1016/j.vaccine.2019.11.040 PMID:31776028

Granda Asencio, L. Y., Espinoza Freire, E. E., & Mayon Espinoza, S. E. (2019). Las TICs como herramientas didácticas del proceso de enseñanza-aprendizaje. *Revista Conrado*, *15*(66), 104–110.

Greyling, T., & Rossouw, S. (2022). Positive attitudes towards COVID-19 vaccines: A cross-country analysis. *PLoS One*, *17*(3), e0264994. doi:10.1371/journal.pone.0264994 PMID:35271637

Gust, D. A., Darling, N., Kennedy, A., & Schwartz, B. (2008). Parents with doubts about vaccines: Which vaccines and reasons why. *Pediatrics*, *122*(4), 718–725. doi:10.1542/peds.2007-0538 PMID:18829793

Hamborsky, J., Kroger, A., & Wolfe, S. (2015). *Epidemiology and prevention of vaccine-preventable diseases.* Centres for Disease Control and Prevention (CDC). https://www.cdc.gov/vaccines/pubs/pinkbook/index.html

Hernández, C. A. H., Prada, R. P., & Espinel, G. A. E. (2022). Competencias TIC del docente en época de no presencialidad: Una mirada desde los estudiantes de comunicación social. *Revista Saber. Ciencia y Libertad*, *17*(1), 432–453. doi:10.18041/2382-3240aber.2022v17n1.7825

Hernández Angulo, J., Garbey Savigne, E., Enríquez O'Farrill, I. J., Hernández Angulo, J., Garbey Savigne, E., & Enríquez O'Farrill, I. J. (2021). Importancia de la perspectiva afectiva e intercultural en la comunicación profesional en tiempos de crisis. *Revista Habanera de Ciencias Médicas*, *20*(1). http://scielo.sld.cu/scielo.php?script=sci_abstract&pid=S1729-519X2021000100018&lng=es&nrm=iso&tlng=es

Jarrett, C., Wilson, R., O'Leary, M., Eckersberger, E., & Larson, H. J. (2015). Strategies for addressing vaccine hesitancy—A systematic review. *Vaccine*, *33*(34), 4180–4190. doi:10.1016/j.vaccine.2015.04.040 PMID:25896377

Jin, Q., Raza, S. H., Yousaf, M., Munawar, R., Shah, A. A., Hassan, S., Shaikh, R. S., & Ogadimma, E. C. (2022). Ingraining polio vaccine acceptance through public service advertisements in the digital era: The moderating role of misinformation, disinformation, fake news, and religious fatalism. *Vaccines*, *10*(10), 10. Advance online publication. doi:10.3390/vaccines10101733 PMID:36298598

Kalimeri, K. G., Beiró, M., Urbinati, A., Bonanomi, A., Rosina, A., & Cattuto, C. (2019). Human values and attitudes towards vaccination in social media. *Companion Proceedings of the 2019 World Wide Web Conference*, 248-254. 10.1145/3308560.3316489

Kapur, R. (2019). *Use of ICT in Improving Quality of Education.* Retrieved from https://www.researchgate.net/publication/333446961_Use_of_ICT_in_Improving_Quality_of_Education

Kreps, S. E., Goldfarb, J. L., Brownstein, J. S., & Kriner, D. L. (2021). The relationship between us adults' misconceptions about COVID-19 vaccines and vaccination preferences. *Vaccines*, *9*(8), 901. doi:10.3390/vaccines9080901 PMID:34452025

Kroger, A., Bahta, L., & Hunter, P. (2022). *General best practice guidelines for immunization. Best practices guidance of the Advisory Committee on Immunization Practices.* ACIP.

Laro-González, E. (2020). Innovar enseñando: La educación del futuro. Las TICs como factor motivador en la enseñanza. *Revista Jurídica de Investigación e Innovación Educativa, 21*, 11–23. doi:10.24310/REJIE.2020.v0i21.7530

Lasco, G., & Yu, V. G. (2021). Communicating COVID-19 vaccines: Lessons from the dengue vaccine controversy in the Philippines. *BMJ Global Health, 005422*(3), e005422. Advance online publication. doi:10.1136/bmjgh-2021-005422 PMID:33653732

Leask, J., Kinnersley, P., Jackson, C., Cheater, F., Bedford, H., & Rowles, G. (2012). Communicating with parents about vaccination: A framework for health professionals. *BMC Pediatrics, 12*(1), 154. doi:10.1186/1471-2431-12-154 PMID:22998654

Ledford, V., & Nan, X. (2023). Boosting vaccination: Accelerating vaccine uptake through communication science. *Metode Science Studies Journal, 0*(13), 103–109. doi:10.7203/metode.13.23759

Leung, S. O. A., Akinwunmi, B., Elias, K. M., & Feldman, S. (2019). Educating healthcare providers to increase Human Papillomavirus (HPV) vaccination rates: A Qualitative Systematic Review. *Vaccine: X, 3*, 100037. doi:10.1016/j.jvacx.2019.100037 PMID:31463471

Li, M., Luo, Y., Watson, R., Zheng, Y., Ren, J., Tang, J., & Chen, Y. (2021). Healthcare workers' (HCWs) attitudes and related factors towards COVID-19 vaccination: A rapid systematic review. *Postgraduate Medical Journal*. . doi:10.1136/postgradmedj-2021-140195

Lin, C., Mullen, J., Smith, D., Kotarba, M., Kaplan, S. J., & Tu, P. (2021). Healthcare providers' vaccine perceptions, hesitancy, and recommendation to patients: A systematic review. *Vaccines, 9*(7), 713. doi:10.3390/vaccines9070713 PMID:34358132

Maatuk, A. M., Elberkawi, E. K., Aljawarneh, S., Rashaideh, H., & Alharbi, H. (2022). The COVID-19 pandemic and E-learning: Challenges and opportunities from the perspective of students and instructors. *Journal of Computing in Higher Education, 34*(1), 21–38. doi:10.100712528-021-09274-2 PMID:33967563

MacIntyre, C. R. (2013). Elderly vaccination—The glass is half full. *Health (Irvine, Calif.), 5*(12), 80–85. doi:10.4236/health.2013.512A011

MacIntyre, C. R., Menzies, R., Kpozehouen, E., Chapman, M., Travaglia, J., Woodward, M., ... Adair, T. (2016). Equity in disease prevention: Vaccines for the older adults–a national workshop, Australia 2014. *Vaccine, 34*(46), 5463–5469. doi:10.1016/j.vaccine.2016.09.039 PMID:27686835

Madewell, Z. J., Chacón-Fuentes, R., Jara, J., Mejía-Santos, H., Molina, I.-B., Alvis-Estrada, J. P., Ortiz, M.-R., Coello-Licona, R., & Montejo, B. (2021). Knowledge, attitudes, and practices of seasonal influenza vaccination in healthcare workers, Honduras. *PLoS One, 16*(2), e0246379. Advance online publication. doi:10.1371/journal.pone.0246379 PMID:33539428

Mohamed, K., Rzymski, P., Islam, M. S., Makuku, R., Mushtaq, A., Khan, A., Ivanovska, M., Makka, S. A., Hashem, F., Marquez, L., Cseprekal, O., Filgueiras, I. S., Fonseca, D. L. M., Mickael, E., Ling, I., Arero, A. G., Cuschieri, S., Minakova, K., Rodríguez-Román, E., ... Rezaei, N. (2022). COVID-19 vaccinations: The unknowns, challenges, and hopes. *Journal of Medical Virology, 94*(4), 1336–1349. doi:10.1002/jmv.27487 PMID:34845731

Nachega, J. B., Sam-Agudu, N. A., Masekela, R., van der Zalm, M. M., Nsanzimana, S., Condo, J., Ntoumi, F., Rabie, H., Kruger, M., Wiysonge, C. S., Ditekemena, J. D., Chirimwami, R. B., Ntakwinja, M., Mukwege, D. M., Noormahomed, E., Paleker, M., Mahomed, H., Tamfum, J.-J. M., Zumla, A., & Suleman, F. (2021). Addressing challenges to rolling out COVID-19 vaccines in African countries. *The Lancet. Global Health*, *9*(6), 746–748. doi:10.1016/S2214-109X(21)00097-8 PMID:33713633

Ochieng, C., Anand, S., Mutwiri, G., Szafron, M., & Alphonsus, K. (2021). Factors associated with COVID-19 vaccine hesitancy among visible minority groups from a global context: A scoping review. *Vaccines*, *9*(12), 1445. doi:10.3390/vaccines9121445 PMID:34960192

Pan American Health Organization. (2020). *Communication on safe vaccination. Guidelines to support health personnel in communication with mothers, fathers, caregivers and patients.* . doi:10.37774/9789275322819

Papadopoulos, I. (2006). *Transcultural health and social care: development of culturally competent practitioners*. Churchill Livingstone Elsevier.

Papadopoulos, I. (2011). *Courage, compassion and cultural competence. In the 13th Anna Reynvaan Lecture*. De Stadsschouwburg - Amsterdam City Theatre, Academic medical Centre, University of Amsterdam. https://www.researchgate.net/publication/303885059_Courage_Compassion_and_Cultural_Competence_The_Anna_Reynvaan_Lecture_2011

Paternotte, E., Scheele, F., Seeleman, C. M., Bank, L., Scherpbier, A. J. J. A., & van Dulmen, S. (2016). Intercultural doctor-patient communication in daily outpatient care: Relevant communication skills. *Perspectives on Medical Education*, *5*(5), 268–275. doi:10.1007/S40037-016-0288-Y PMID:27638395

Paterson, P., Meurice, F., Stanberry, L. R., Glismann, S., Rosenthal, S. L., & Larson, H. J. (2016). Vaccine hesitancy and healthcare providers. *Vaccine*, *34*(52), 6700–6706. doi:10.1016/j.vaccine.2016.10.042 PMID:27810314

Pfattheicher, S., Petersen, M. B., & Böhm, R. (2022). Information about herd immunity through vaccination and empathy promote COVID-19 vaccination intentions. *Health Psychology*, *41*(2), 85–93. doi:10.1037/hea0001096 PMID:34570535

Pinaka, O., Spanou, I., Papadouli, V., Papanikolaou, E., Gioulekas, F., & Mouchtouri, V. A. (2021). The role of local primary healthcare units in increasing immunization uptake among children in vulnerable social groups through vaccination campaigns. *Public Health in Practice (Oxford, England)*, *2*, 100185. doi:10.1016/j.puhip.2021.100185 PMID:36101591

Poveda-Pineda, D. F., Cifuentes-Medina, J. E., Poveda-Pineda, D. F., & Cifuentes-Medina, J. E. (2020). Incorporation of information and communication technologies (ICT) during the learning process in higher education. *Formación Universitaria*, *13*(6), 95–104. doi:10.4067/S0718-50062020000600095

Ree, M. E. B. van der, & Lozano, J. A. M. (2019). Competencias claves de los estudiantes universitarios para el uso de las TIC. *Revista de Comunicación de la SEECI*, 43-72. doi:10.15198/seeci.2019.50.43-72

Ridda, I., Motbey, C., Lam, L., Lindley, I. R., McIntyre, P. B., & Macintyre, C. R. (2008). Factors associated with pneumococcal immunisation among hospitalised elderly persons: A survey of patient's perception, attitude, and knowledge. *Vaccine*, *26*(2), 234–240. doi:10.1016/j.vaccine.2007.10.067 PMID:18054818

Rocamora-Pérez, P., López-Liria, R., Padilla-Góngora, D., Morales-Montoya, A., Díaz-López, P., & Aguilar-Parra, J. M. (2018). Training needs for immigrant caregivers of older people in the home in Europe: IENE7 Project. *Revista INFAD de Psicología. International Journal of Developmental and Educational Psychology.*, *4*(1), 1. Advance online publication. doi:10.17060/ijodaep.2018.n1.v4.1307

Rodrigo, M. (n.d.). *La comunicación intercultural.* Portal de comunicación. https://www.educacion.navarra.es/documents/57308/57732/2011_comintercultural.pdf/8bcd2117-0621-4ad8-973b-8d13f83ce0f0

Ross, D., & Magris, M. (2012). The role of communication and knowledge management as evidenced by HCP vaccination programs in the Netherlands, Germany and Italy: possible suggestions for medical translators. *Linguistica Antverpiensia, New Series – Themes in Translation Studies, 11.* doi:10.52034/lanstts.v11i.300

Salomoni, M. G., Di Valerio, Z., Gabrielli, E., Montalti, M., Tedesco, D., Guaraldi, F., & Gori, D. (2021). Hesitant or not hesitant? A systematic review on global COVID-19 vaccine acceptance in different populations. *Vaccines*, *9*(8), 873. doi:10.3390/vaccines9080873 PMID:34452000

Samad, L., Butler, N., Peckham, C., & Bedford, H. (2006). Incomplete immunisation uptake in infancy: Maternal reasons. *Vaccine*, *24*(47), 6823–6829. doi:10.1016/j.vaccine.2006.06.039 PMID:17050048

Sharpe, R., & Benfield, G. (2016). The student experience of e-learning in higher education: A review of the literature. *Brookes EJournal of Learning and Teaching*, 1.

Siciliani, L., Wild, C., McKee, M., Kringos, D., Barry, M. M., Barros, P. P., De Maeseneer, J., Murauskiene, L., & Ricciardi, W. (2020). Strengthening vaccination programmes and health systems in the European Union: A framework for action. *Health Policy (Amsterdam)*, *124*(5), 511–518. doi:10.1016/j.healthpol.2020.02.015 PMID:32276852

Smith, P. J., Chu, S. Y., & Barker, L. E. (2004). Children who have received no vaccines: Who are they and where do they live? *Pediatrics*, *114*(1), 187–195. doi:10.1542/peds.114.1.187 PMID:15231927

Statista. (n.d.). *Salud e Industria farmacéutica.* https://es.statista.com/sectores/1174/salud-e-industria-farmaceutica/

Tversky, A., & Kahneman, D. (1981). The framing of decisions and the psychology of choice. *Science*, *211*(4481), 453–458. doi:10.1126cience.7455683 PMID:7455683

UNICEF. (2019). *United Nations children's fund; participants' manual: interpersonal communication for immunization package.* https://ipc.unicef.org/sites/ipcfi/files/2019-06/UNICEF_Participant_Manual_FINAL_2.pdf

UNICEF. (n.d.) *Interpersonal Communication for Immunization Initiative.* https://ipc.unicef.org/

Valverde-Martínez, M. Á., López-Liria, R., Martínez-Cal, J., Benzo-Iglesias, M. J., Torres-Álamo, L., & Rocamora-Pérez, P. (2023). Telerehabilitation, a viable option in patients with persistent post-COVID syndrome: A systematic review. *Health Care*, *11*(2), 2. Advance online publication. doi:10.3390/healthcare11020187 PMID:36673555

Venegas-Ramos, L., Luzardo-Martínez, H. J., & Pereira-Santana, A. (2020). Conocimiento, formación y uso de herramientas TIC aplicadas a la Educación Superior por el profesorado de la Universidad Miguel de Cervantes. *Edutec. revista electrónica de tecnología educativa, 71*, 35-52. . doi:10.21556/edutec.2020.71.1405

Voudoukis, N., & Pagiatakis, G. (2022). Massive Open Online Courses (MOOCs): Practices, trends, and challenges for the higher education. *European Journal of Education and Pedagogy*, *3*(3), 288–295. doi:10.24018/ejedu.2022.3.3.365

Washington State Department of Health. (n.d.). *Immunization Action Coalition Washington.* Immunity Community. https://immunitycommunitywa.org/iacw/

WHO. (2020b). *Coronavirus disease (Covid-19) outbreak.* https://www.who.int/es/emergencies/diseases/novel-coronavirus-2019/advice-for-public

WHO. (2022). *Implementation guide for vaccination of health workers.* https://apps.who.int/iris/rest/bitstreams/1453340/retrieve

World Health Organization WHO. (2020a). *Global Vaccine Action Plan 2011-2020.* https://www.who.int/immunization/global_vaccine_action_plan/en/

Writers, S. (2022, December 20). *Vaccine myths debunked.* https://www.publichealth.org/public-awareness/understanding-vaccines/vaccine-myths-debunked/

Zaremohzzabieh, Z., Roslan, S., Mohamad, Z., Ismail, I. A., Ab Jalil, H., & Ahrari, S. (2022). Influencing factors in MOOCs adoption in higher education: A meta-analytic path analysis. *Sustainability (Basel), 14*(14), 14. Advance online publication. doi:10.3390u14148268

Zonneveld, M., Patomella, A.-H., Asaba, E., & Guidetti, S. (2020). The use of information and communication technology in healthcare to improve participation in everyday life: A scoping review. *Disability and Rehabilitation*, *42*(23), 3416–3423. doi:10.1080/09638288.2019.1592246 PMID:30966833

KEY TERMS AND DEFINITIONS

Culturally Competent Compassion: The human quality of understanding the suffering of others and having the desire to take action, using culturally appropriate and acceptable nursing interventions. This approach acknowledges and respects the cultural backgrounds of both patients and caregivers, considering the unique circumstances in which care is provided.

Equity: The principle of equity in health is connected to the notion of fairness in society as proposed by Rawls' theory of justice. It emphasizes the need for healthcare services and interventions to be distributed based on individual needs, aiming to reduce social, environmental, and economic disparities. Ensuring equal access to immunization is a crucial aspect of the right to health.

Herd Immunity: Herd immunity occurs when a large number of people in a community become immune to a disease. This prevents the virus or bacteria from spreading easily, protecting those who cannot be vaccinated or who have immune systems. By getting vaccinated, we help build this herd immunity and prevent the spread of disease, providing protection for the entire community.

Immunity: Immunity is the body's ability to protect itself from disease. Our immune system fights against germs and bacteria that can make us sick. When our body encounters a germ or bacteria, it produces special defenses called antibodies to fight it, which help destroy the germs and prevent us from getting sick. There are two types of immunity, active, which is generated by our own body, and passive which is the protection against disease through antibodies produced by another human or animal (like maternal antibodies).

Side Effect: Any health problem that happens after vaccination is considered an adverse event and it can be a true adverse reaction that is related to the vaccine, or a coincidental event that happened following vaccination.

Vaccination: It is a simple, safe and effective way to protect ourselves against harmful diseases, taking advantage of our body's natural defences to develop resistance against different germs and strengthen our immune system. This term is often used interchangeably with immunization.

Vaccine Hesitancy: Vaccine hesitancy is when someone is left in limbo and cannot decide whether or not to get vaccinated, even though vaccines are available. This includes those who refuse to be vaccinated, those who are late in receiving vaccines, those who accept them but are undecided about their use, or those who get vaccinated with some but not others. Vaccine reluctance is somewhat complicated and context dependent, and can vary by time, place, and specific vaccines.

Vaccines: A way to train immune system to create antibodies, from only killed or weakened forms of germs like viruses or bacteria, that cannot cause the disease or put us at risk of its complications.

Vulnerable Groups Vaccination: A group formed by pregnant women, premature babies, individuals with chronic and immunodeficient diseases, older adults and people from developing countries are considered vulnerable.

Chapter 6
Predicting University Students' Mobile Learning Attitudes in Language Classrooms:
The Role of Digital Literacy

Ali Orhan
https://orcid.org/0000-0003-1234-3919
Zonguldak Bülent Ecevit University, Turkey

ABSTRACT

This study aimed to examine the predictive role of digital literacy skills on mobile English learning attitudes. It was conducted with 135 university students studying in the school of foreign languages of a state university in Turkey, and the data were gathered using digital literacy scale and English mobile learning attitude scale. This study showed that university students' digital literacy skills and mobile English learning attitudes were high and their digital literacy skills have a moderate and positive relationship with their mobile English learning attitudes. Besides, it was concluded that university students' digital literacy skills significantly predicted their mobile English learning attitudes and they together explained 50% of the total variance on their mobile English learning attitudes. Also, university students' scores on the attitude subscale of the digital literacy scale presented the largest effect on their mobile English learning attitudes.

INTRODUCTION

With the quick advancement in mobile technology, integration of mobile technologies in teaching and learning processes has become a vital factor to enhance the quality of education (Almaiah, 2018; Sophonhiranrak, 2021), and hence, mobile learning (m-learning) has gained importance, especially in higher education. Thanks to m-learning, learning, and teaching activities can be extended beyond a

DOI: 10.4018/978-1-6684-8861-4.ch006

traditional classroom because most people have portable devices today (Wu, 2014). Potential benefits of m-learning make it an attractive topic for many researchers, and it has become to be used in many different fields (Almaiah et al., 2020).

Language learning is one of these fields because m-learning makes these portable devices potentially beneficial tools in language-learning environments. Researchers have agreed on the positive effect of m-learning on language learning (Darmi & Albion, 2014; Bano et al., 2018; Elaish et al., 2019; Tommerdahl et al., 2022; Viberg & Grönlund, 2012; Kim & Kwon, 2012). A plethora of studies has revealed that mobile language learning is effective to facilitate various language skills like listening, writing, reading, etc. and language components like grammar and pronunciation (Kondo et al., 2012; Çavuş & İbrahim, 2017; Hwang & Fu, 2019; Hwang et al., 2014; Li, 2022; Read et al., 2021; Özer & Kılıç, 2018; Shadiev et al., 2018; Yu et al., 2023). Also, previous meta-analysis and systematic review studies have revealed that m-learning significantly enhances academic achievement in language learning environments (Chen et al., 2020; Sung et al., 2015; Mihaylova et al. 2022; Mahdi, 2018; Cho et al., 2018; Taj et al., 2016).

It can be said that successful implementation and use of m-learning largely depend on teachers' and learners' willingness and acceptance to use m-learning and their beliefs on how this technology would be beneficial in education processes for them (Al-Shehri, 2014; Alowayr, 2022; Al-Rahmi et al., 2022; Al-Seghayer, 2014; Aytekin et al., 2022; Padmanathan & Jogulu, 2018; Su & Chao, 2022; Abu-Al-Aish & Love, 2013). Although there are a lot of studies concluding the positive effect of m-learning on learning outcomes in language classrooms, little is known about the possible variables that can affect learners' and teachers' willingness to adopt m-learning and their attitudes towards it (Akour, 2009; Almaiah et al., 2016; Tsuei & Hsu, 2019). Several studies have been conducted to investigate the factors that can be effective on willingness to adopt m-learning and attitudes towards it and they concluded that some variables like parental involvement and parents' attitudes (Kong, 2018; Tsuei & Hsu, 2019), usefulness and the effort needed to learn how to use m-learning (Montrieux et al., 2014; Odiakaosa et al., 2017; Sarrab et al., 2016), perceived enjoyment and compatibility of m-learning (Sánchez-Prieto et al., 2019), ICT literacy (Mac Callum et al., 2014b) play a vital role on accepting m-learning and attitudes towards it.

Also, digital literacy (DL) is another important factor that can influence individuals' attitudes towards m-learning. DL is a vital requirement for effective m-learning environments during language learning (Kukulska-Hulme & Traxler, 2005) and it stands at the heart of m-learning. There is a sound theoretical base regarding the significant effect of DL on attitudes towards m-learning. Also, there are some studies indicating DL is an important factor that can be effective on attitudes towards m-learning (Özcan, 2022; Puebla et al., 2022; Mac Callum et al., 2014a; Moya & Camacho, 2021; Krasulia & Saks, 2020; Dashtestani, 2016; Saad et al., 2021). However, there has not been much research on the predictive role of DL on attitudes towards m-learning in language classrooms, and hence, empirical evidence regarding this predictive role is limited. Therefore, this chapter aimed to investigate the predictive role of DL levels of university students on their attitudes towards m-learning in language classrooms. To this end, the following question was sought:

1. Do university students' scores on subscales of the Digital Literacy Scale (attitude, technical, cognitive, and social) significantly predict their m-learning attitudes?

LITERATURE REVIEW

Mobile technology can be defined as any portable and personal devices (Naismith et al., 2014) and these devices' light weight and portability (Kukulska-Hulme, 2009) make them effective tools to be used for different purposes in different fields such as tourism, entertainment, and banking, etc. (Hew, 2009; Lakhal et al., 2007). In addition to these different fields, these devices have started to be used for teaching/learning purposes recently (Mayer, 2020). They have become one of the critical components of learning environments because they can connect teachers, learners, and content independent of space and time (Wu et al., 2012). Therefore, m-learning has gained importance in education in the last two decades and it is used widely for teaching and learning purposes, especially in higher education (Danish & Hmelo-Silver, 2020; Rysbayeva et al., 2022).

M-learning can be simply defined as learning in mobile environments using handheld mobile devices (Seppala & Alamaki, 2003). It refers to the usage of technology based on mobile tools like tablets, e-readers, smartphones, or portable computers to improve learners' performance, enhance their knowledge and skills, and support their learning (Clark & Mayer, 2008). It allows learners to access the content and materials without time and place constraints thanks to mobile devices and internet (Lan & Sie, 2010) and it enhances active learning, productivity, and performance of the learners by providing them with real-time interaction, discussion opportunities among learners and teachers, and instant feedback (Keskin & Metcalf, 2011; Vasiliou & Economides, 2007). Therefore, we can say that m-learning gives learners a flexible learning environment where they can access the learning content whenever, wherever, and however they want, and hence, it enhances the learner autonomy and learners take control of their own learning (Clarke & Svanaes, 2014). M-learning not only provides students with online learning space but also enables them to access learning materials and activities quickly without time and place constraints, increases the students-students and students-teacher interaction by creating extra channels for communication, and creates opportunities for innovative learning (Chavoshi & Hamidi, 2019; Hamidi & Jahanshaheefard, 2019). Thanks to m-learning, learners can share their ideas among each other not only inside the school but also outside the school (Al-Emran et al., 2016) resulting in bridging the gap between formal and informal learning environments (Johnson et al., 2011). Learners can access the knowledge not only in the class via their teachers and peers but also outside the school using their mobile devices and this affects their learning capability and increases the possibility of meaningful learning (Almaiah & Alismaiel, 2019; Kumar & Chand, 2019). Therefore, m-learning provides learners with productive, collaborative, adaptive, and communicative learning environments in remote locations (Laurillard & Pachler, 2007) and learners can access richly diversified learning materials and activities (Huang et al., 2007).

Mobile technologies are also seen as effective tools to be used in language learning contexts and m-learning has started to be used widely in language learning environments (Levy & Kennedy, 2005), and hence, mobile language learning has recently gained interest from researchers and teachers (Godwin-Jones, 2008). Mobile language learning which refers to the usage of mobile technologies to enhance and assist the language learning process offers diverse advantages for language learning environments (Kukulska-Hulme, 2012). Mobile language learning can enable interactive and spontaneous learning content and activities across diverse contexts. Mobile language learning mostly occurs in informal learning environments such as while commuting between school and home instead of sitting in the classroom and it transfers language learning from formal learning environments to informal ones. Therefore, mobile language learning offers a flexible learning environment for learners and language learning goes beyond the traditional four walls of the school thanks to m-learning (Beres, 2011). Mobile language learning increases the access of learners to contextual, situated, and authentic materials and extralinguistic input, the possibility of interaction within and beyond the learning environment, and opportunities for additional language practice and interaction outside the classroom (Hoven & Palalas, 2011; Comas-Quinn et al., 2009). Besides, it transforms the language learning process into a more learner-centered and collaborative one by providing learners with the chance of accessing the content whenever or wherever they want and communicating with each other when they need help (Nah et al., 2008).

Digital Literacy and Mobile Learning Attitudes

DL can be simply defined as the mixture of different literacies related to the usage of digital technologies (Ng, 2012). It is about not only using technology in the teaching and learning process but also finding, creating, sharing, and consuming information effectively in digital environments (Hamutoğlu et al., 2017; Ng, 2012). Therefore, it can be said that DL means a mixture of technical proficiency and meaningful engagement in the digital world (Lankshear & Knobel, 2008; Madigan et al., 2007). Digital literate individuals have the ability to communicate, collaborate, solve problems, create and share content, and manage information in digital environments (Gilster, 1997). DL is about the enhancement of the awareness and attitude level of individuals regarding digital technologies. DL provides individuals with the necessary skills to access digital resources, evaluate, interpret, analyze, and synthesize them, reveal new information, create new channels for communication, and transfer these skills to daily life (Markauskaite, 2007; Martin, 2008). In other words, individuals can adapt to the digital world and stay there thanks to their DL skills (Manovich, 2001). Therefore, DL is really important in today's world which is full of digital and internet technologies integrated into the social, work, and education life of individuals.

DL is a concept which stands at the heart of m-learning because incorporation of m-learning into education requires the improvement of a fundamental understanding of technology and some abilities which are necessary to use technology and new interactive media. DL is related to individuals' adaptability and tendency towards using new technologies (Ng, 2012). Therefore, individuals are expected to be digital literate for effective m-learning environments and high DL skills can enhance the attitudes towards using mobile technologies for educational purposes (Ng, 2015). During mobile language learning, individuals have to use diverse mobile gadgets, internet, and different language-learning apps which makes DL a vital requirement to be successful in m-learning environments during language learning (Kukulska-Hulme & Traxler, 2005; Özcan, 2022) because DL is a prerequisite to use these gadgets and tools effectively (Baran, 2014). According to Brown and Mbati (2015), low DL skills are an important challenge for both teachers and learners during m-learning and high DL skills are a must for effective m-learning. In other words, DL is crucial to maximize the potential of m-learning and it is essential to realize the learning potential of m-learning (Terras & Ramsay, 2012; Puebla et al., 2022).

In his study aiming to examine the predictive role of DL on prospective teachers' m-learning attitudes, Özcan (2022) concluded that DL levels of 443 university students significantly predicted their m-learning attitudes. In a similar vein, in their study aiming to investigate the factors affecting adults' adoption and use of m-learning and their attitude towards it while learning a second foreign language, Puebla et al. (2022) found that DL level of adults significantly predicted their intention, attitude, and use of language learning apps. Also, Mac Callum et al. (2014a) carried out a study with 175 lecturers to examine the effect of DL along with some other variables on m-learning acceptance and they concluded that DL has an important effect on the lecturers' behavioral intention to use m-learning. Similarly, in their study aiming to investigate the moderating effect of DL on behavioral intention of teachers towards m-learning, Saad et al. (2021) concluded that DL plays an important role on teachers' behavioral intention towards m-learning. In their study aiming to identify the critical factors that can affect the adoption and use of m-learning, Moya & Camacho (2021) reviewed 26 studies and identified five categories. They stated that DL is the third most important factor that has a significant effect on the adoption and use of m-learning. In their study aiming to examine students' perceptions towards mobile language learning, Krasulia and Saks (2020) identified some challenges faced by students during mobile language learning and they categorized these challenges as technical and DL constraints. They stated that lack of DL skills is an important barrier against the success of m-learning in language classrooms. In a similar vein, Dashtestani (2016) concluded that low DL is one of the important challenges that might discourage learners from m-learning, and hence, hinders the effect of it in his study.

METHOD

A cross-sectional survey design was employed in this chapter. While university students' scores on attitude, technical, cognitive, and social subscales of the Digital Literacy Scale were predictor variables, their m-learning attitude was the dependent variable.

Study Group

The study group consists of 135 university students (74 female and 61 male) studying in the School of Foreign Languages of a state university in Türkiye. This school offers one-year English preparatory education for students before they start to study in their departments. All of the students are in their preparatory year, and they are learning English as a foreign language. The mean age of the students was 19.15 (SD=0.85) and their ages ranged from 18 to 22. A-priori power analysis which was conducted before the study using G*Power 3 software program (Faul et al., 2007) revealed that a minimum of 129 data were required to conduct a linear multiple regression analysis (alpha=0.05; power=0.95; 4 predictor variables) for a medium effect size (f^2=0.15). Hence, it can be said that the sample size of 135 was adequate.

Data Collection Tools

Digital Literacy Scale (DLC)

The DLC developed by Ng (2012) and adapted into Turkish by Hamutoğlu et al. (2017) was used to determine university students' DL levels. The scale has 17 items and four subscales, namely, attitude, technical, cognitive, and social. The attitude subscale of DLC aims to investigate individuals' attitudes towards the usage of digital and mobile technologies for learning while the technical subscale of DLC refers to having the technical abilities which are required to use digital and mobile technologies both for learning and in daily life. Also, the cognitive subscale of DLC targets to measure individuals' ability to think critically while producing, consuming, and handling digital information and it also refers to having the necessary skills to select, use, and evaluate appropriate software programs for learning or doing a task while the social subscale is about using the internet in a responsible way to communicate, socialize, and learn. Turkish adaptation study of DLC which was carried out with 395 university students indicated that the factor structure of Turkish DLC was consistent with the original scale. Cronbach Alpha values of the subscales ranged between 0.70 and 0.89 while it was 0.93 for the total scale in the Turkish adaptation study. In this chapter, the reliability coefficients were found to be 0.76, 0.80, 0.68, and 0.66 for attitude, technical, cognitive, and social subscales, respectively while it was calculated as 0.87 for the total scale.

English Mobile Learning Attitude Scale (EMLAS)

EMLAS developed by Liu (2017) and adapted into Turkish by Önal & Önal (2019) was used to investigate university students' m-learning attitudes. The scale has 21 items and four subscales, namely, perceived usefulness, affection, perceived control, and behavior. Turkish adaptation study of EMLAS which was carried out with 309 university students indicated that the four-factor structure of Turkish EMLAS was consistent with the original scale. Cronbach Alpha values of the subscales ranged between 0.70 and 0.88 while it was 0.90 for the total scale in the Turkish adaptation study. In this chapter, the reliability coefficient was found to be 0.85 for the total scale.

Data Collection

Ethical committee approval to conduct this study was obtained from ZBEU and the data were collected in 2022-2023 academic year. Before implementation of instruments, students were informed about all privacy and confidentiality issues, aim of the study, and their right to withdraw from the study whenever they want. The instruments were completed in about 15-20 minutes.

Data Analysis

Before starting to analyze the data, firstly all of the variables were checked to ensure that the dataset did not include any missing data. After that, investigation of skewness and kurtosis values revealed that the dataset presented normal distribution (see Table 1). Then, possible outliers per variable and multivariate outliers were checked using Z transformation and Mahalanobis Distance values and these values showed that the dataset did not include any influential outliers. Also, examination of Pearson correlation, CI, VIF, and tolerance values showed that there was no high correlation among the variables. The data were analyzed using descriptive statistics, Pearson correlation, and multiple linear regression with enter method.

RESULTS

Descriptive statistics for the DLC and EMLAS can be seen in Table 1.

As shown in Table 1, university students presented high mobile English learning attitudes ($\overline{X}$=3.67). Also, they had high scores regarding the subscales of DLC indicating that they had high DL skills. Besides, they had the highest mean score in the cognitive subscale ($\overline{X}$=3.98) and it was followed by the attitude ($\overline{X}$=3.69), technical ($\overline{X}$=3.68), and social ($\overline{X}$=3.64) subscales.

Table 1. Descriptive statistics for the DLC and EMLAS and Skewness and Kurtosis values for each instrument

	Skewness		Kurtosis			
	Statistic	**Std. Error**	**Statistic**	**Std. Error**	$\bar{X}$	**sd**
Attitude subscale	-0.312	0.209	-0.001	0.414	3.69	0.66
Technical subscale	-0.362	0.209	-0.663	0.414	3.68	0.74
Cognitive subscale	-0.675	0.209	0.060	0.414	3.98	0.75
Social subscale	-0.550	0.209	-0.070	0.414	3.64	0.88
Mobile English learning attitude	-0.211	0.209	0.004	0.414	3.67	0.53

Table 2. Inter-correlations among DL skills and mobile English learning attitudes

	Technical Subscale	Cognitive Subscale	Social Subscale	Mobile L. A.
1. Attitude subscale	0.574**	0.545**	0.413**	0.616**
2. Technical subscale	-	0.405**	0.471**	0.570**
3. Cognitive subscale	-	-	0.519**	0.525**
4. Social subscale			-	0.492**
Note: **correlation is significant at p< 0.01; Mobile L. A. means mobile English learning attitudes				

As shown in Table 2, university students' scores regarding the attitude (r = 0.616), technical (r = 0.570), cognitive (r = 0.525), and social (r = 0.492) subscales of DLC had a moderate and positive relationship with their mobile English learning attitudes. Also, subscales of DLC were found to be weakly to moderately correlated to each other.

Table 3. Multiple linear regression results regarding the predictive role of DL skills on mobile English learning attitudes

	B	Std. Error	β	t	p
Constant	1.242	0.217	-	5.715	0.00
Attitude subscale	0.257	0.067	0.317	3.813	0.00
Technical subscale	0.177	0.057	0.245	3.078	0.00
Cognitive subscale	0.122	0.057	0.171	2.142	0.03
Social subscale	0.094	0.046	0.156	2.030	0.04
$R=0.708$, $R^2=0.502$, $F_{(4, 134)}= 32.723$, $p<0.01$					

As shown in Table 3, multiple linear regression analysis results showed that university students' scores on the attitude subscale ($\beta=0.317$ $t_{(135)=}3.813$, $p<0.05$), technical subscale ($\beta=0.245$, $t_{(135)=}3.078$, $p<0.05$), cognitive subscale ($\beta=0.171$, $t_{(135)=}2.142$, $p<0.05$), and social subscale ($\beta=0.156$, $t_{(135)=}2.030$, $p<0.05$)

significantly predicted their mobile English learning attitudes ($R=0.708$, $R^2=0.502$, $p<0.01$). Also, one-way ANOVA test results indicated that the established regression model was significant ($F_{(4,134)}=32.723$, $p<0.01$). Therefore, it can be said that university students' DL skills are a significant predictor for their m-learning attitudes in language classrooms and they together explained 50% of the total variance on their m-learning attitudes. Also, university students' scores regarding the attitude subscale ($\beta=0.317$) had the largest effect on m-learning attitudes and it was followed by the technical ($\beta=0.245$), cognitive ($\beta=0.171$), and social ($\beta=0.156$) subscale scores, respectively.

DISCUSSION AND RECOMMENDATIONS

This chapter aimed to examine the predictive role of university students' DL skills on their mobile English learning attitudes. It was found that university students' DL skills and mobile English learning attitudes were high. Another result obtained in this chapter indicated that university students' DL skills had a moderate and positive relationship with their mobile English learning attitudes. Also, the subscales of DLC were found to be weakly to moderately correlated to each other. Most of the previous studies aiming to investigate university students' DL skills (Akman, 2021; Bay, 2021; Kaya-Özgül et al., 2023; Ocak & Karakuş, 2019) and m-learning attitudes (Turgut & Akdemir, 2019; Yağan, 2023; Yılmaz & Çakır, 2022) in Türkiye revealed that university students had high DL skills and m-learning attitudes. Therefore, it can be said that most of the previous literature, including this chapter, have revealed that university students have high DL skills and m-learning attitudes in Türkiye which is an important result that shows Turkish university students' DL skills and m-learning attitudes, which are really important in this century, are not far from the desired level.

This chapter also showed that university students' scores on subscales of DLC significantly predicted their mobile English learning attitudes and they together explained 50% of the total variance on their m-learning attitudes. Özcan (2022) found that university students' DL levels significantly predicted their m-learning attitudes in his study aiming to investigate the predictive role of DL on 443 prospective teachers' m-learning attitudes. Also, Puebla et al. (2022) carried out a study to examine the factors affecting adults' adoption and use of m-learning and their attitude towards it in the language learning process and they concluded that DL level of adults significantly predicted their intention, attitude, and use of language learning apps which indicates that DL is an important factor on m-learning attitudes. Similarly, in their study aiming to examine the effect of different variables including DL on lecturers' m-learning acceptance, Mac Callum et al. (2014a) concluded that DL is an important factor that affects the lecturers' behavioral intention to use m-learning. Similarly, Saad et al. (2021) found that DL plays an important role on teachers' behavioral intention towards m-learning in their study aiming to investigate moderating effect of DL on behavioral intention of teachers towards m-learning. In their study, Moya & Camacho (2021) reviewed 26 studies to find out the critical factors which can be effective on the usage of m-learning, and they concluded that the critical factors can be categorized into five groups and DL is the third most vital factor which can affect the adoption and use of m-learning significantly. Besides, Krasulia and Saks (2020) and Dashtestani (2016) tried to investigate students' perceptions towards m-learning in language classrooms and they concluded that lack of DL skills is one of the important challenges that can hinder the effect of m-learning and discourage learners from it. In short, a huge body of previous studies, including this chapter, has shown that DL is an effective factor on students' m-learning attitudes. Therefore, it can be said that previous literature confirmed the results of this chapter.

As DL has a strong relation with individuals' adaptability and tendency towards using new technologies for learning and teaching purposes (Ng, 2012), it is an important concept which stands at the heart of m-learning and is a vital requirement for effective m-learning environments during language learning (Kukulska-Hulme & Traxler, 2005). Mobile language learning process includes using diverse mobile gadgets, language learning apps, and the internet. Therefore, it can be said that students need to have the necessary skills and competencies to use new interactive media, mobile technologies, and internet which makes DL an important requirement for effective m-learning environments (Kukulska-Hulme & Traxler, 2005) because DL is a prerequisite to use these technologies in an effective and correct way (Baran, 2014). Individuals are required to be digital literate for effective m-learning environments (Ng, 2015) because low DL skills are an important problem for both teachers and learners in m-learning environments (Brown & Mbati, 2015). In other words, high DL skills can increase the attitudes towards using mobile technologies in education (Ng, 2015) and maximize the potential of m-learning (Terras & Ramsay, 2012). Therefore, it can be said that the result regarding the significant predictive role of DL on university students' mobile English learning attitudes coincides with the theoretical background.

Besides, this chapter showed that the attitude subscale had the largest effect on m-learning attitudes, and it was followed by the technical, cognitive, and social subscales, respectively. Ng (2012) states that the attitude subscale of DLC aims to investigate individuals' attitudes towards the usage of digital and mobile technologies for learning. Individuals with high scores regarding the attitude subscale of DLC think that digital and mobile technologies motivate them to learn and learning with them is fun. Also, they think that learning with digital and mobile technologies makes learning more interesting, enhances their learning capacity, and allows them to learn independently (Ng, 2012). Therefore, it is not surprising that the attitude subscale had the largest effect on m-learning attitudes because it is expected that the individuals with high scores regarding the attitude subscale of DLC also have positive attitudes towards m-learning. Besides, Ng (2012) states that the technical subscale of DLC refers to possessing the technical and operational skills which are needed to use digital and mobile technologies both for learning and in everyday activities. Individuals with high DL in terms of the technical subscale can use peripheral devices like mobile phones, earphones, speakers, etc. understand file structures, manage data transfer, download and install applications, set up and use communication and social networking tools (Ng, 2012). The cognitive subscale of DLC targets to measure individuals' ability to think critically while producing, consuming, and handling digital information. It also refers to having the necessary skills to select, use, and evaluate appropriate software programs for learning or doing a task. Individuals with high DL in terms of the cognitive subscale are aware of the ethical, moral, and legal issues related to mobile and digital environments. The social subscale is about using the internet in a responsible way to communicate, socialize, and learn. Individuals with high DL in terms of the social subscale have netiquette, use appropriate language and words in order not to cause any misinterpretation and misunderstanding, respect other people, and can keep their personal information private and protect their safety and privacy. During the mobile language learning process, learners need to use diverse mobile and internet technologies like earphones, speakers, language learning apps, or internet, etc. Therefore, students should have the necessary skills and competencies to use new interactive media, mobile technologies, and the internet. Also, it can be expected that when they see themselves as competent to use these mobile and internet technologies, they will be more willing to adopt and use m-learning. Therefore, this can be shown as the possible explanation for the result of this chapter which indicated that the technical subscale is the second strongest predictor for m-learning attitudes because the technical subscale of DLC is directly related to these technical and operational skills which are necessary for m-learning environments.

LIMITATIONS AND IMPLICATIONS FOR FURTHER RESEARCH

This chapter provides some important results regarding the predictive role of DL on m-learning. However, it has several limitations. The study group is the first limitation because this study was conducted with university students recruited from a single university. Therefore, the predictive role of DL on m-learning can be investigated with other sample groups consisting of students from various educational levels. Second, this chapter focused on the predictive role of DL on m-learning in language learning context. Therefore, this can be shown another limitation and it would be a good idea to investigate the predictive relation between DL and m-learning in other educational settings. Third, the data for this chapter were collected using self-report quantitative tools which can be affected by social desirability, and this can be shown as a limitation. Therefore, conducting other studies using qualitative or mixed methods would be a good idea to better understand the predictive role of DL on m-learning.

CONCLUSION

In short, this chapter showed that university students' DL skills and mobile English learning attitudes were high. Also, their scores on attitude, technical, cognitive, and social subscales significantly predicted their mobile English learning attitudes. This result which coincides with the theoretical background and is confirmed by previous literature shows the significant role of DL on m-learning attitudes. Based on this result, it can be concluded that high DL skills are needed for effective m-learning environments. As successful implementation of m-learning largely depends on the learners' and teachers' acceptance and willingness to adopt m-learning and their attitudes towards it (Al-Seghayer, 2014; Al-Shehri, 2014; Liaw & Huang, 2012; Padmanathan & Jogulu, 2018), it can be said that this result is important because it reveals an important factor which is effective on individuals' m-learning attitudes. Therefore, DL skills should be implemented during the effort of maximizing the potential of m-learning. It can be said that this chapter has contributed to the literature and extended it because it presents evidence regarding the predictive role of DL skills on university students' m-learning attitudes in language learning context. Also, this chapter has contributed to the literature by investigating the predictive role of subscales of DLC comparatively because previous research lacks enough evidence regarding this.

REFERENCES

Abu-Al-Aish, A., & Love, S. (2013). Factors influencing students' acceptance of M-learning: An investigation in higher education. *International Review of Research in Open and Distance Learning*, *14*(5), 82–107. doi:10.19173/irrodl.v14i5.1631

Akman, Y. (2021). Dijital okuryazarlık, çevrim içi öğrenme ve akademik isteklilik arasındaki ilişkinin incelenmesi. *The Journal of Turkish Educational Sciences*, *19*(2), 1012–1036. doi:10.37217/tebd.982846

Akour, H. (2009). *Determinants of mobile learning acceptance: an empirical investigation in higher education* [Unpublished PhD dissertation]. Oklahoma State University.

Al-Emran, M., Elsherif, H. M., & Shaalan, K. (2016). Investigating attitudes towards the use of mobile learning in higher education. *Computers in Human Behavior*, *56*, 93–102. doi:10.1016/j.chb.2015.11.033

Al-Rahmi, A. M., Al-Rahmi, W. M., Alturki, U., Aldraiweesh, A., Almutairy, S., & Al-Adwan, A. S. (2022). Acceptance of mobile technologies and M-learning by university students: An empirical investigation in higher education. *Education and Information Technologies*, *27*(6), 7805–7826. doi:10.100710639-022-10934-8

Al-Seghayer, K. (2014). The four most common constraints affecting English teaching in Saudi Arabia. *International Journal of English Linguistics*, *4*(5), 17–26. doi:10.5539/ijel.v4n5p17

Al-Shehri, S. (2014). Mobile learning in the Arab world: Contemporary and future implications. In X. Xu (Ed.), *Interdisciplinary mobile media and communications: social, political, and economic implications* (pp. 48–62). IGI Global. doi:10.4018/978-1-4666-6166-0.ch003

Almaiah, M. A. (2018). Acceptance and usage of a mobile information system services in University of Jordan. *Education and Information Technologies*, *23*(5), 1873–1895. doi:10.100710639-018-9694-6

Almaiah, M. A., Al-Khasawneh, A., & Althunibat, A. (2020). Exploring the critical challenges and factors influencing the e-learning system usage during COVID-19 pandemic. *Education and Information Technologies*, *25*(6), 5261–5280. doi:10.100710639-020-10219-y PMID:32837229

Almaiah, M. A., & Alismaiel, O. A. (2019). Examination of factors influencing the use of mobile learning system: An empirical study. *Education and Information Technologies*, *24*(1), 885–909. doi:10.100710639-018-9810-7

Almaiah, M. A., Jalil, M. A., & Man, M. (2016). Extending the TAM to examine the effects of quality features on mobile learning acceptance. *Journal of Computers in Education*, *3*(4), 453–485. doi:10.100740692-016-0074-1

Alowayr, A. (2022). Determinants of mobile learning adoption: Extending the unified theory of acceptance and use of technology (UTAUT). *The International Journal of Information and Learning Technology*, *39*(1), 1–12. doi:10.1108/IJILT-05-2021-0070

Aytekin, A., Özköse, H., & Ayaz, A. (2022). Unified theory of acceptance and use of technology (UTAUT) in mobile learning adoption: Systematic literature review and bibliometric analysis. *COLLNET Journal of Scientometrics and Information Management*, *16*(1), 75–116. doi:10.1080/09737766.2021.2007037

Bano, M., Zowghi, D., Kearney, M., Schuck, S., & Aubusson, P. (2018). Mobile learning for science and mathematics school education: A systematic review of empirical evidence. *Computers & Education*, *121*, 30–58. doi:10.1016/j.compedu.2018.02.006

Baran, E. (2014). A review of research on mobile learning in teacher education. *Journal of Educational Technology & Society*, *17*(4), 17–32.

Bay, D. N. (2021). Digital literacy levels of preschool teacher candidates. *Mustafa Kemal University Journal of the Faculty of Education*, *5*(7), 172–187.

Beres, D. (2011). Mobile-assisted language learning from the student perspective: Encouraging effective language learning strategies outside of the classroom. In B. R. Facer & M. Abdous (Eds.), *Academic podcasting and mobile assisted language learning: applications and outcomes* (pp. 93–110). IGI Global. doi:10.4018/978-1-60960-141-6.ch006

Brown, T. H., & Mbati, L. S. (2015). Mobile learning: Moving past the myths and embracing the opportunities. *International Review of Research in Open and Distance Learning*, *16*(2), 115–135. doi:10.19173/irrodl.v16i2.2071

Çavuş, N., & İbrahim, D. (2017). Learning English using children's stories in mobile devices: Children's stories in mobile devices. *British Journal of Educational Technology*, *48*(2), 625–641. doi:10.1111/bjet.12427

Chavoshi, A., & Hamidi, H. (2019). Social, individual, technological and pedagogical factors influencing mobile learning acceptance in higher education: A case from Iran. *Telematics and Informatics*, *38*, 133–165. doi:10.1016/j.tele.2018.09.007

Chen, Z., Chen, W., Jia, J., & An, H. (2020). The effects of using mobile devices on language learning: A meta-analysis. *Educational Technology Research and Development*, *68*(4), 1769–1789. doi:10.100711423-020-09801-5

Cho, K., Lee, S., Joo, M., & Becker, B. (2018). The effects of using mobile devices on student achievement in language learning: A meta-analysis. *Education Sciences*, *8*(3), 1–16. doi:10.3390/educsci8030105

Clark, R. C., & Mayer, R. E. (2008). *E-learning and the science of instruction: Proven guidelines for consumers and designers of multimedia learning*. Wiley.

Clarke, B., & Svanaes, S. (2014). *An updated literature review on the use of tablets in education*. Family Kids and Youth.

Comas-Quinn, A., Mardomingo, R., & Valentine, C. (2009). Mobile blogs in language learning: Making the most of informal and situated learning opportunities. *ReCALL*, *21*(1), 96–112. doi:10.1017/S0958344009000032

Danish, J., & Hmelo-Silver, C. E. (2020). On activities and affordances for mobile learning. *Contemporary Educational Psychology*, *60*(101829), 1–5. doi:10.1016/j.cedpsych.2019.101829

Darmi, R., & Albion, P. (2014, February). A review of integrating mobile phones for language learning. *International Conference on Mobile Learning*, Madrid, Spain.

Dashtestani, R. (2016). Moving bravely towards mobile learning: Iranian students' use of mobile devices for learning English as a foreign language. *Computer Assisted Language Learning*, *29*(4), 815–832. do i:10.1080/09588221.2015.1069360

Elaish, M. M., Shuib, L., Ghani, N. A., & Yadegaridehkordi, E. (2019). Mobile english language learning (MELL): A literature review. *Educational Review*, *71*(2), 257–276. doi:10.1080/00131911.2017.1382445

Faul, F., Erdfelder, E., Lang, A. G., & Buchner, A. (2007). G* power 3: A flexible statistical power analysis program for the social, behavioral, and biomedical sciences. *Behavior Research Methods*, *39*(2), 175–191. doi:10.3758/BF03193146 PMID:17695343

Gilster, P. (1997). *Digital literacy*. John Wiley.

Godwin-Jones, R. (2008). Emerging technologies: Web-writing 2.0: Enabling, documenting, and assessing writing online. *Language Learning & Technology*, *12*(2), 7–13.

Hamidi, H., & Jahanshaheefard, M. (2019). Essential factors for the application of education information system using mobile learning: A case study of students of the university of technology. *Telematics and Informatics*, *38*, 207–224. doi:10.1016/j.tele.2018.10.002

Hamutoğlu, N. B., Canan Güngören, Ö., Kaya Uyanık, G., & Gür Erdoğan, D. (2017). Adapting digital literacy scale into Turkish. *Ege Journal of Education*, *18*(1), 408–429.

Hew, K. (2009). Use of audio podcast in K-12 and higher education: A review of research topics and methodologies. *Educational Technology Research and Development*, *57*(3), 333–357. doi:10.100711423-008-9108-3

Hoven, D., & Palalas, A. (2011). (Re)conceptualizing design approaches for mobile language learning. *CALICO Journal*, *28*(3), 699–720. doi:10.11139/cj.28.3.699-720

Huang, J. H., Lin, Y. R., & Chuang, S. T. (2007). Elucidating user behavior of mobile learning: A perspective of the extended technology acceptance model. *The Electronic Library*, *25*(5), 585–598. doi:10.1108/02640470710829569

Hwang, G., & Fu, Q. (2019). Trends in the research design and application of mobile language learning: A review of 2007-2016 publications in selected SSCI journals. *Interactive Learning Environments*, *27*(4), 567–581. doi:10.1080/10494820.2018.1486861

Hwang, W., Chen, H., Shadiev, R., Huang, Y., & Chen, C. (2014). Improving English as a foreign language writing in elementary schools using mobile devices in familiar situational contexts. *Computer Assisted Language Learning*, *27*(5), 359–378. doi:10.1080/09588221.2012.733711

Johnson, L., Smith, R., Willis, H., Levine, A., & Haywood, K. (2011). *The 2011 horizon report*. The New Media Consortium.

Kaya-Özgül, B., Aktaş, N., & Çetinkaya-Özdemir, E. (2023). Examination of digital literacy levels of primary school teachers and primary school teacher candidates according to various variables. *Cumhuriyet International Journal of Education*, *12*(1), 204–221.

Keskin, N. O., & Metcalf, D. (2011). The current perspectives, theories and practices of mobile learning. *The Turkish Online Journal of Educational Technology*, *10*(2), 202–208.

Kim, H., & Kwon, Y. (2012). Exploring smartphone applications for effective mobile-assisted language learning. *Multimedia-Assisted Language Learning*, *15*(1), 31–57. doi:10.15702/mall.2012.15.1.31

Kondo, M., Ishikawa, Y., Smith, C., Sakamoto, K., Shimomura, H., & Wada, N. (2012). Mobile Assisted Language Learning in university EFL courses in Japan: Developing attitudes and skills for self-regulated learning. *ReCALL*, *24*(2), 169–187. doi:10.1017/S0958344012000055

Kong, S. C. (2018). Parents' perceptions of e-learning in school education: Implications for the partnership between schools and parents. *Technology, Pedagogy and Education*, *27*(1), 15–31. doi:10.1080/1475939X.2017.1317659

Krasulia, A., & Saks, K. (2020, July). Students' perceptions towards mobile learning in an English as a foreign language class. *20th International Conference on Advanced Learning Technologies*. 10.1109/ICALT49669.2020.00078

Kukulska-Hulme, A. (2009). Will mobile learning change language learning? *ReCALL*, *21*(2), 157–165. doi:10.1017/S0958344009000202

Kukulska-Hulme, A. (2012). Smart devices or people? A mobile learning quandary. *International Journal of Learning and Media*, *4*(3-4), 73–77. doi:10.1162/IJLM_a_00105

Kukulska-Hulme, A., & Traxler, J. (2005). *Mobile learning: A handbook for educators and trainers*. Routledge.

Kumar, B. A., & Chand, S. (2019). Mobile learning adoption: A systematic review. *Education and Information Technologies*, *24*(1), 471–487. doi:10.100710639-018-9783-6

Lakhal, S., Khechine, H., & Pascot, D. (2007). Evaluation of the effectiveness of podcasting in teaching and learning. In G. Richards (Ed.), *Proceedings of World Conference on E-Learning in Corporate, Government, Healthcare, and Higher Education* (pp. 6181-6188). AACE.

Lan, Y. F., & Sie, Y. S. (2010). Using RSS to support mobile learning based on media richness theory. *Computers & Education*, *55*(2), 723–732. doi:10.1016/j.compedu.2010.03.005

Lankshear, C., & Knobel, M. (2008). Introduction: Digital literacies - Concepts, policies and practices. In C. Lankshear & M. Knobel (Eds.), *Digital literacies: concepts, policy and practice* (pp. 1–17). Peter Lang.

Laurillard, D., & Pachler, N. (2007). Pedagogical forms of mobile learning: Framing. In N. Pachler (Ed.), Mobile learning: Towards a research agenda (pp. 33-54). WLE Centre, IoE.

Levy, M., & Kennedy, C. (2005). Learning Italian via Mobile SMS. In A. Kukulska-Hulme & J. Traxler (Eds.), *Mobile learning: a handbook for educators and trainers* (pp. 76–83). Routledge.

Li, R. (2022). Effects of mobile-assisted language learning on EFL/ESL reading comprehension. *Journal of Educational Technology & Society*, *25*(3), 15–29.

Liaw, S. S., & Huang, H. M. (2012). A case of study of investigating users' acceptance toward mobile learning. In F. L. Gaol (Ed.), *Recent progress in data engineering and internet technology* (pp. 299–306). Springer. doi:10.1007/978-3-642-28798-5_41

Liu, T. Y. (2017). Developing an English mobile learning attitude scale for adult learners. *Journal of Educational Technology Systems*, *45*(3), 424–435. doi:10.1177/0047239516658448

Mac Callum, K., Jeffrey, L., & Na, K. (2014a). Factors impacting teachers' adoption of mobile learning. *Journal of Information Technology Education*, *13*, 141–162. doi:10.28945/1970

Mac Callum, K. M., Jeffrey, L., & Kinshuk. (2014b). Comparing the role of ICT literacy and anxiety in the adoption of mobile learning. *Computers in Human Behavior*, *39*, 8–19. doi:10.1016/j.chb.2014.05.024

Madigan, E. M., Goodfellow, M., & Stone, J. A. (2007). Gender, perceptions, and reality: Technological literacy among first-year students. *SIGCSE Bulletin*, *39*(1), 410–414. doi:10.1145/1227504.1227453

Mahdi, H. S. (2018). Effectiveness of mobile devices on vocabulary learning: A meta-analysis. *Journal of Educational Computing Research*, *56*(1), 134–154. doi:10.1177/0735633117698826

Manovich, L. (2001). *The language of new media.* The MIT Press.

Markauskaite, L. (2007). Exploring the structure of trainee teachers' ICT literacy: The main components of, and relationships between, general cognitive and technical capabilities. *Educational Technology Research and Development*, *55*(6), 547–572. doi:10.100711423-007-9043-8

Martin, A. (2008). Digital literacy and the "digital society". In C. Lankshear & M. Knobel (Eds.), *Digital literacies: Concepts, policies, and practices* (pp. 151–176). Peter Lang.

Mayer, R. E. (2020). Where is the learning in mobile technologies for learning? *Contemporary Educational Psychology*, *60*(101824), 1–3. doi:10.1016/j.cedpsych.2019.101824

Mihaylova, M., Gorin, S., Reber, T. P., & Rothen, N. (2022). A meta-analysis on mobile assisted language learning applications: Benefits and risks. *Psychologica Belgica*, *62*(1), 252–271. doi:10.5334/pb.1146 PMID:36186897

Montrieux, H., Grove, F. D., & Schellens, T. (2014). Mobile learning in secondary education: Teachers' and students' perceptions and acceptance of tablet computers. *International Journal of Mobile and Blended Learning*, *6*(1), 26–40. doi:10.4018/ijmbl.2014040103

Moya, S., & Camacho, M. (2021). Identifying the key success factors for the adoption of mobile learning. *Education and Information Technologies*, *26*(4), 3917–3945. doi:10.100710639-021-10447-w

Nah, K. C., White, P., & Sussex, R. (2008). The potential of using a mobile phone to access the Internet for learning EFL listening skills within a Korean context. *ReCALL*, *20*(3), 331–347. doi:10.1017/S0958344008000633

Naismith, L., Lonsdale, P., Vavoula, G., & Sharples, M. (2004). *Literature review in mobile technologies and learning: Report 11.* Futurelab Series.

Ng, W. (2012). Can we teach digital natives digital literacy? *Computers & Education*, *59*(3), 1065–1078. doi:10.1016/j.compedu.2012.04.016

Ng, W. (2015). Sustaining mobile learning at a personal level: Mobile digital literacy. In W, Ng, & T. M. Cumming (Eds.), Sustaining mobile learning: theory, research and practice (pp. 84-107). Routledge.

Ocak, G., & Karakuş, G. (2019). An investigation of digital literacy self-efficacy skills of pre-service teachers in terms of different variables. *Afyon Kocatepe University Journal of Social Sciences*, *21*(1), 129–147.

Odiakaosa, O. J., Dlodlo, N., & Jere, N. (2017). Teacher and learner perceptions on mobile learning technology: A case of Namibian high schools from the hardap region. *Higher Educator-An International Journal*, *1*(1), 13–41. doi:10.26762/he.2017.30000002

Önal, N., & Önal, N. T. (2019). Adaptation of English mobile learning attitude scale into Turkish: Validity and reliability study. *Journal of Social Sciences of Mus Alparslan University*, *7*(1), 283–289.

Özcan, M. (2022). Evaluation of prospective teachers' digital literacy levels and mobile learning attitudes. *Journal of Educational Technology & Online Learning*, *5*(2), 367–378. doi:10.31681/jetol.1020586

Özer, O., & Kılıç, F. (2018). The effect of mobile-assisted language learning environment on EFL students' academic achievement, cognitive load and acceptance of mobile learning tools. *Eurasia Journal of Mathematics, Science and Technology Education*, *14*(7), 2915–2928. doi:10.29333/ejmste/90992

Padmanathan, Y., & Jogulu, L. N. (2018). Mobile learning readiness among Malaysian Polytechnic students. *Journal of Information Systems and Technology Management*, *3*, 113–125.

Puebla, C., Fievet, T., Tsopanidi, M., & Clahsen, H. (2022). Mobile-assisted language learning in older adults: Chances and challenges. *ReCALL*, *34*(2), 169–184. doi:10.1017/S0958344021000276

Read, T., Kukulska-Hulme, A., & Barcena, E. (2021). Supporting listening comprehension by social network-based interaction in mobile assisted language learning. *Porta Linguarum Revista Interuniversitaria De Didáctica De Las Lenguas Extranjeras*, *35*(35), 295–309. doi:10.30827/portalin.v0i35.15341

Rysbayeva, G., Berdaliyeva, A., Kuralbayeva, A., Baiseitova, N., Uspabayeva, A., Zhapparbergenova, E., & Poshayeva, G. (2022). Students' attitudes towards mobile learning. *International Journal of Engineering Pedagogy*, *12*(2), 129–140. doi:10.3991/ijep.v12i2.29325

Saad, A. M., Mohamad, M. B., & Tsong, C. K. (2021). Behavioural intention of lecturers towards mobile learning and the moderating effect of digital literacy in Saudi Arabian universities. *International Transaction Journal of Engineering Management & Applied Sciences & Technologies*, *12*(1), 1–9.

Sánchez-Prieto, J. C., Hernández-García, A., García-Peñalvoa, F. J., Chaparro-Peláez, J., & Olmos-Miguelañez, S. (2019). Break the walls! Second-order barriers and the acceptance of mLearning by first-year pre-service teachers. *Computers in Human Behavior*, *95*, 158–167. doi:10.1016/j.chb.2019.01.019

Sarrab, M., Shibli, I. A., & Badursha, N. (2016). An empirical study of factors driving the adoption of mobile learning in Omani higher education. *International Review of Research in Open and Distance Learning*, *17*(4), 331–349. doi:10.19173/irrodl.v17i4.2614

Seppala, P., & Alamaki, H. (2003). Mobile Learning in teacher training. *Journal of Computer Assisted Learning*, *19*(3), 330–335. doi:10.1046/j.0266-4909.2003.00034.x

Shadiev, R., Hwang, W., & Liu, T. (2018). Investigating the effectiveness of a learning activity supported by a mobile multimedia learning system to enhance autonomous EFL learning in authentic contexts. *Educational Technology Research and Development*, *66*(4), 893–912. doi:10.100711423-018-9590-1

Sophonhiranrak, S. (2021). Features, barriers, and influencing factors of mobile learning in higher education: A systematic review. *Heliyon*, *7*(4), 1–10. doi:10.1016/j.heliyon.2021.e06696 PMID:33869873

Su, C. Y., & Chao, C. M. (2022). Investigating factors influencing nurses' behavioral intention to use mobile learning: Using a modified unified theory of acceptance and use of technology model. *Frontiers in Psychology*, *13*, 1–10. doi:10.3389/fpsyg.2022.673350 PMID:35651564

Sung, Y., Chang, K., & Yang, J. (2015). How effective are mobile devices for language learning? A meta-analysis. *Educational Research Review*, *16*, 68–84. doi:10.1016/j.edurev.2015.09.001

Taj, I., Sulan, N., Sipra, M., & Ahmad, W. (2016). Impact of mobile assisted language learning (MALL) on EFL: A meta-analysis. *Advances in Language and Literary Studies*, *7*(2), 76–83.

Terras, M. M., & Ramsay, J. (2012). The five central psychological challenges facing effective mobile learning. *British Journal of Educational Technology*, *43*(5), 820–832. doi:10.1111/j.1467-8535.2012.01362.x

Tommerdahl, J. M., Dragonflame, C. S., & Olsen, A. A. (2022). Online first). A systematic review examining the efficacy of commercially available foreign language learning mobile apps. *Computer Assisted Language Learning*, 1–30. doi:10.1080/09588221.2022.2035401

Tsuei, M., & Hsu, Y. Y. (2019). Parents' acceptance of participation in the integration of technology into children's instruction. *The Asia-Pacific Education Researcher*, *28*(5), 457–467. doi:10.100740299-019-00447-3

Turgut, K., & Akdemir, E. (2019). The attitude of prospective teachers towards mobile learning. *International Journal of Society Researchers*, *13*(19), 500–526.

Vasiliou, A., & Economides, A. A. (2007). Mobile collaborative learning using multicast MANETs. *International Journal of Mobile Communications*, *5*(4), 423–444. doi:10.1504/IJMC.2007.012789

Viberg, O., & Grönlund, Å. (2012, October). Mobile assisted language learning: A literature review. *11th world conference on mobile and contextual learning*.

Wu, Q. (2014). Learning ESL vocabulary with smartphones. *Procedia: Social and Behavioral Sciences*, *143*, 302–307. doi:10.1016/j.sbspro.2014.07.409

Wu, W. H., Wu, Y. C. J., Chen, C. Y., Kao, H. Y., Lin, C. H., & Huang, S. H. (2012). Review of trends from mobile learning studies: A meta-analysis. *Computers & Education*, *59*(2), 817–827. doi:10.1016/j.compedu.2012.03.016

Yağan, S. A. (2023). Mobile learning in higher education: A mixed research synthesis. Erzincan University Journal of Education Faculty, 25(1), 108-125.

Yılmaz, N., & Çakır, R. (2022). Preservice teachers' digital literacy and attitudes towards mobile learning. *Turkish Journal of Educational Studies*, *9*(2), 291–325.

Yu, Z., Xu, W., & Sukjairungwattana, P. (2023). Motivation, learning strategies, and outcomes in mobile English language learning. *The Asia-Pacific Education Researcher*, *32*(4), 545–560. doi:10.100740299-022-00675-0

KEY TERMS AND DEFINITIONS

Digital Literacy: A mixture of technical proficiency and meaningful engagement in the digital world.

Mobile Learning: A learning in mobile environments using handheld mobile devices like tablets, e-readers, smartphones, or portable computers.

Mobile Technology: Any portable and personal devices.

Chapter 7
Online Learning in Times of Pandemic:
A Systematic Review

Ana María Martínez-Martínez
Universidad de Almería, Spain

María del Pilar Díaz-López
https://orcid.org/0000-0002-0888-925X
Universidad de Almería, Spain

Christian Roith
https://orcid.org/0000-0003-4830-2756
Universidad de Almería, Spain

Noelia Gómez Navarro
Universidad de Almería, Spain

ABSTRACT

The present study deals with a systematic review on virtual education in higher education during the COVID-19 health crisis to know the digital competence of teachers and student perception of online learning. The scientific literature was analyzed during the years 2020-2022 based on empirical and theoretical studies in the Spanish university environment. The results show a certain weakness of telematic teaching with respect to face-to-face teaching. Technology is a powerful tool that requires planning, an aspect that was not adequately developed in times of pandemic. COVID-19 has revealed technological deficiencies in teachers, which implies that we are still developing in the digital framework and their use is reduced to the minimum expression of all the potential they have. In general, university students feel satisfied with virtual teaching. They have recognized the potential that ICT has in the "learning to learn" competition, but have emphasized an increase in academic stress, backed by the psychological impact of confinement.

DOI: 10.4018/978-1-6684-8861-4.ch007

INTRODUCTION

On March 11, 2020, the World Health Organization (WHO, 2020) reports on a new disease caused by an unprecedented type of coronavirus, the SARSCoV-2 virus, which causes the disease COVID-19 with alarming levels of transmission, qualifying this fact as a pandemic in a global world (García Peñalvo & Corell, 2020).

This situation required drastic measures for citizens, by the Centers for Disease Control and Prevention (ODC, 2020), focused on closing borders and social isolation as a preventive measure for the spread of the virus. Each country proposed different strategies, guidelines, and action protocols supported by the WHO recommendations. This devastating framework affected the normal functioning of societies in all spheres, economic, social, health and educational (Grande de Prado et al., 2021). The right to education was seriously affected, a right included in the Sustainable Development Goals of the United Nations Organization (UN, 2019).

The Ministry of Education and Vocational Training (MEFP) exceptionally ordered the postponement and/or suspension of school activities in all educational areas. Education had to resort to Information and Communication Technology (ICT) to guarantee the teaching-learning processes of the students.

To make training effective, the MEFP enabled in the different autonomous communities, digital educational resources hosted in Procomn (network of open educational digital resources, where you can search, view, and download learning material), in the Educational, Digital, Innovative Project (EDIA) to promote the improvement of learning through digital dynamics and linked to active methodology, and the one promoted by the Center for the Innovation and Development of Distance Education (CIDEAD), a format with experience in teaching the distance modality, all of them for pre-university teaching.

However, it is the universities that have shown their great potential with respect to other educational levels, offering synchronous and asynchronous classes to adapt to the new educational challenges in order to ensure academic continuity (García-Peñalvo & Corell, 2020).

In Spain, Blackboard users make up 35% of universities. It is an e-learning tool that allows access to quality education at any time and place, both in public and private universities. Another online teaching university manager is Moodle, with features similar to the previous one.

The necessary resources were available to meet the educational challenge, but teachers urgently demanded the need for training to be able to implement classes with the highest quality and educational guarantee, as until then ICT had basically been used as complementary support and help (Fernández Torres et al., 2020). Now they stood as the only possible tool for teaching and evaluation (Hodges et al., 2020; Suárez-Guerrero et al., 2021). The immediate transition to digital platforms meant a radical, unnatural change, which evidenced the need to generate training courses and adaptation to the virtual environment (Baladrón et al., 2020; García-Peñalvo and Corell 2020).

An added problem of this health crisis, as reported by the Conference of University Rectors (CRUE), was the large percentage of students who could not continue classes normally due to technical problems, an aspect that further aggravated the situation (Gil Villa et al., 2020; Lozano et al., 2020).

GOALS

Know and analyze studies on virtual education in higher education during the health crisis of COVID-19 to know the digital competence of teachers and student perception of online learning.

MATERIALS AND METHODS

The study corresponds to the performance of a systematic review of a descriptive nature, on empirical studies, and supported by the methodology proposed by Okoli and Schabram (2010). The systematic review is a research work addressing the existing bibliographic sources, with the purpose of extracting methodically and objectively studies published in databases of greater recognition and scientific repercussion, with the purpose of knowing the current state of the matter.

For this, two search engines have been consulted: Dialnet and Google Scholar following a strict rigor in the search for information. The implementation of these works was focused from 2020-2022, in light of the techno-pedagogical impact caused by the pandemic and the educational repercussions. The keywords that guided the review were associated with the "ICT" "University", "COVID", "Pandemic", "Virtual Education", using the Boolean operator AND in order to limit the information necessary for the body of study. The inclusion and exclusion criteria are related to the following:

- Studies published between 2020 and 2022.
- Empirical and/or theoretical works.
- Articles published in English and Spanish.
- Open access and full text.
- Direct relationship with the search elements; those whose study population did not correspond to Spanish university students and in turn did not contemplate at least two criteria were excluded.

The establishment of these criteria allowed for the refinement of the search, yielding sufficient information to be able to respond to the proposed objective.

After performing the search in Dialnet and Google Scholar and defining the search terms, the results obtained were 577 articles. Table 1 shows the findings of the searches carried out.

Table 1. Descriptors and search criteria used

Databases and Search Terms	Results
Academic Google	
1st. "Spanish university" and "virtual education" and "covid-19"	168 results
2nd. "pandemic" and "digital competence" and "Spanish university"	205 results
3rd. "covid-19" and "higher education in Spain" and "ICT"	115 results
Dialnet	
1st. Spanish university "and" virtual education "and" covid-19	41 results
2nd. pandemic "and" digital competence "and" Spanish university	36 results
3rd. covid-19 "and" higher education in Spain "and" ICT	12 results

The information was filtered in the Dialnet and Google databases. Duplicate documents have been removed. The titles and abstracts of the same were addressed, which considerably narrowed the studies that would be part of the study by eliminating all those whose population was students of secondary education or of non-Spanish university origin.

Of the 83 articles chosen, a second screening was performed, in which empirical and/or theoretical essays were selected that contemplated online learning and assessment in times of pandemic. A critical reading of the entire text was carried out, excluding 68 of them because they did not have relevant information for the study. The conclusive documents were represented by 15 papers. Figure 2 shows the flowchart with the search and selection process.

Figure 1. Flowchart for the choice of documents for the present study. elaboration of our own

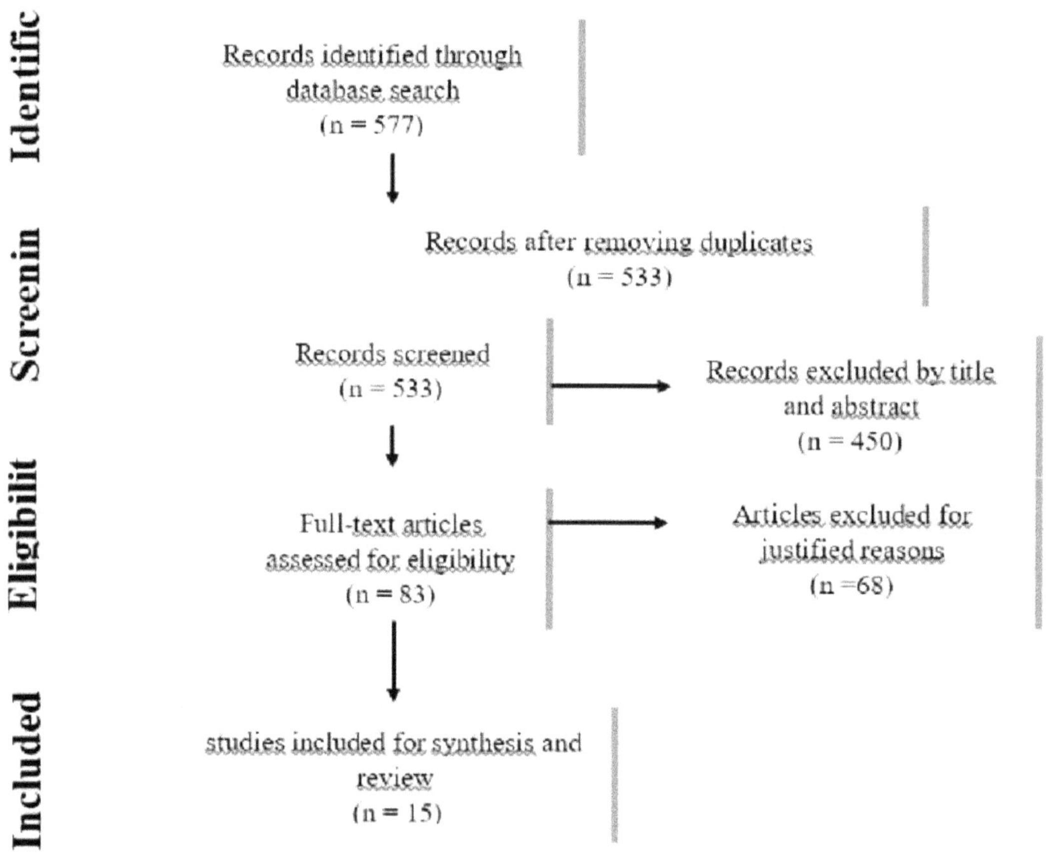

RESULTS

Based on previous review studies (Martínez-Baena and Faus-Boscá, 2018), the results of the selected works have been briefly included in a table (Table 3).

Quantitative Analysis

The extracted results are shown below, indicating the title of the article, reference authors, year of publication, type of methodology used in the research and a brief summary of the main findings.

The 15 articles come from various journals; 2 of them have been published in the Education in the Knowledge Society journal; 2 in the Latino Magazine of Social Communication; 2 belong to open access MDPI, the rest being of miscellaneous origin. With respect to authors, there is a plural distribution of authorship. The author who repeats publication with other authors is García Peñalvo, due to his extensive and extensive proliferation of studies on this subject.

Regarding the reference universities, they all correspond to the national territory with private and public representation, with concurrence in the 17 autonomous communities of the Spanish State. Degrees point to various qualifications: Degree in Early Childhood and Primary Education, Social Education, Pedagogy, Communication, Master's in teacher training, Journalism, Audiovisual Communication and Advertising. The Faculty of Arts and Humanities is also represented in the studies, but does not specify the degrees under study.

Research Methodology Used

Of the 15 selected studies, 66.66% respond to quantitative empirical research, 6.6% to mixed and qualitative empirical research, and 20% to theoretical studies.

To respond to the formulated objective, we have proceeded to answer three questions: how has the student perception of online education been? How has the evaluation been carried out? And how has the digital competence of university teaching staff been in times of pandemic? These questions lead us to establish 3 blocks that are identified in Table 3.

In order to establish the most detailed information, attention will be paid to developing the information according to the generated blocks, although some of them may be connected to other mentioned topics. The information for each of the categories is presented below:

Student Perception of Online Education in the Times of COVID-19

This category refers to the perception of Spanish higher education students about online learning in times of pandemic, weaknesses, and strengths—and degree of satisfaction.

Technological advances implemented in classrooms in recent decades have allowed the use of ICT to become the pedagogical tool par excellence, offering important possibilities for sharing, processing, and transmitting information through its potential educational channel. Studies indicate that online education has been a complex task in times of pandemic at the university level. All the universities promoted institutional challenges to provide immediate responses to immediate problems, which was a lifeline for teachers and students, given the uncertainty of the new addition to face-to-face (Baladrón et al., 2020; Díaz-Noguera, et al., 2022; Rodicio-García et al., 2020).

Table 3. The documents selected in the search

	Qualification	Authors/Year	Methodology	Goals	Main findings
Student perception of online education in times of COVID	Student perception of online education in times of COVID-19: University of Almería (Spain)	Castellano Gil et al. (2021)	Quantitative empirical research	To investigate the perception of a group of university students about the new modality of virtual emerging teaching implemented due to the COVID-19 health crisis.	The absence of barriers to access, use, and skills in relation to technological devices is verified. The evaluation system is more demanding than the face-to-face one.
	The Digital Divide in Spanish Students in the face of the COVID-19 crisis	Rodicio-García et al. (2020)	Quantitative empirical research	Find out how students live the non-face-to-face teaching imposed by the pandemic	University students consider that they do not have sufficient technological skills to respond to educational demands. Difficulty coping with technological changes
	Autonomy, motivation, and digital pedagogy are key factors in the perception of Spanish higher education students about online learning during the COVID-19 pandemic	Diaz-Noguera et al. (2022)	Quantitative empirical research	To know the adaptability of students to the digital transformation in university teaching through three constructs: motivations, digital pedagogy, and student autonomy.	It has been shown that the factors that favor the adaptation of students to the digital transformation were professional collaboration, digital pedagogy, student empowerment, self-learning, and the promotion of student initiative.
	Perceptions and expectations in university students from the adaptation to non-face-to-face teaching motivated by the COVID-19 pandemic	Gil Villa et al. (2020)	Quantitative empirical research	Expose the disadvantages of the telematic system adopted during confinement with respect to equal opportunities, well-being and satisfaction, and teaching quality.	University students prefer face-to-face when considering a meager university adaptation in times of pandemic. They consider poor institutional communication, both in communications with students and in listening to them.
	The digital transformation of university teaching in communication during the COVID-19 crisis in Spain: an approach from the perspective of the student body.	Balazon et al. (2020)	Quantitative empirical research	To know how university students adapted in the transition from face-to-face to virtuality.	Students from private universities approve of the management of their centers. The main resources are focused on uploading documents to the platform and holding video conferences. However, the pandemic has exposed the technological limitations of universities.
	Educational image of technology in COVID-19. The look of the Teaching students with an ICT mention	Suárez-Guerrero et al. (2021)	Qualitative empirical research	To know the educational perception of digital technology in the COVID-19 crisis in Teaching students	Students evaluated the depth of the digital response during COVID-19, considering the training requirement of teachers in digital competence.
Methodological Challenges in Higher Education in the COVID-19 pandemic	Higher education and COVID-19: methodological adaptation and online assessment at two universities in Barcelona	Ortega et al. (2021)	theoretical research	Highlight the consequences generated by the pandemic and the transformation and adaptation of the teaching-learning processes.	The adaptations of the teaching methodologies and the evaluation procedures have contributed to solving an adverse situation in a complex scenario.
	Online assessment in Higher Education in times of COVID-19	Garcia Penalvo et al. (2020)	theoretical research	Help university teachers organize the online assessment	Put at the service of the greatest number of teacher recommendations to carry out the online evaluation satisfactorily
	Online Higher Education in times of coronavirus. What do the students think?	Diez-Gutierrez and Gajardo. (2021)	Quantitative empirical research	Describe the assessment of university students towards evaluation systems and strategies, as well as online teaching	Students demand greater flexibility in online assessment. They demand evaluation practices oriented toward learning, but they reject peer evaluation.

continues on following page

Table 3. Continued

	Qualification	Authors/Year	Methodology	Goals	Main findings
Digital competence in Higher Education	Qualitative analysis of the implementation of e-learning and hybrid teaching during the COVID-19 pandemic in Spanish Universities	Bustamante et al. (2022)	mixed empirical research	Analyze the perceptions of students and teachers about how they have adapted to virtual spaces during and after confinement.	The results show a greater preference for face-to-face teaching over virtual teaching.
	University and pandemic: communication in distance education	Fernandez-Torres et al. (2021)	Quantitative empirical research	Know the new scenario configured by the pandemic and communication between teacher and student	There is evidence of difficulty on the part of teachers and students when faced with virtual teaching. Impoverishment of didactic communication.
	Impact of the remote emergency teaching and learning process on digital competence and on the state of mind of teacher training	Gomez-Gomez et al. (2021)	Quantitative empirical research	Analyze whether during remote emergency teaching and learning there have been significant changes in the state of mind and digital competence of teachers and students	Changes in mood and stress have been shown to have a significant increase in mean values. Need for digital pedagogical training for the educational community.
	Perception of digital teaching competence of bilingual teachers in training: Comparative study between face-to-face teaching and emergency distance learning	Palacios-Hidalgo et al. (2022)	Quantitative empirical research	The perceptions of mastery of teacher digital competence are compared between two groups of bilingual teachers in training.	Digital competence perceived as positive. Students prefer face-to-face instead of emergency distance learning.
	COVID-19: an enzyme for the digital transformation of teaching or a reflection of a methodological and competence crisis in higher education?	Garcia-Peñalvo and Corell (2020)	theoretical research	Reflect on online teaching and assessment in times of pandemic	It highlights a deep competence gap both in digital aspects and in teaching methodologies in the case of Spanish universities.
	Digital competence of university students and academic performance in times of COVID-19	Garcia-Prieto et al. (2022)	Quantitative empirical research	Analyze the incidence of digital competence on the academic performance of university students in times of pandemic.	It is shown that the higher the levels of digital competency, the higher the academic performance of the students.

COVID-19 made visible the digital divide between Spanish students and the factors associated with it. On the one hand, access problems, taking into account that many of the students were confined in rural areas with meager coverage, and on the other hand, and possibly the most serious, lack of digital skills. The INE (2019) warned that only 40% of Internet users have advanced digital skills, and the rest are basically users of social networks (Instagram, Twitter, Facebook, or YouTube), with the group with the greatest skills being those between 16 and 24 years. This same study indicates that 91.4% of households have an Internet connection and 80.9% have some technological equipment.

In this sense, the descriptive study carried out by Rodicio et al. (2020) refers to that it is one thing to have technology in homes and quite another thing is that students have the necessary skills to meet educational demands. University students are not in a better situation compared to other groups studied in digital competence. They also address the digital divide by gender, arguing that the results do not show statistically significant differences between men and women, which infers that ICTs are used equally by both genders. What stands out is that women use it differently than men. Girls are more active and determined when making video calls, creating material to upload it to social networks, and men have greater technical specialization, such as managing programs, compressing files, installing devices, or using programming languages. This infers that there is a low representation of women in strategic sectors related to ICT.

Considering the degree of satisfaction of university students, the study by Gil Villa et al. (2020) considers that the university has not properly adapted to virtual teaching or evaluations, there being great deficiencies with respect to face-to-face teaching. A large number of students report that this fact will considerably affect their grades with lower results than pre-COVID promotions, which could affect their job prospects. A large part of the problems is due to ineffective communication and listening by teachers, since not all students have the same possibilities of accessing new technologies, nor would dramatic situations have happened in the family environment, which would be an aggravating factor compared to other colleagues. In general terms, students are little or not satisfied with the virtual education received and especially with the evaluation process of the different subjects. According to the students, the teaching suffered because all the weight fell on the predisposition, skills, and digital competences of the teachers, and not all the teachers were up to the task, since some cases focused on proposing readings focused on the exam, leaving in demonstrates the training needs of teachers. It is palpable that the educational quality of the students suffered and not all the teachers were up to the task, since some cases focused on proposing readings focused on the exam, revealing the teachers' training needs. The educational quality of the students suffered and not all the teachers were up to the task, since some cases focused on proposing readings focused on the exam, revealing the teachers' training needs.

These findings contrast with the research carried out by Castellano Gil et al. (2021). From his research, it is clear that the student body makes a positive assessment of the technological infrastructure and connectivity to move from face-to-face to virtuality. The problem is that not everyone has a desktop computer to take the online classes, making up for this inconvenience with the mobile phone. 63% of those surveyed have approached this change with positivity, making a positive assessment of this academic period. However, they point out that although the on-line modality favors autonomous work, it interferes with group or collaborative work, noting, in addition, the degree of demand, volume of tasks carried out, and the little time granted for taking exams. These aspects confer a greater effort and study on the tasks with respect to face-to-face. The advantage of this teaching modality is that it favors the participation of the students in the dynamization of the classes. The study also presents negative characteristics related to inexperience in the virtual classroom, both teaching and learning, connection problems, as well as less demanding teaching work. Faced with a situation such as the pandemic, the students stated that they had not received any type of training from the university for the use of the virtual platform Blackboard Collaborate, which they took advantage of to appeal to the university about these training needs. Although all demand to implement the training, the majority manifested a good domain of execution.

Baladron et al. (2020) also infer a high degree of satisfaction to virtual teaching, using the university's virtual classroom for this purpose, although sometimes they could combine other supports, such as Microsoft Teams, social networks, Google Meet, etc. 58% of the teachers have used at least three different resources in the development of their sessions, such as uploading subject content, videoconferences, and designing an activity. These aspects have been widely valued by the students, together with self-assessment tests. The actions most valued by students are the resolution of doubts, keeping in direct contact with the teacher, and promoting content for the subject.

Students, on the other hand, according to the study carried out by Suárez-Guerrero et al. (2022) confer benefits to online classes in relation to the personalization of learning, allowing better attention to diversity. They consider the need to rethink the methodologies for working with technology. The positive side that students attribute to the pandemic is the fact that teacher training in the face of the growth of technologies is an indication of university quality. The technological mistakes of the pandemic are summarized in the following terms: use the same model as in face-to-face education; slow or no adaptation; not considering the emotional problems generated by confinement; thinking that having access is having digital skills; the fraudulent act of evaluations. The weaknesses that students give to the virtual educational system are generalized anxiety due to the high-class load, adding to the forced social isolation, sometimes perceiving virtual classes as overwhelming. Other studies highlight the difficulty of assimilating and understanding certain terms, poor internet connection, loneliness, modesty when interacting through audio and cameras, as well as the lack of space to be able to carry out classes correctly (Fernández Torres et al., 2020), sometimes perceiving virtual classes as overwhelming. Other studies highlight the difficulty of assimilating and understanding certain terms, poor internet connection, loneliness, modesty when interacting through audio and cameras, as well as the lack of space to be able to carry out classes correctly (Fernández Torres et al, 2020), sometimes perceiving virtual classes as overwhelming.

Diaz Noguera et al. (2022) pointed out that the factors that favored student adaptation to digital transformation were through professional collaboration, digital pedagogy, and student empowerment and self-learning.

Methodological Challenges in Higher Education in the COVID-19 Pandemic

This section aims to shed light on the methodological, teaching, and evaluation challenges, as well as the assessment that university students make during the confinement of COVID-19.

Once the government of Spain decreed the state of alarm on April 14, 2020, the universities had to urgently improvise not only the online educational activities but also the evaluation system, adapting it to the technological platforms. That is why the CRUE, on March 30, urged universities to have alternative evaluation tools available to deal with this exceptional situation. On the other hand, the Spanish Network of University Quality Agencies (REACU, 2020a) required universities to adopt evaluation methodologies with quality criteria as dictated by the European Higher Education Area. Thus, alternative methods to single evaluation were proposed, opting for other evaluation measures, such as continuous evaluation with individual or group work, activities that encourage reflection, making the evaluation result have less weight on the final grade (Ministry of Universities, 2020). In the words of Ibarra-Saiz and Rodríguez-Gómez (2010), this alludes to an evaluation focused on learning. This was a challenge that universities had to overcome since to date they had focused almost exclusively on the exam culture.

In this sense, Ortega-Ortigoza et al. (2021) point out the adaptive effort that universities had to make. His work would come to exemplify some adaptations carried out in the Degree of Social Education and Pedagogy, focused on the methodological adaptations of the learning processes, online evaluation, and tutorial accompaniment, in order to be able to meet the objectives set in the teaching guides. Garcia-Peñalvo et al. (2020) developed a guide with recommendations to help teachers and universities in this process.

Based on the premises set by the Ministry of Universities, the rectors of each university proceed by means of Instructions, to build the general guidelines for teaching and evaluation operation for the 2019-2020 school year. Thus, by means of addendums, the modifications of the learning methodological processes and the contributions aimed at reformulating the criteria for the single and continuous evaluation and the tutoring of the students are detailed.

Ortega-Ortigosa et al. (2021) point out that the greatest difficulty centered on the limited room for maneuver in teaching replanning once the total suspension of face-to-face teaching was decreed. Added to this problem were the precarious digital skills to carry out the virtual modality (Moreno-Rodríguez, 2020). These authors show that the adaptation of teachers with less digital competence was reflected in overloading students with activities, readings, and complementary work in their subject. Teachers confused online teaching with the total availability that students should have because they were at home all day. In turn, the students made the same mistake, considering that the teachers should not have been disconnected from work either. For the implementation of virtual classes, the university's own platform, BBCollaborate, was used, which allowed synchronous classes to be held. One of the major drawbacks, according to the authors, was the poor connection and participation of the students. At the same time, e-learning tools were used to clarify curricular content, which favors asynchronous interaction. Online, group, and individual tutoring was the methodological strategy that grew exponentially to resolve doubts about the subject itself and, above all, doubts regarding the exams. The tutoring was an emotional support approach in turbulent times. The exams underwent methodological changes: replacement of the classic exam by reflective questions, resolution of practical cases, among others.

The work carried out by Díez-Gutiérrez et al. (2021) provides relevant information in this regard. Most of the students indicate that the teaching activity and the online exams must be performed until it is possible to resume face-to-face. To do this, they request that universities grant technological access to students who need it, facilitating the loan of computers and Internet access. The vast majority of students consider that a fair form of evaluation would be through the preparation of papers, projects, or portfolios. Others prefer multiple choice questions or short questions. In addition, to a lesser extent, they contemplate oral exams through videoconferences. Definitely, what they request is a continuous evaluation that uses a diversity of instruments that allow the evaluation of progress and that the weight of the qualification does not fall entirely on the final exam. Regarding the use of programs that certify authorship or surveillance, the answers are varied. There are those who doubt its reliability to those who consider them ethically questionable. The greatest consensus among the student body is in the use by the teachers of anti-plagiarism programs for the works. On the other hand, they claim the possibility of repeating face-to-face exams for those who could not take them electronically. Without supporting the general approval, they request greater flexibility to be able to recover if they fail and formalize enrollment in the following course without having exceeded the minimum number of credits required. They reject peer review, because they are not considered competent for it and allege that this work belongs to the teacher.

Digital Competence in Higher Education

The digitization of society implies the need to promote the acquisition of key skills for optimal development and operation. Teachers in the classroom must have the necessary skills to guide the learning process of these skills in students. In response to the recommendations of UNESCO (2011) on the mastery of ICT by teachers, he clarifies that it is not enough for teachers to be digitally competent to teach students. The teacher must master the use of technological and digital tools, which allow the student to function as an autonomous citizen in today's society. ICTs have changed the way we see society and solving the problems of today.

Hodges et al. (2020) argue that the adaptation from face-to-face to virtual in times of social isolation does not meet the teaching quality criteria, since online education requires exhaustive planning, an aspect that did not take place during COVID. -19. In this sense, Fernández-Torres et al. (2020) address how universities have adapted to the new scenario. From their study, it can be deduced that 61% of teachers considered that their level of digital literacy was optimal. However, both students and teachers agree to point out the great difficulty in adapting to the new scenario, an aspect that is not without attention, taking into account the digital skills they claim to possess. In this sense, it is possible to think that the difficulty to which they refer is associated more, to constant improvisation in response to the changes demanded by the pandemic, than to virtual teaching itself. Along the same lines, COVID-19 has revealed that although university teachers used resources and/or platforms in face-to-face teaching, they were not fully aware of their great potential, which infers their occasional use. Emergency non-face-to-face teaching led teachers to recycle and explore tools that, although they were previously aware of their existence, they did not use in their teaching practice. Fernandez Torres et al. (2020) are emphatic in stating that the rush to act in the new scenario evidenced the need for better adaptation, greater use, and knowledge by teachers in the face of the amalgam of possibilities offered by new technologies. However, they found some advantages in distance education: greater involvement of the students, greater dynamization of the classes by the teacher, and greater feedback from the teaching staff with the students. The most used resources were videoconferences and emails, leaving other tools such as blogs, wikis, and social networks relegated to the background. Network connection problems were a common denominator in many universities. In this sense, Gómez-Gómez et al. (2021) pointed out that the internet suffered a decrease in connection quality, an aspect that hindered the educational process, although in general they had sufficient devices to face the teaching-learning processes (Rodicio-García et al., 2020). Furthermore, they observed greater involvement of the students, greater dynamization of the classes by the teacher and greater feedback from the teaching staff with the students.

Gomez-Gomez et al. (2021), also question the need to improve digital competence, especially by teachers to teach virtual classes with greater autonomy, avoiding overloading students with work (Moreno-Rodríguez, 2020). As indicated by Palacios-Hidalgo et al. (2022), digital teaching competence must begin with future teachers. To do this, they must provide teachers in training with strategies that allow them to develop and evaluate their digital competence, be able to experiment in teaching contexts, and offer quality-training models from the universities themselves.

DISCUSSION

New technologies provide us with the necessary resources and tools for the conception of new social learning spaces.

Despite the good results shown by the latest reports on the use of ICT in Spanish universities, COVID-19 has shown certain shortcomings. The hasty adaptation in the virtuality of teaching has only fueled the idea of weakness of this modality, with respect to face-to-face teaching (Fernández-Torres et al., 2020). Many universities have improvised, they have arrived late in training, or, in the worst case, they have used obsolete learning models in person. There is no doubt that technology is a powerful tool, but it requires planning, an aspect that could not be developed adequately in the turbulent times of the pandemic. Teachers are aware that online learning is not the same,

This controversial environment has evidenced the technological deficiencies of many teachers, which implies that we are still in the process of developing in the digital framework, and their use is reduced to the minimum expression of all the potential they have (Fernández-Torres et al., 2020). The need to promote and design more educational platforms and resources has been emphasized. It is essential to meet the training needs of teachers in virtual spaces, considering the different learning styles of the students (Suárez-Guerrero).

In the academic field, students entering compulsory secondary education have been observed to have poor digital training, showing a low level of ICT literacy. This lack has become an even bigger problem, as students have found it necessary to quickly learn about these technologies to continue their education. Both teachers and students face a gap in their digital skills that must be addressed. Digital education is here to stay, with the use of teaching methods in hybrid spaces, which demands the implementation of effective action plans to improve the situation (Bustamante et al., 2022).

Scientific evidence shows interesting and contradictory results regarding the impact of COVID on university students. There is documented evidence that online learning has had great benefits among the student body compared to the previous year. The enormous potential that ICTs have in the "learning-to-learn" competition has been recognized, promoting autonomy and creativity in students (González et al., 2020).

However, there are also studies that have shown an increase in academic stress, which infers a greater decrease in motivation and academic performance (Gómez-Gómez et al., 2021; Suarez-Guerrero et al., 2020). These results are supported by the psychological impact of confinement, anxiety, life satisfaction, family concerns, and resilience, among others (Luque et al., 2021). With the pandemic, the levels of stress and alteration in the states of mind suffered by teachers and students have also increased, caused by the time dedicated to the teaching task, workload, and personal factors (Gómez-Gómez, 2021).

It is important to consider that technological resources are not by themselves a guarantee of success. It is necessary to recognize their potential and the learning opportunities that they offer students in general, but planning, organization, and flexibility are essential elements to take advantage of ICT and in this sense, there is still a long way to go (Fernández Torres et al., 2020). However, taking into account that the adaptation had to be done at a forced pace and with limited technological support, the degree of satisfaction of the educational agents-teachers and students-has been satisfactory. What was possible was done with the resources available (Díaz-Noguera et al., 2022; Baladrón et al., 2020), but both teachers and students have found more drawbacks than advantages to online teaching (Fernández Torres et al., 2020). In fact, students prefer face-to-face to virtual by emphasizing the inherent benefits of interpersonal relationships (Bustamante et al., 2022).

It has been shown that digital teaching competence not only consists of extensive technical use, it also requires competencies, knowledge, skills, and attitudes on how to educate with technology. Online learning has validated its usefulness and has established itself as a permanent fixture in the educational landscape. Teachers and students have experienced the benefits and opportunities that online learning provides, leading to increased interest and confidence in this educational approach. In addition, hybrid modalities have also been proven to be equally important. Many educational institutions are evaluating its implementation, which combines face-to-face with virtual education. This combination allows flexibility and adaptability in the learning process, as well as greater access to resources and improvement in the digital skills necessary for insertion into the labor market. In this sense, it has become a powerful and efficient tool for continuous learning in an increasingly digitized world (Moreno-Oliver et al., 2022).

Below is a table detailing the strengths and weaknesses that online learning has revealed.

Table 4. Strengths and weaknesses of online learning

Strengths of Online Learning	Weaknesses of Online Learning
● Accessibility to education from anywhere as long as you have internet access. ● Flexible scheduling. The student sets the pace of learning. ● Variety of resources and tools, such as videos, interactive presentations, and online collaboration platforms, which can facilitate the understanding of concepts. ● Virtual interaction through discussion forums, chats, and videoconferences.	● Digital Gap ● Lack of face-to-face interaction. ● Self-discipline and motivation. ● Limited access to specific resources. Some disciplines require access to laboratories, specialized equipment, or other offline resources, limiting students' ability to engage in hands-on experiments and activities.

The present study could be improved by other ways of research related to the effectiveness of virtual learning. For this, both pedagogical modalities should be considered -face-to-face and virtual- to conclude which of the two provides a greater retention of knowledge and, consequently, a greater academic performance.

Another possible line of research may be related to the development of an integrated methodology that combines tools and strategies in learning platforms, to guarantee active and autonomous learning, with the purpose of providing relevant feedback to teachers and beneficial to them.

The main limitation of this study lies in its temporary and specific focus, since online teaching has only been investigated in depth during a particular and singular period, due to the confinement imposed on the population due to the COVID-19 pandemic. This has evidenced the role each university has played in virtual teaching and the difficulties experienced in each of them.

Another weakness of this work has been not contemplating the post-COVID panorama, above all, the hybrid modality that has come to stay in the educational panorama. This modality allows a certain degree of flexibility for both students and teachers since they can alternate between face-to-face classes and virtual activities as appropriate.

CONCLUSION

COVID-19 has had a significant impact on online learning. Some important implications of the pandemic for education are:

Abrupt transition to online learning as educational institutions were forced to close their physical facilities. This created challenges for students, teachers, and parents, as they had to quickly adapt to new learning environments and technologies.

The pandemic highlighted the existing digital divide in different regions and communities. Not all students had equal access to electronic devices, internet connectivity, and the digital skills necessary to fully participate in online learning. Teachers also needed immediate training to be able to teach classes online. This has caused an imbalance in educational quality. Online learning can present additional challenges in terms of student engagement and motivation. The lack of social interaction, the difficulty of maintaining attention for long periods, can alter the motivation and educational commitment of students. Online learning requires students to be self-disciplined and capable of managing their time affectively. Students must be responsible for organizing their schedules and studies, meeting deadlines, and staying motivated without direct supervision from the teacher.

Online learning can limit opportunities for interaction and collaboration between students and teachers. The lack of face-to-face interaction can affect the quality of discussions, the exchange of ideas, and social learning, which are important aspects of educational development.

The pandemic and the transition to online learning have had a significant impact on the mental health and well-being of both students and faculty. Social isolation, pandemic-related stress, and academic challenges can contribute to higher levels of anxiety, depression, and burnout in the educational community.

REFERENCES

Aguilera-Hermida, A. P. (2020). Uso y aceptación de estudiantes universitarios del aprendizaje en línea de emergencia debido a la COVID-19. *Revista internacional de investigación educativa abierta*. doi:10.1016/j.ijedro.2020.100011

Baladrón Pazos, A. J., Correyero Ruiz, B., & Manchado Pérez, B. (2020). La transformación digital de la docencia universitaria en comunicación durante la crisis de la COVID-19 en España: una aproximación desde la perspectiva del alumnado. *Revista latina de comunicación social, 78*, 265-287. https://www.doi.org/10.4185/RLCS-2020-1477

Bustamante, J. C., Segura-Berges, M., Lizalde-Gil, M., & Peñarrubia-Lozano, C. (2022). análisis cualitativos de la implementación del e-learning y la enseñanza híbrida durante la pandemia del COVID-19 en las universidades españolas. *Sostenibilidad, 14*(19), 1–19. doi:10.3390u141912003

Castellano Gil, J. M., Almagro, J., & Fajardo, Á. B. (2021). Percepción estudiantil sobre la educación online en tiempos de COVID-19: Universidad de Almería (España). *Revista Scientific, 6*(19), 185–207. doi:10.29394/Scientific.issn.2542-2987.2021.6.19.9.185-207

CRUE Universidades Españolas. (2020). *Comunicado de la CRUE a los rectores de las universidades españolas de 30 de marzo de 2020*. CRUE Universidades Españolas.

Díaz-Noguera, M. D, Hervás-Gómez, C., De la Calle-Cabrera, A. M. y López-Meneses, E. (2022). La autonomía, la motivación y la pedagogía digital son factores clave en las percepciones de los estudiantes españoles de educación superior sobre el aprendizaje en línea durante la pandemia de COVID-19. *Revista internacional de investigación ambiental y salud pública, 19*(2), 654. doi:10.3390/ijerph19020654

Díez-Gutiérrez, E. J., & Gajardo K. (2021). Evaluación online en educación superior en tiempos de coronavirus ¿Qué piensan los estudiantes? *Bordón: Revista de pedagogía, 73*(1), 39-58.

Fernández Torres, M. J., Chamizo Sánchez, R. & Sánchez Villarrubia, R. (2021). Universidad y pandemia: la comunicación en la educación a distancia. *Revista internacional de comunicación, 52*, 156-174. . doi:10.12795/Ambitos.2021.i52.10

García-Peñalvo, F. J. y Corell, A. (2020). La COVID-19: ¿enzima de la transformación digital de la docencia o reflejo de una crisis metodológica y competencial en la educación superior? *Campus virtuales, 9*(2), 83-98.

García Peñalvo, F. J., Corell, A., Abella García, V. y Grande de Prado, M. (2020). La evaluación online en la educación superior en tiempos de la COVID-19. *Education in the Knowledge Society: EKS,* 21. . doi:10.14201/eks.23013

García Prieto, F. J., López-Aguilar, D. & Delgado-García, M. (2022). Competencia digital del alumnado universitario y rendimiento académico en tiempos de COVID-19. *Revista de medios y educación Pixel-Bil, 64*, 1-41 doi:10.12795/pixelbit.91862

Gil-Villa, F., Urchaga, J. D., & Sánchez-Fernández, A. (2020). Proceso de digitalización y adaptación a la enseñanza no presencial motivada por la pandemia de COVID-19: análisis de la percepción y repercusiones en la comunidad universitaria. *Revista latina de comunicación social, 78*, 99-119. https://www.doi.org/10.4185/RLCS-2020-1470

Gómez-Gómez, M., Hijón-Neira, R., Santacruz-Valencia, L., & Pérez-Marín, D. (2022). Impacto del proceso de enseñanza y aprendizaje remoto de emergencia en la competencia digital y el estado de ánimo en la formación del profesorado. *Education in the Knowledge Society, 23*. Advance online publication. doi:10.14201/eks.27037

González, T., de la Rubia, M., Hincz, K., López, M. C., Subirats, L., Fort, S., & Sacha, G. M. (2020). Influencia del confinamiento por COVID-19 en el desempeño de los estudiantes en la educación superior. *PLoS One, 15*(10). Advance online publication. doi:10.1371/journal.pone.0239490 PMID:33035228

Grande de Prado, M., García-Peñalvo, F. J., Corell, A. & Abella García, V. (2021). Evaluación en Educación Superior durante la pandemia de la COVID-19. *Campus virtuales, 1*(10), 49-58.

Hodges, C., Moore, S., Lockee, B., Trust, T., & Bond, A. (2020). La diferencia entre la enseñanza remota de emergencia y el aprendizaje en línea. *EDUCAUSE Review, 27*, 1–12. https://www.educacionyfp.gob.es/prensa/actualidad/2020/04/20200415-sectorial.html

Ibarra Sáiz, M. S. & Rodríguez-Gómez, G. (2010). Los procedimientos de evaluación como elementos de desarrollo de la función orientadora en la universidad. *Revista española de orientación y psicopedagogía, 21*(2), 443-461. . doi:10.5944/reop.vol.21.num.2.2010.11558

INE. (2019). *Encuesta sobre equipamiento y uso de la tecnología de la información y comunicación en los hogares*. INE. https://www.ine.es/dyngs/INEbase/es/operacion.htm?c=estadistica_C&cid=1254736176741&menu=ultiDatos&idp=1254735976608

Liberati, A., Altman, D. G., Tetzlaff, J., Mulrow, C., Gøtzsche, P. C., & Loannidis, J. P. (2009). La declaración Prisma para informar revisiones sistemáticas y metaanálisis de estudios que evalúan las intervenciones de atención médica: Explicación y elaboración. *PLoS Medicine*, *6*(7), 1–28. PMID:19621070

Lozano-Díaz, A., Fernández-Prados, J. S., Figueredo Canosa, V., & Martínez-Martínez, A. M. (2020). Impactos del confinamiento por el COVID-19 entre universitarios: Satisfacción vital, resiliencia y capital social online. *RISE*, *9*(1), 79–104. doi:10.17583/rise.2020.5925

Luque Vilca, O. M., Bolivar Espinoza, N., Achahui Ugarte, V. E., & Gallegos Ramos, J. R. (2022). Estrés académico en estudiantes universitarios frente a la educación virtual asociada al covid-19. *Puriq*, *4*, e200. doi:10.37073/puriq.4.1.200

Martínez-Baena, A., & Faus-Boscá, J. (2018). Acoso escolar y Educación Física: Una revisión sistemática. *Retos*, *34*, 338–345.

Ministerio de Educación y Formación Profesional. (2020). *El Ministerio de Educación y Formación Profesional y las CCAA acuerdan mantener la duración del curso escolar hasta junio.* https://www.educacionyfp.gob.es/prensa/actualidad/2020/04/20200415-sectorial.html

Moreno Oliver, V., Hernández-Leo, D., & Peig Olivé, E. (2022). Impacto pre-post pandemia en escenarios educativos online e híbridos: El caso de la Escuela de Ingeniería de la Universitat Pompeu Fabra. *Revista Actas de las Jenui*, *7*, 223–229.

Moreno-Rodríguez, R. (2020) Reflexiones en torno al Impacto del Covid-19 sobre la Educación Universitaria: Aspectos a Considerar acerca de los Estudiantes con Discapacidad. *Revista internacional de educación para la justicia social,* *9*(3), 1-6. https://revistas.uam.es/ riejs/article/view/1222

OkoliC.SchabramK. (2010). Una guía para realizar una revisión sistemática de la literatura sobre la investigación de sistemas de información. *Revista Electrónica* SSRN, 10 (26) doi:10.2139/ssrn.1954824

ONU. (2019). *Página principal de la Naciones Unidas*. https://www.un.org/sustainabledevelopment/es/objetivosde-desarrollo-sostenible/

Organización Mundial de la Salud. (2020). *Coronavirus*. Consultado el 18 de enero desde https://www.who.int/docs/default-source/coronaviruse/situation-reports/20200213-sitrep-24-covid-19.pdf?sfvrsn=9a7406a4_4

Ortega Ortigoza, D., Rodríguez-Rodríguez, J. & Mateos Inchaurrondo, A. (2021). Educación superior y la COVID-19: adaptación metodológica y evaluación online en dos universidades de Barcelona. *Revista digital de investigación en docencia universitaria,* *15*(1), 1-13. http://dx.doi.org/. doi:10.19083/ridu.2021.1275

Palacios-Hidalgo, F. J., Huertas-Abril, C. A. y Gómez-Parra, M. E. (2022). Competencia digital docente percibida del profesorado bilingüe en formación: Estudio comparativo entre enseñanza presencial y enseñanza a distancia de emergencia. *Revista de educación a distancia (RED),* *22*(69), 1-15. . doi:10.6018/red.466771

REACU. (2020a). *Acuerdo de REACU de 3 de abril de 2020, ante la situación de excepción provocada por el COVID19.* https://bit.ly/2wuHhqK

Rodicio-García, M. L., Ríos-de-Deus, M. P., Mosquera-González, M. J. y Penado Abilleira, M. (2020). La brecha digital en estudiantes españoles ante la Crisis de la Covid-19. *Revista internacional de educación para la justicia social, 9*(3), 103-125. . doi:10.15366/riejs2020.9.3.006

Suárez-Guerrero C., Lloret-Catalá C. y Lizandra J. (2022). Imagen educativa de la tecnología en la COVID-19. La mirada del alumnado de Magisterio con mención TIC. *Teknokultura. Revista de cultura digital y movimientos sociales, 19*(2), 177-188. doi:10.5209/tekn.77951

UNESCO. (2011). *UNESCO ICT competency framework for teachers*. UNESCO. https://bit.ly/3g2Fj2s

Chapter 8
Becoming a Successful Pre-Service English as a Foreign Language Lecturer in the Digital Divide

Gülin Zeybek
https://orcid.org/0000-0002-6863-7169
Isparta University of Applied Sciences, Turkey

ABSTRACT

The ephemeral nature of life is such that what was present in our lives yesterday may undergo a transformation or cease to exist altogether, as if it had never been. Technology has emerged as a significant factor in mitigating the challenges posed by various crises, particularly during recent global pandemics, natural calamities, and mass migrations. Undoubtedly, the incorporation of technology in the realm of education is a significant phenomenon. Currently, attention is directed towards academic institutions that offer programs for teacher education. Numerous responsibilities are assigned to the faculty members who provide instruction to individuals pursuing a career in teaching. The education of pre-service teachers regarding technology necessitates a distinct emphasis and instruction. The primary subject matter of this chapter pertains to the instruction of technology integration for pre-service English teachers, with a particular emphasis on the digital divide.

INTRODUCTION

The pervasive presence of technology in contemporary society has generated novel prospects and demands for educators and learners alike. However, the utilization of technology in the instruction of foreign languages has been relatively restricted (Kessler & Hubbard, 2017). The integration of the latest technological advancements in language education is not a straightforward process, as it cannot be assumed that personal and social practices can be seamlessly transferred into the classroom. While

DOI: 10.4018/978-1-6684-8861-4.ch008

language teachers and students are the primary beneficiaries of these advancements, their effective integration requires careful consideration and planning (Winke & Goertler, 2008). Utilizing technological tools in the realm of education necessitates a distinct level of awareness and proficiency, despite their implementation (Al-Jarrah, Talafhah & Al-Jarrah, 2019). It is imperative to differentiate the utilization of technology for the purpose of learning rather than solely for the sake of technology (Koehler & Mishra, 2008). Consequently, it is imperative to provide foreign language instructors with the requisite expertise and competencies for choosing and integrating novel technological innovations into pedagogical approaches within the classroom setting (Kessler & Hubbard, 2017).

When examining the ability and inclination of university professors, who serve as pre-service teacher trainers, to incorporate technology into their instructional practices, the findings are not particularly encouraging. According to Duart's (2011) analysis of the incorporation of the Internet into the teaching and learning process within universities, it has been confirmed that there is a low level of utilization by instructors, with only 50% of them employing it for pedagogical purposes. Marcelo, Yot, and Mayor (2015) corroborate the findings of Kedrova and Potemkin (2015) regarding the limited use of digital technologies by university instructors. Specifically, 44% of teachers reportedly use such technologies sparingly in their pedagogical approach, and when they do, they tend to rely on a narrow range of tools that primarily support lecture-based instruction rather than student-centered activities. Mercader and Gairín (2017) conducted a study examining the tools utilized by university instructors. Their findings indicate that visual presentations and virtual platforms are the most commonly used tools to support lecture sessions. In contrast, tools that require active participation from students, such as social networks, blogs, and augmented reality, are less frequently utilized and considered less important by instructors.

Given the widespread recognition among university educators, irrespective of their academic field, of the potential of digital technologies for facilitating learning (Hue & Jalil, 2013; Kurt, 2011), it is pertinent to inquire why these educators do not employ a wider range of digital technologies in their instructional practices. The significance of teacher education in technology integration is underscored by Pegrum (2014), who accentuates the advantages of this approach in terms of knowledge acquisition and collaborative learning. According to Kukulska-Hulme et al. (2017), individuals who have received professional training in classroom education possess robust pre-existing competencies and are motivated to refine and augment them in response to the technological advancements in learners' communication practices that have emerged in the 21st century.

The provision of technology-based courses in English Language Teaching (ELT) programs, such as those in Turkey, tends to be limited and typically delivered by instructors who possess a personal interest or expertise in the subject matter. The availability of such courses is subject to variation across different universities, with a majority of them lacking a technology course that is tailored to specific content. Additionally, compulsory courses on fundamental computer skills are offered by all universities. According to Aydın (2013), Başal (2015), and Kuru Gönen (2019), it has been observed that English language teachers are graduating from their respective departments without being adequately equipped with the requisite field-specific knowledge pertaining to technological practices in English language teaching.

Pre-service teacher education programs do not provide specialized training on technology integration for English language teaching. To clarify, it appears that Pre-Service Teachers (hereafter referred to as PSTs) do not receive explicit instruction regarding the selection, timing, and implementation of technology in the creation or modification of foreign language learning tasks. Research on language teacher education and technology integration has emphasized the limited availability and importance of such courses within English language teacher training programs. It has been recommended that pre-service English teachers receive training on the utilization of technology in language teaching, as indicated by Aydın (2013), Başal (2015), and Gülbahar (2008).

Research conducted in Turkey indicates that both pre-service teachers (PSTs) and in-service teachers are not meeting the expected standards in regard to integrating technology into their teaching practices (Akbulut, Odabaşı & Kuzu, 2011; Yılmaz, 2007). Insufficient proficiency in technology and inadequate comprehension of the implementation of educational technologies, such as interactive whiteboards and portable electronic devices like tablets and smartphones, in pedagogy, are among the limitations (Tezci, 2011; Yavuz-Konokman, Yanpar-Yelken & Sancar-Tokma, 2013). According to previous research conducted by Özel and Arıkan (2015) and Şahin Kızıl (2017), language teachers have demonstrated a lack of proficiency in utilizing contemporary technologies for the purpose of generating language teaching materials. Additionally, teachers possess insufficient knowledge regarding how to facilitate their students' use of these technological tools and how to integrate them effectively within diverse classroom settings (Hockly, 2013). According to Warschauer and Meskill's (2000) observation, the effective utilization of technology in language teaching is dependent on "humanware" rather than hardware or software. This implies that the success of technology integration in education is largely determined by the teacher's ability to plan, design, and execute educational activities effectively.

THEORETICAL GROUNDS FOR EDUCATING PRE-SERVICE TEACHERS

The scholarly discourse on language teacher education and technology primarily draws upon constructivist and socio-cultural perspectives, with a focus on equipping language teachers with the necessary skills and knowledge to effectively utilize technology-enhanced language learning (TELL) and computer technologies. Despite the abundance of literature on the utilization of various technologies for English language instruction, a noticeable dearth of scholarly works or theoretical directives in teacher training pertaining to the pedagogical approach for incorporating technology into language education is evident. Consequently, the theoretical perspectives and frameworks pertaining to technology and language teacher education discussed herein are contingent upon the comprehensive integration of technology, with particular emphasis on the incorporation of technology in teacher education.

Constructivist Views in PST Education for Technology Integration

The theory of constructivism has been widely recognized as a significant framework for elucidating the process of teacher education (Kosnik, Menna, Dharamshi & Beck, 2018). Additionally, it provides valuable insights into the training of teacher candidates with regard to the utilization of technology. Pre-service The key components of this perspective involve the construction of experiential knowledge by teachers, social learning, the integration of all aspects of an individual, and the establishment of an inclusive and equitable learning community. Henceforth, these teachers will be referred to as PSTs (Beck & Kosnik, 2006). Consequently, in order to achieve successful learning outcomes, it is imperative that prospective teachers (PSTs) experience a sense of safety and inclusivity within their classroom environment, fostering a sense of community (Beck & Kosnik, 2006). To facilitate the acquisition of technological integration skills, it is anticipated that pre-service teachers (PSTs) will be afforded practical experiences that enable them to collaboratively construct knowledge in this area. Furthermore, this perspective proposes the establishment of a secure atmosphere that promotes the sense of safety among prospective teachers when they employ technology for instructional purposes. The aforementioned experiences during the learning process have been deemed to have a significant impact on the manner in which individuals perceive and integrate technology within their prospective classrooms (Hart, 2002).

Various models have been developed for the purpose of training teachers on the implementation of technology in foreign language education, drawing from the features of the constructivist approach. The ECCR model of teacher development in technology integration, which was introduced by Son in 2018, is among the latest models in this field. This model comprises four key components, namely exploration, communication, collaboration, and reflection, which serve as the foundation for the model proposed in the current study. The following section presents a detailed overview of the ECCR model.

The ECCR model, which is derived from the terms Exploration, Communication, Collaboration, and Reflection, is founded on the theoretical framework of constructivism. The suggested elements for training foreign language teachers in this model are interconnected and each element represents a distinct learning activity (Son, 2018).

As per this theoretical framework, the engagement in active exploration, interactive communication, professional collaboration, and critical reflections are instrumental in facilitating the development of skills and knowledge among prospective teachers (Son, 2004). Thus, prospective teachers experience a sense of professional efficacy and personal assurance when employing technology within the educational setting. In accordance with the constructivist perspective, prospective teachers are motivated to engage in the process of knowledge discovery and adopt a practical approach to learning (Son, 2018).

The cultivation of language educators who are proficient in technology necessitates a process of inquiry that serves as a foundation for their professional development. Son (2018) suggests that in order for teachers to effectively utilize technology in their teaching, they must engage in careful observation and exploration of available technologies and resources. This involves assessing the availability of technologies, considering various options, enrolling in technology-based courses, reading relevant publications, attending conferences related to technology, and implementing one technology at a time to ensure efficient handling. Preliminary teacher candidates (PSTs) and practicing language educators are anticipated to engage in a process of self-exploration and exhibit characteristics of autonomous social learners (Son, 2014).

Communication encompasses engaging in exchanges with individuals who are acquiring a new language, professional associates, mentors, and scholars (Son, 2018). Computer-mediated communication (CMC) has been utilized as a means of facilitating communication, collaboration, and reflection through various channels such as video conferencing, email, blogs, and discussion forums. The aforementioned tools are deemed imperative for the professional growth of educators, as research has established a strong correlation between the efficacy of teachers and their proficiency in interpersonal communication (Meskill, 2009). Moreover, these tools are regarded as valuable assets to educators' practical knowledge, encompassing their professional and personal viewpoints on the incorporation of technology in their pedagogy, as well as their capacity to motivate themselves and their pupils to acquire proficiency in utilizing novel tools (Haines, 2016).

Collaboration necessitates educators to engage in collective efforts to exchange their resources, concepts, and encounters. It is recommended that teacher education programs incorporate online collaboration practices to facilitate collaborative experiences within a virtual setting (Ernest et al., 2013). Therefore, educators have the opportunity to acquire competencies such as devising, formulating, and overseeing virtual collaborative endeavors, as well as providing lucid directives, scaffolding, and motivation to their pupils. In addition, utilizing the aforementioned computer-mediated communication (CMC) tools, educators have the ability to collectively exchange concepts and materials with both their peers and pupils (Son, 2006).

The act of reflecting is widely acknowledged as a fundamental element in the realm of teacher education, as described by Son (2018), and is defined as the systematic analysis and evaluation of one's own experiences (Richards & Farrell, 2005, p. 7). Reflective practice is considered a fundamental concept that allows educators to engage in critical self-reflection regarding their experiences with teaching and learning (Son, 2004). Richards and Farrell (2005) suggest that incorporating technological tools into teaching can be improved through various strategies such as self-monitoring, peer observation, maintaining teaching diaries and portfolios, conducting action research, and engaging in team teaching.

In summary, the constructivist perspective posits that the instruction of pre-service teachers (PSTs) in technology integration involves the examination of educational technologies, utilization of said technologies for communication, creation of collaborative educational settings with communication tools, cooperation with peers in resource sharing, and introspection on their developmental processes as they integrate technology. The prescribed procedures outlined in this theoretical framework ought to be adhered to within the context of pre-service teacher training. The utilization of constructivist education in technology-focused teacher training models is expected to yield advantageous results.

Socio-cultural Views in PST Education for Technology Integration

The socio-cultural theory has evolved into a multifaceted theoretical framework for examining the process of foreign language teacher learning (Nguyen, 2019). Johnson and Golombek (2020) assert that the socio-cultural perspective of teacher education presents an all-encompassing cognitive framework that can inform and revolutionize the notions and practices in research on language teacher education. Viewed through this lens, the recognition of teachers as learners of teaching underscores the significance of the social contexts they encounter in their professional growth, as well as the expertise required to navigate these contexts, in fostering their cognitive advancement (Johnson, 2009). Moreover, as noted by Nguyen (2019), it is imperative to comprehend a PST as a person who embodies diverse cultural and historical backgrounds and is confronted with a plethora of challenges spanning across the past, present, and future. Research has revealed that the personal backgrounds of teachers, including their cultural heritage (Nguyen, 2017), previous educational and professional experiences (Nguyen & Brown, 2016), and beliefs (Yuan & Lee, 2014), have a significant impact on the learning experiences and outcomes of pre-service teachers. Thus, it is crucial to consider the beliefs and acceptance of existing technologies, as well as the preferences of pre-service teachers, when training them for technology integration. The manner in which individuals formulate their designs for future EFL lessons that incorporate technology is influenced by the beliefs and choices they bring to the training process. Furthermore, within the socio-cultural framework, educators are acknowledged as individuals who are both impacted by and have an impact on social circumstances. (Johnson & Golombek, 2018). Therefore, facilitating opportunities for pre-service teachers (PSTs) to collaborate and communicate their perspectives on the integration of technology is crucial. Drawing upon the socio-cultural perspective, scholars (Bax, 2003, 2011; Chao, 2015; Torsani, 2016; Guichon & Hauck, 2011; Guichon, 2012) have examined the topic of pre-service teacher (PST) education for technology integration through the lens of various theoretical constructs, including normalization, transfer, and reconstruction.

Bax (2003) proposed the concept of normalization, which refers to the point at which technology has become integrated into individuals' everyday routines. Bax (2011) presents a redefinition of normalization in his scholarly article, drawing on the principles of Neo-Vygotskian Theory. The author references the following sources to support his argument. The process of learning and development is not solely influenced by culture, but rather is culturally based. Additionally, this process is social in nature and not individualized. Communication plays a crucial role in the development of learning. It is important to note that understandings are constructed within culturally formed settings. The provision of assistance or instruction is a common and significant aspect of human mental development. It is possible to expand an individual's learning or problem-solving abilities by providing appropriate assistance or instruction. (Mercer & Fisher, 1997, pp. 13-16, in Bax, 2011, p. 7)

Bax (2011) posits that the integration of new technology in language teaching should not be viewed as an isolated endeavor, but rather as a socially-mediated process that is shaped by cultural contexts, communication, and support. The process of normalization is influenced by various social and cultural factors, in addition to the technological attributes or any other individual agent. These elements operate in a complex manner to shape the normalization process. Stated differently, the English instructor is not an autonomous entity operating in seclusion, but rather a societal entity impacted by their cultural environment.

Bax (2011) argued that conducting a "Needs Audit" is a necessary step prior to implementing technology in language instruction. The term 'Needs Audit' pertains to the critical evaluation of the relevance and efficacy of the technology under consideration for implementation. This approach prioritizes learning by taking into account the more cost-effective and pragmatic options. Hence, it is recommended that the notion of 'Needs Audit' be introduced to prospective language teachers (PSTs) as a means of assessing the suitability and efficacy of technological tools at their disposal for use in their specific teaching contexts, rather than simply exposing them to technology integration.

The capacity to transfer is an essential notion within the socio-cultural framework of teacher education. The term "transfer" refers to the act of language educators utilizing the skills acquired during their technology-related coursework in their instructional practices. This phenomenon holds significant value in determining the nature and degree of knowledge acquisition by teachers through technology-oriented courses (Torsani, 2016). As per Chao's (2015) findings, the term denotes the utilization of a particular technology in its original form, as taught in a technology-oriented course, without any modifications. In the event that educators are instructed on the development of an oral communication assignment utilizing a designated technological instrument during their technology course, successful implementation necessitates the utilization of said instrument for the same purpose in their practical teaching encounters. Chao (2015) examines the distinction between in-service and pre-service teachers, highlighting the variance in their utilization of technology within distinct contexts and their respective identities. Chao (2015) proposes the concept of long-term transition, drawing from Beach (2003)'s 'consequential transitions' and recognizing the importance of socio-cultural contexts in facilitating transfer. This perspective posits that it is not sufficient for teachers to merely apply the knowledge acquired in their technology course. Instead, teacher trainers should facilitate a more profound reflection among prospective teachers regarding their perspectives and dispositions towards technology utilization. This is because, by the time these prospective teachers enter the classroom, novel technologies may have surfaced. Through contemplation of technological tools, their applications, and their impact, educators can develop proficiencies in implementing technology in diverse settings, thereby fostering ingenuity in the utilization of novel technologies (Torsani, 2016).

RESEARCH IN PST EDUCATION ON TECHNOLOGY INTEGRATION

Contemporary developments in the field of education underscore the imperative of incorporating digital technologies into established pedagogical settings with the aim of revolutionizing and augmenting the learning experience (Anderson, 2009). The utilization of contemporary Information and Communication Technologies (ICTs) and the Internet affords learners and educators the potential to enhance their academic and social development through the provision of novel assessment paradigms, collaborative platforms, visual aids, and innovative learning methodologies (Meyer, 2015). The literature highlights that educators of foreign languages, including those who are in-service and pre-service, exhibit favorable dispositions towards integrating information and communication technologies (ICTs) and the Internet into their language classes, irrespective of the pedagogical approach (blended or traditional) implemented in educational institutions (Aydın, 2013; Barnes, 2018; Mei, Brown & Teo, 2018; Yokuş, 2016). Nevertheless, it has been observed that foreign language instructors may experience a deficit in cognitive and technical proficiency, which could impede their ability to effectively utilize information and communication technologies readily available to them (Lozano & Izquierdo, 2019). Despite the numerous expectations of competencies from teachers by various organizations and institutions, there remains a lack of clarity regarding how teacher training should facilitate the acquisition of these competencies for the purpose of integrating technology into teaching (Aydın, 2017).

Certain scholars have posited that the inadequate and unsuitable training provided to instructors of foreign languages is one of the contributing factors to the unfavorable digital divide observed among educators (Hsu, 2016; Lord & Lomicka, 2011). According to Kuure et al. (2016), foreign language instructors must possess the knowledge and skills to effectively integrate information and communication technologies (ICTs) into their teaching practices. Additionally, they must receive training on how to design language learning tasks that incorporate the latest technological advancements. Hsu (2016) highlighted the deficiency in foreign language teacher education and the integration of information and communication technology (ICT) in language teaching. The author suggested that teacher training programs should provide courses that enable educators to maximize the efficient utilization of ICTs and the Internet in accordance with the curriculum standards. According to Lord and Lomicka's (2011) assertion, foreign language instructors should possess a diverse range of contemporary technological resources. Lozano and Izquierdo (2019), on the other hand, emphasized that foreign language educators must not only acquire the most recent technologies but also acquire the knowledge of how to integrate them effectively in the classroom from a pedagogical perspective.

In 2018, Yeh conducted a quantitative research study that examined the level of technology integration and preparation among Teachers of English to Speakers of Other Languages (TESOL) programs in the United States. The data was gathered by the researcher from pre-service teachers (PSTs) who were currently enrolled in thirteen Teaching English to Speakers of Other Languages (TESOL) programs. The TPACK survey was utilized by the individual in question to ascertain the perceptions of prospective teachers of English to Speakers of Other Languages (TESOL) regarding the integration of technology in their teacher preparation programs. The findings of this research indicate that the prospective teachers (PSTs) did not experience any instances of technology integration as a means of facilitating English as a Second Language (ESL) instructional practices. Moreover, the individual expressed that there was a lack of assistance for the incorporation of technology into their instructional practices during their practicum experience. According to the participants' accounts, their exposure to their professors' utilization of technology was limited to their coursework in the TESOL program. The participants reported a lack of technology integration with their practicum instructors. The findings of this study indicate that the participants' lack of exposure to training and positive exemplars during their educational experiences has left them ill-equipped to effectively incorporate technology into their teaching practices.

Kuru Gönen (2019) conducted a qualitative research study in Turkey to address the insufficient training and practical experience of pre-service English language teaching (ELT) teachers. The study involved eight pre-service teachers and ninety-five English language learners and lasted for twelve weeks. The research utilized an experimental design to explore the perceptions of both the PSTs and foreign language students regarding the integration of technology into foreign language classrooms during a situated practicum experience. The author proposed a framework for the incorporation of technology in second language (L2) classrooms, which comprises five sequential stages: (1) Acquisition of essential competencies and expertise, (2) Examination of the educational institution and classroom setting, (3) Identification of suitable technological resources, (4) Assimilation of technology into the curriculum, and (5) Evaluation of the entire process. The data was gathered by the researcher utilizing reflective diaries and focus-group interviews. The study findings indicated that the integration of technology had a positive impact on language skills, specifically in the areas of oral proficiency, vocabulary acquisition, and retention rates. Additionally, technology integration fostered a positive classroom environment that was characterized by increased enjoyment, motivation, interactivity, and cooperation during lessons. Furthermore, technology integration facilitated teacher engagement and efficiency by accommodating diverse learning styles and enabling the identification of students' individual levels. The findings of the study have also brought to light certain limitations, including technical issues, time constraints for students with lower proficiency levels, disinterest or boredom, and the impact of class size (Kuru Gönen, 2019). The study's findings indicate that it is feasible to provide training to prospective teachers (PSTs) on technology integration using particular models, given the appropriate opportunities. Nevertheless, the quantity of such models is restricted.

Aydın (2017) has proposed an alternative model that offers a distinct perspective on the matter. This model proposes a collaborative approach between pre-service teachers (PSTs) and mentor teachers at practicum schools to develop technology-enhanced language teaching materials, as opposed to solely training PSTs. In 2017, Aydın proposed the implementation of the 'Reverse Mentoring Model' and a course titled 'Digital Material Development' for pre-service teacher education. The purpose of this proposal was to facilitate the integration of technology into language teaching. The author examined the challenges associated with in-service and pre-service training programs and emphasized that such programs fail to produce the intended results in terms of enhancing the technological integration skills of both in-service and pre-service educators. As a proposed solution, the researcher has suggested the implementation of the 'Reverse Mentoring Model'. This model involves providing instruction on technology integration to the Pre-Service Teachers (PSTs) and assigning them the task of preparing materials from a shared pool. Subsequently, prospective teachers will be requested to utilize the electronic resources in order to formulate instructional designs for their teaching practicum. The in-service teachers (mentors) will observe the presentations of the pre-service teachers (PSTs) who will instruct their mentors on the utilization and preparation of digital tools for their respective lessons. The mentor teachers also equip resources for the collective use. The aforementioned swimming pool is intended for utilization by both pre-service teachers and currently employed educators across the nation. The course provided by Aydın (2017) requires that prospective teachers (PSTs) adhere to the prescribed model, which entails observing one another during teaching practice and improving their utilization of digital tools through constructive feedback. The proposed model and corresponding course are expected to yield advantageous results for pre- and in-service teachers with respect to their ability to create digital materials. However, it is worth noting that the model does not encompass the training component of technology integration and lesson preparation for language teachers. As such, there remains a need for a distinct model that can effectively address this aspect of language teacher education.

In their 2017 study, Sert and Li utilized qualitative methods to investigate the CALL knowledge possessed by 111 EFL pre-service teachers in Turkey. The researchers sought to examine the participants' abilities in designing and developing activities based on CALL. Over a period of 14 weeks, the participants underwent training on the creation of computer-assisted language learning (CALL) materials, including online teaching tools and Web 2.0 tools. Reflective diaries were gathered from the participants during this time. The findings of the study indicate that knowledge of Computer-Assisted Language Learning (CALL) is grounded in three key areas: disciplinary knowledge, encompassing an understanding of the advantages and functions of technology; instructional knowledge, which involves designing activities and technical aspects; and contextual knowledge, which takes into account the availability of technology and the needs of students. The researchers put forth a novel model for teacher training in Computer-Assisted Language Learning (CALL) pedagogy, as per the results of their investigation. The authors of this novel framework posited that the preparation of English as a Foreign Language instructors ought to encompass the phases of experiential learning, critical thinking, practical application, and introspection. The authors emphasized the significance of providing guidance to prospective science teachers (PSTs) regarding the development of practical knowledge processes, augmenting their decision-making abilities, and engaging in reflective practices concerning the implementation of computer-assisted language learning (CALL). Consequently, the outcomes and theoretical framework presented in the conclusion of this research indicate a requirement for structured instruction on the implementation of technology in the classroom for English as a Foreign Language Pre-Service Teachers.

The aforementioned studies indicate that in order to effectively integrate technology into teaching, teacher education programs should offer comprehensive and structured training that encompasses experiential learning and reflective practices. Nonetheless, these investigations also underscore a significant deficiency in language instructor training with respect to the incorporation of technology in their instructional practices. The majority of studies approach technology holistically when instructing pre-service teachers (PSTs) and do not provide targeted pedagogical approaches for contemporary technologies.

DIGITAL DIVIDE AND THE PST EFL LECTURER

The integration of technology in foreign language education necessitates the examination of the distinct differences and disparities among students and educators in relation to the digital divide. Insufficient user expertise or resources can significantly impact the outcomes of technology assimilation and education. The successful integration of technology in the classrooms of Pre-Service EFL Lecturers can be achieved through a comprehensive understanding of the implications of the digital divide and appropriate action taken in response.

The term "digital divide" typically denotes inequalities in the ability to access and effectively utilize contemporary Information and Communication Technologies (ICTs) (van Dijk, 2006; Scheerder, Deursen & van Dijk, 2017). The resolution of these disparities is of utmost importance in the assimilation of these technologies in educational settings, as being positioned on either the advantageous or disadvantageous end could potentially impact the results of Pre-service English as a Foreign Language (EFL) instruction.

The occurrence of recent unprecedented incidents such as pandemics, large-scale fires, and natural calamities has brought to the forefront the existence of the digital divide. This refers to the situation where a significant number of individuals are deprived of access to information and knowledge. The ability to consider the potential consequences of a shortage of technical resources or a limited understanding of certain technologies has become increasingly critical for Pre-service EFL instructors as they design their technology-based courses, as noted by Köksal (2004).

The digital divide in terms of information and communication technologies (ICTs) has predominantly been observed with regards to the utilization and accessibility of the Internet (Castells, 2002). The digital divide became more widespread, especially with the increase in personal computer ownership and Internet accessibility (Eastin, Cicchirillo & Mabry, 2015). According to Bucy and Newhagen (2004), individuals who have access to the Internet are considered to be situated on the favorable end of the divide, which is regarded as the initial stage of the digital divide. Conversely, the correlation between possessing Internet connectivity and occupying a privileged position within the digital divide has recently been subject to scrutiny. In several Western countries, individuals who encountered no difficulties in accessing the Internet and enjoyed unfettered connectivity were unable to exhibit the requisite proficiencies (Scheerder et al., 2017). Hargittai (2002) observed a shift from the digital divide to the skills divide, referred to as the second-level digital divide. This highlights the importance of possessing technical skills to effectively utilize the Internet (Mossberger, Tolbert & Stansbury, 2003; van Deursen, Helsper & Eynon 2016). The researchers have recently classified the existing gap as a tertiary level digital divide and have introduced a prerequisite for favorable outcomes in internet usage (Scheerder, et al., 2017; Wei, Teo, Chan & Tan, 2011). The third phase of the digital divide pertains to instances where the acquisition of technological competencies does not yield favorable outcomes. According to Brandtzaeg, Heim, and Karahasanovic's (2011) argument, the variation in consequences can be attributed to "user types." This implies that factors such as an individual's disposition and inclination towards technology, in addition to their utilization and aptitude for creating meaning, play a significant role in establishing this novel form of division. The researchers established a typology of Internet users by conducting a survey on a sample of 12,666 Europeans aged between 16 and 74 years. Brandtzg et al. (2011) classified users into five distinct categories, namely non-users, sporadic users, entertainment users, instrumental users, and advanced users. It was determined that the instrumental and advanced categories of internet users are more likely to derive greater benefits from the internet, such as acquiring new skills and achieving better outcomes.

The possession and accessibility of necessary technological tools significantly contribute to determining an individual's position within the digital divide. Van Deursen and van Dijk (2019) have stated that the aforementioned equipment consists of devices that have the capability to access the Internet, including but not limited to desktops, laptops, tablets, smartphones, and smart televisions. Additionally, auxiliary devices such as printers, scanners, and hard drives are also included in this category. The investigation focuses on the analysis of three key factors that contribute to the inequality in "material access". The factors under consideration are: (1) disparities in device accessibility; (2) discrepancies in device and peripheral assortment; and (3) dissimilarities in device and peripheral upkeep expenses. (van Deursen & van Dijk, 2019, p. 256). Individuals who utilize a blend of mobile devices, such as smartphones and tablets, in conjunction with desktops and laptops, derive greater advantages from their Internet usage in relation to "device opportunities." This is due to the fact that the integration of both types of devices provides a diverse range of potential opportunities. The utilization of diverse peripheral devices, including supplementary displays, online printed resources, and supplementary hard drives, serves to augment the overall Internet experience. Finally, it is worth noting that the expenses associated with maintaining the devices may exceed their initial purchase price, as a result of supplementary costs such as software licensing and repair fees. (van Deursen & van Dijk, 2019). Hence, the capacity to fulfill all of these prerequisites for material accessibility can be regarded as advantageous in laying strong groundwork for the implementation of technology in educational institutions.

In order to effectively navigate the digital divide, it is imperative to not only have access to necessary resources, but also to possess the requisite technological competencies. The technical capabilities can be classified into four distinct categories, namely operational skills, information navigation skills, social skills, and creative skills. (van Deursen et al., 2016). Operational skills have been identified by researchers as the primary and essential abilities required for utilizing the Internet. The concept of information navigation skills encompasses the aptitude to effectively explore, identify, choose, and evaluate digital resources. Social skills refer to the ability to effectively utilize communication and interpersonal connections to comprehend and exchange information, while simultaneously accumulating social capital. The production of diverse and superior online content for dissemination to a wider audience necessitated the possession of innovative skills. (van Deursen et al., 2016). Consequently, it is imperative that both English as a Foreign Language (EFL) learners and instructors possess the necessary competencies when incorporating technology into the educational process.

Scheerder et al. (2017) conducted an analysis of 948 papers pertaining to the digital divide and identified seven distinct categories of factors that contribute to this phenomenon, beyond the fundamental requirements of resource access and skill acquisition. As per the analysis of these scholars, the digital divide is influenced by various factors such as socio-demographic, economic, social, cultural, individual, material, and motivational aspects. According to the proponents of the concept of the digital divide, the inequalities observed in this divide cannot be solely attributed to the age of users. Additionally, the group with the most advanced online skills is not necessarily the so-called digital generation, but rather those individuals possessing the aforementioned characteristics (Harigatti, 2002; van Deursen & van Dijk, 2019). The impact of generational disparities has been observed in various aspects of technology utilization, including general technology usage (Olson, O'Brien, Rogers & Charness, 2011), internet usage (Van Volkom, Stapley & Amaturo, 2014), mobile phone usage (Ling, 2010), social media usage (Van Volkom et al., 2014), and the emotional responses of users (Van Volkom et al., 2014). According to the definition provided by Çapar and Kecik (2016), a generation refers to a cohort of individuals who share the same age and exist within the same historical period. Prensky (2001) has classified the younger generation as "digital natives" and the older generations as "digital immigrants." This categorization may have an impact on the digital divide, as the abilities and motivations of these generations in the digital realm could play a role in determining their positions. It is imperative to consider the age discrepancies between Pre-service EFL instructors and their pupils when integrating contemporary teaching and learning technologies.

As stated in the introduction, it is noteworthy that the ELT departments' curriculum in Turkey does not include dedicated courses aimed at the integration of technology in the context of EFL instruction. Despite the recent decision by the Council of Higher Education in Turkey to delegate the responsibility of curriculum development to individual universities, the prevailing trend among Faculties of Education is to continue utilizing the previous curriculum established by the Council of Higher Education. Consequently, a significant number of prospective teachers are completing their studies without possessing the requisite proficiency in utilizing contemporary technologies to augment their pedagogical practices in the domain of English language instruction. Regarding this matter, it is possible to incorporate technology-focused courses into the English Language Teaching (ELT) curriculum. These courses would provide Pre-Service Teachers (PSTs) with information and guidance on how to effectively integrate the latest technologies. Given the dynamic nature of technological progress, it remains uncertain what novel advancements may emerge with each successive improvement. It is recommended that teacher candidates be instructed and provided with courses on the implementation of technologies. In addition to the courses that may be incorporated into the English Language Teaching (ELT) curriculum, prospective teachers (PSTs) have the opportunity to participate in or host seminars, conferences, and workshops that provide guidance on the integration of technology. Prospective teachers (PSTs) have the option to organize events and engage in discussions to exchange their personal experiences regarding the latest technologies through a Community of Practice (CoP). Alternatively, they may be motivated to participate in events hosted by academic institutions. As per Hanson-Smith's (2016) assertion, individuals who belong to a Community of Practice (CoP) engage in the exchange of knowledge, provision of guidance and mentorship to novices, and collaborative efforts towards honing the requisite skills within the specific domain or environment where the CoP has been established. Thus, the establishment of the groundwork for the perpetuation of career advancement subsequent to tertiary education will also be initiated. An additional implication pertaining to the education of prospective teachers may involve the incorporation of technology-based prerequisites within the curricula that facilitate the preparation of micro-teaching demonstrations by said individuals. As a requirement, Teaching English to Young Learners or Teaching Language Skills courses may prompt PSTs to incorporate technology into their lesson plans. Consequently, prospective teachers (PSTs) are afforded an opportunity to engage in the application of technology in content-specific domains of English Language Teaching (ELT) and will be subjected to a greater degree of exposure to technology implementation in comparison to a solitary technology integration course. ELT department professors can utilize various technologies to conduct lessons with PSTs. By adopting this approach, prospective teachers can engage in the process of observing their educators and acquiring knowledge about the methods employed for the purpose of integration. In addition, pre-service teachers (PSTs) have the ability to engage in reflective practices regarding their learning outcomes through the utilization of technology. This includes evaluating the efficacy of technological tools, identifying areas of proficiency and deficiency, and subsequently modifying their pedagogical approaches to incorporate these technologies in the future.

REFERENCES

Akbulut, Y., Odabaşı, H., & Kuzu, A. (2011). Perceptions of preservice teachers regarding the integration of information and communication technologies in Turkish education faculties. *The Turkish Online Journal of Educational Technology*, *10*(3), 175–184.

Al-Jarrah, J. M., Talafhah, R. H., & Al-Jarrah, T. M. (2019). ESL teacher perceptions of using educational mobile applications to develop the language skills of ESL elementary school students. *European Journal of Foreign Language Teaching.*, *4*(1), 65–86.

Anderson, N. (2009). *Equity and information communication technology (ICT) in education*. Peter Lang.

Aydın, B. (2017). Three birds with a stone: Technology integration in language education with reverse mentoring model. *Journal of Teacher Education and Educators*, *6*(2), 177–190.

Aydın, S. (2013). Teachers' perceptions about the use of computers in EFL teaching and learning: The case of Turkey. *Computer Assisted Language Learning*, *26*(3), 214–233. doi:10.1080/09588221.2012.654495

Barnes, J. A., Jr. (2018). *Teachers' perceptions of implementing m-learning using pedagogical approaches* [Unpublished doctoral dissertation]. The University of Memphis, Memphis, TN.

Başal, A. (2015). English language teachers and technology education. *Journal of Theory and Practice in Education.*, *1*(4), 1496–1511.

Bax, S. (2003). CALL—Past, present and future. *System*, *31*(1), 13–28. doi:10.1016/S0346-251X(02)00071-4

Bax, S. (2011). Normalisation revisited: The effective use of technology in language education. *International Journal of Computer-Assisted Language Learning and Teaching*, *1*(2), 1–15. doi:10.4018/ijcallt.2011040101

Beach, K. (2003). Consequential transitions: A developmental view of knowledge propagation through social organizations. In T. Tuomi-Gröhn & Y. Engeström (Eds.), *Between school and work: New perspectives on transfer and boundary-crossing* (pp. 39–62). Earli.

Beck, C., & Kosnik, C. (2006). *Innovations in teacher education: a social constructivist approach.* SUNY Press. doi:10.1353/book5061

Brandtzæg, P. B., Heim, J., & Karahasanović, A. (2011). Understanding the new digital divide—A typology of Internet users in Europe. *International Journal of Human-Computer Studies*, *69*(3), 123–138. doi:10.1016/j.ijhcs.2010.11.004

Bucy, E. P., & Newhagen, J. E. (2004). *Media access: Social and psychological dimensions of new technology use*. Psychology Press.

Çapar, M., & Keçik, İ. (2017). Generation types view and perspectives of EFL instructors on their expectations related to professional development programs. *ELT Research Journal, 5*(2).

Castells, M. (2002). *The Internet galaxy: Reflections on the Internet, business, and society*. Oxford University Press. doi:10.1093/acprof:oso/9780199255771.001.0001

Chao, C. C. (2015). Rethinking transfer: Learning from CALL teacher education as consequential transition. *Language Learning & Technology*, *19*(1), 102–118.

Darling-Hamond, L. & Bransford, J. (2005). *Preparing Teachers Education for a Changing World.* Jossey-Bass Publishing Co.

Duart, J. M. (2011). La Red en los procesos de enseñanza de la Universidad. *Comunicar*, *37*(19), 10–13. doi:10.3916/C37-2011-02-00

Eastin, M. S., Cicchirillo, V., & Mabry, A. (2015). Extending the digital divide conversation: Examining the knowledge gap through media expectancies. *Journal of Broadcasting & Electronic Media*, *59*(3), 416–437. doi:10.1080/08838151.2015.1054994

Ernest, P., Catasús, M. G., Hampel, R., Heiser, S., Hopkins, J., Murphy, L., & Stickler, U. (2013). Online teacher development: Collaborating in a virtual learning environment. *Computer Assisted Language Learning, 26*(4), 311–333. doi:10.1080/09588221.2012.667814

Guichon, N. (2012). *Versl'intégration des TIC dansl'enseignement des langues.* Didier.

Gülbahar, Y. (2008). ICT usage in higher education: A case study on preservice teachers and instructions. *The Turkish Online Journal of Educational Technology, 7*(1), 32–36.

Haines, K. (2016). Expanding the knowledge base of teachers' use of communication tools for language learning. *System, 62*, 102–112. doi:10.1016/j.system.2016.07.008

Hanson-Smith, E. (2016). Teacher education and technology. The Routledge handbook of language learning and technology, 210-222.

Hargittai, E. (2002). Second-level digital divide: Differences in people's online skills. *First Monday, 7*(4). Advance online publication. doi:10.5210/fm.v7i4.942

Hart, L. C. (2002). Preservice teachers' beliefs and practice after participating in an integrated content/methods course. *School Science and Mathematics, 102*(1), 4–15. doi:10.1111/j.1949-8594.2002.tb18191.x

Hockly, N. (2013). Mobile learning. *ELT Journal, 67*(1), 80–84. doi:10.1093/elt/ccs064 PMID:23923690

Hsu, L. (2016). Examining EFL teachers' technological pedagogical content knowledge and the adoption of mobile-assisted language learning: A partial least square approach. *Computer Assisted Language Learning, 29*(8), 1287–1297. doi:10.1080/09588221.2016.1278024

Hue, L. T., & Ab Jalil, H. (2013). Attitudes towards ICT integration into curriculum and usage among university lecturers in Vietnam. *International Journal of Instruction, 6*(2), 53–66.

Johnson, K. E. (2009). *Second language teacher education: a sociocultural perspective.* Routledge. doi:10.4324/9780203878033

Johnson, K. E., & Golombek, P. R. (2018). Making L2 teacher education matter through Vygotskian-inspired pedagogy and research. In The Routledge handbook of sociocultural theory and second language development (pp. 443-456). Taylor and Francis. doi:10.4324/9781315624747-28

Johnson, K. E., & Golombek, P. R. (2020). Informing and transforming language teacher education pedagogy. *Language Teaching Research, 24*(1), 116–127. doi:10.1177/1362168818777539

Kedrova, G., & Potemkin, S. (2015, October). New trends in implementation of ICT in higher education. In *2015 9th International Conference on Application of Information and Communication Technologies (AICT) (pp. 531-535).* IEEE. 10.1109/ICAICT.2015.7338617

Kessler, G., & Hubbard, P. (2017). Language Teacher Education and Technology. In C. A. Chapelle & S. Sauro (Eds.), *The handbook of technology and second language teaching and learning* (pp. 278–292). Wiley. doi:10.1002/9781118914069.ch19

Koehler, M. J., Mishra, P., & Yahya, K. (2007). Tracing the development of teacher knowledge in a design seminar: Integrating content, pedagogy and technology. *Computers & Education, 49*(3), 740–762. doi:10.1016/j.compedu.2005.11.012

Köksal, D. (2004). To Kill the Blackboard? Technology in Language Teaching and Learning. *Turkish Online Journal of Educational Technology-TOJET*, *3*(3), 62–72.

Kosnik, C., Menna, L., Dharamshi, P., & Beck, C. (2018). Constructivism as a framework for literacy teacher education courses: The cases of six literacy teacher educators. *European Journal of Teacher Education*, *41*(1), 105–119. doi:10.1080/02619768.2017.1372742

Kukulska-Hulme, A., Lee, H., & Norris, L. (2017). Mobile learning revolution: Implications for language pedagogy. In C. A. Chapelle & S. Sauro (Eds.), *The handbook of technology and second language teaching and learning* (pp. 217–233). Wiley. doi:10.1002/9781118914069.ch15

Kurt, S. (2011). Use of constructivist approach in architectural education. *Procedia: Social and Behavioral Sciences*, *15*, 3980–3988. doi:10.1016/j.sbspro.2011.04.402

Kuru Gönen, S. İ. (2019). A qualitative study on a situated experience of technology integration: Reflections from pre-service teachers and students. *Computer Assisted Language Learning*, *32*(3), 163–189. doi:10.1080/09588221.2018.1552974

Ling, R. (2010). Texting as a life phase medium. *Journal of Computer-Mediated Communication*, *15*(2), 277–292. doi:10.1111/j.1083-6101.2010.01520.x

Lord, G., & Lomicka, L. (2011). Calling on educators: Paving the way for the future of technology and CALL. In N. Arnold & L. Ducate (Eds.), *Present and future promises of CALL: From theory and research to new directions in language teaching* (pp. 441–470). CALICO.

Lozano, A. A., & Izquierdo, J. (2019). Technology in Second Language Education: Overcoming the Digital Divide. *Emerging Trends in Education*, *2*(3), 52–70. doi:10.19136/etie.a2n3.3250

Marcelo, C., Yot, C., & Mayor, C. (2015). Enseñar con tecnologías digitales en la Universidad. *Comunicar*, *45*(23), 117–124.

Mei, B., Brown, G. T., & Teo, T. (2018). Toward an understanding of preservice English as a Foreign Language teachers' acceptance of computer-assisted language learning 2.0 in the People's Republic of China. *Journal of Educational Computing Research*, *56*(1), 74–104. doi:10.1177/0735633117700144

Mercader, C., & Gairín, J. (2017). ¿Cómo utiliza el profesorado universitario las tecnologías digitales en sus aulas? *Revista de Docencia Universitaria*, *15*(2), 257–274. doi:10.4995/redu.2017.7635

Mercer, N., & Fisher, E. (1997). The importance of talk. In R. Wegerif & P. Scrimshaw (Eds.), *Computers and talk in the primary classroom* (pp. 13–21). Multilingual Matters.

Meskill, C. (2009). CMC in language teacher education: Learning with and through instructional conversations. *Innovation in Language Learning and Teaching*, *3*(1), 51–63. doi:10.1080/17501220802655474

Meyer, B. (2015). Learning through telepresence with iPads: Placing schools in local/global communities. *Interactive Technology and Smart Education*, *12*(4), 270–284. doi:10.1108/ITSE-09-2015-0027

Mossberger, K., Tolbert, C. J., & Stansbury, M. (2003). *Virtual inequality: beyond the digital divide*. Georgetown University Press.

Nguyen, M. H. (2017). Negotiating contradictions in developing teacher identity during the EAL practicum in Australia. *Asia-Pacific Journal of Teacher Education*, *45*(4), 399–415. doi:10.1080/1359866X.2017.1295132

Nguyen, M. H. (2019). *English language teacher education: a sociocultural perspective on preservice teachers' learning in the professional experience*. Springer. doi:10.1007/978-981-13-9761-5

Nguyen, M. H., & Brown, J. (2016). Influences on preservice writing instruction during the secondary English as an additional language practicum in Australia. *The Australian Journal of Teacher Education*, *41*(8), 84–101. doi:10.14221/ajte.2016v41n8.5

Olson, K. E., O'Brien, M. A., Rogers, W. A., & Charness, N. (2011). Diffusion of technology: Frequency of use for younger and older adults. *Ageing International*, *36*(1), 123–145. doi:10.100712126-010-9077-9 PMID:22685360

Özel, A. G., & Arikan, A. (2015). The use of the internet and web 2.0 tools among EFL instructors. *Mediterranean Journal of Humanities*, *5*(1), 313–325. doi:10.13114/MJH.2015111386

Pegrum, M. (2014). *Mobile learning: languages, literacies and cultures*. Palgrave Macmillan. doi:10.1057/9781137309815

Prensky, M. (2001). Digital natives, digital immigrants. *On the Horizon*, *9*(5), 1–6. doi:10.1108/10748120110424816

Richards, J. C., & Farrell, T. S. C. (2005). *Professional development for language teachers: Strategies for teacher learning*. Cambridge University Press. doi:10.1017/CBO9780511667237

Şahin Kızıl, A. (2017). EFL learners in the digital age: An investigation into personal and educational digital engagement. *RELC Journal*, *48*(3), 373–388. doi:10.1177/0033688216684285

Scheerder, A., van Deursen, A., & van Dijk, J. (2017). Determinants of Internet skills, uses and outcomes: A systematic review of the second-and third-level digital divide. *Telematics and Informatics*, *34*(8), 1607–1624. doi:10.1016/j.tele.2017.07.007

Sert, O., & Li, L. (2019). A qualitative study on CALL knowledge and materials design: Insights from pre-service EFL teachers. In M. Khosrow (Ed.), *Computer-Assisted language learning: concepts, methodologies, tools, and applications* (pp. 2036–2051). IGI Global.

Son, J.-B. (2004). Teacher development in e-learning environments. In J.-B. Son (Ed.), Computer-assisted language learning: concepts, contexts and practices (pp. 107-122). iUniverse.

Son, J.-B. (2006). Using online discussion groups in a CALL teacher training course. *RELC Journal*, *37*(1), 123–135. doi:10.1177/0033688206063478

Son, J.-B. (2014). Learning about computer-assisted language learning: Online tools and professional development. In J.-B. Son (Ed.), APACALL Book Series: Vol. 3. *Computer-assisted language learning: Learners, teachers and tools* (pp. 173–186). Cambridge Scholars Publishing.

Son, J. B. (2018). *Teacher development in technology-enhanced language teaching*. Springer International Publishing. doi:10.1007/978-3-319-75711-7

Torsani, S. (2016). *CALL teacher education: language teachers and technology integration.* Springer. doi:10.1007/978-94-6300-477-0

van Deursen, A. J., Helsper, E. J., & Eynon, R. (2016). Development and validation of the Internet Skills Scale (ISS). *Information Communication and Society, 19*(6), 804–823. doi:10.1080/1369118X.2015.1078834

van Deursen, A. J., & van Dijk, J. A. (2019). The first-level digital divide shifts from inequalities in physical access to inequalities in material access. *New Media & Society, 21*(2), 354–375. doi:10.1177/1461444818797082 PMID:30886536

van Dijk, J. A. (2006). Digital divide research, achievements and shortcomings. *Poetics, 34*(4-5), 221–235. doi:10.1016/j.poetic.2006.05.004

Van Volkom, M., Stapley, J. C., & Amaturo, V. (2014). Revisiting the digital divide: Generational differences in technology use in everyday life. *North American Journal of Psychology, 16*(3), 557–574.

Warschauer, M., & Meskill, C. (2000). Technology and second language learning. In J. Rosenthal (Ed.), *Handbook of undergraduate second language education* (pp. 303–318). Lawrence Erlbaum.

Winke, P., & Goertler, S. (2008). Did we forget someone? Students' computer access and literacy for CALL. *CALICO Journal, 25*(3), 482–509. doi:10.1558/cj.v25i3.482-509

Yavuz-Konokman, G., Yanpar-Yelken, T., & Sancar-Tokmak, H. (2013). An investigation of primary school pre-service teachers' perception on their Tpack in terms of a variety factors: Mersin university case. *Kastamonu Eğitim Dergisi., 21*(2), 665–684.

Yeh, C. Y. (2018). *Investigating technology integration in United States TESOL teacher preparation programs* [Doctoral dissertation]. University of Delaware.

Yılmaz, M. (2007). Instructional technology in training primary school teacher. *Gazi Eğitim Fakültesi Dergisi., 27*(1), 155–167.

Yokuş, G. (2016). *Eğitim fakültesi öğrencilerinin mobil öğrenmeye ilişkin görüşlerinin incelenmesi ve eğitim bilimleri alanına yönelik mobil uygulama geliştirme çalışması: mobil akademi* [Unpublished master's thesis]. Mersin University.

Yuan, R., & Lee, I. (2014). Pre-service teachers' changing beliefs in the teaching practicum: Three cases in an EFL context. *System, 44*, 1–12. doi:10.1016/j.system.2014.02.002

KEY TERMS AND DEFINITIONS

Constructivism: It is an epistemology, a learning or meaning-making theory, that offers an explanation of the nature of knowledge and how human beings learn. It maintains that individuals create or construct their own new understandings or knowledge through the interaction of what they already know and believe and the ideas, events, and activities with which they come in contact.

Digital Divide: It commonly refers to the inequalities in accessing and using new forms of Information and Communication Technologies.

Lecturer: Specifically, as a job in a university or college of higher education, it is a person who delivers lectures.

Pre-Service Teacher (PST): This word refers to those individuals who are now enrolled in a teacher preparation program and are actively pursuing their certification as teachers. The trainees engage in supervised field-based teaching experiences, receiving guidance and mentorship from university faculty members and cooperating teachers in K-12 settings. These individuals are students who have received formal education from institutions of higher learning with the aim of becoming professional educators. Individuals with limited or negligible prior teaching experience in a classroom setting.

Socio-Cultural Theory: This theory asserts that human mental functioning is organized by cultural concepts, activities and artifacts which is mainly a mediated process and individuals are conceived as utilizing their own cultural artifacts and forming new ones to regulate their behavioral and biological activities. Learning results from the dynamic interaction among individuals and cultural artifacts contributing the social formation of the mind and realization of social goals

Technology: It refers to computers or other electronic equipment that connects or interacts with computers or other electronic equipment e.g., LCD projector, interactive whiteboard, digital camera, tablet, smartphone, student response system, et cetera). Technology also encompasses networks (e.g., Internet, LAN (local area network)) and computer software or applications.

Technology Integration: This is a general term covering the stages of technology use by teachers and may refer to the level, intensity, or frequency of technology used.

Chapter 9
Using PBL to Teach English Language and Culture at the Tertiary Level

Madalina Armie
University of Almería, Spain

María Enriqueta Cortés de los Ríos
University of Almería, Spain

Ángeles Jordán Soriano
https://orcid.org/0000-0003-0140-8115
University of Almería, Spain

Nuria del Mar Torres López
University of Almería, Spain

María del Mar Sánchez Pérez
https://orcid.org/0000-0002-6550-9030
University of Almería, Spain

ABSTRACT

The aim of this research is to display how project-based learning (PBL), one of the most innovative methodologies in the contemporary educational panorama, can be beneficial in teaching culture in English language classes at the tertiary level. Students will be due to produce a final project that examines cultural elements in several English-speaking countries combining various disciplines, such as geography, history, literature, and economics. The results will be drawn from two questionnaires: a pre-task survey that will assess the expectations and motivation levels of tertiary students in connection to this pedagogical approach and a post-task survey which will—in this case—assess the effects of these final results in students' motivation and the potential changes concerning their views on PBL.

DOI: 10.4018/978-1-6684-8861-4.ch009

INTRODUCTION

This study's purpose is to incorporate culture into tertiary English language instruction, for which a project-based learning (PBL) approach was applied in courses from various fields. PBL is a pedagogical framework that comprises teaching a subject through having students complete a project (Thomas, 2000). In English language learning, "this approach involves students completing a project by choosing a topic, solving problems, and generating products in the form of meaningful solutions" (Sanatullova-Allison, 2015, p. 7). Alternatively, Suzie and Krauss (2014) provide the following definition of PBL:

[PBL]—powered by contemporary technologies—is a strategy certain to turn traditional classrooms upside down. When students learn by engaging in real-world projects, nearly every aspect of their experience changes. The teacher's role shifts. He or she is no longer the content expert, doling out information in bite-sized pieces. Student behavior also changes. Instead of following the teacher's lead, learners pursue their own questions to create their own meaning. Even the boundaries of the classroom change. Teachers usually design the project as the framework for learning, but students have more agency and may wind up using technology to access and analyze information from all corners of the globe. Connections among learners and experts can happen in real time. That means new kinds of learning communities can come together to discuss, debate, and exchange ideas. (p. 1)

PBL has its roots in Dewy's work (1938), which claims that humans learn by doing. Therefore, with PBL, "students will be driving their own inquiry, working collaboratively to seek viable solutions, using advanced technology, and communicating their findings to the related audience" (Bell, 2010, p. 40). Thus, it is not surprising that this innovative methodology may be more inspiring and engaging for English language learners than more conventional ones since this is a learner-centered approach in which students develop critical thinking, cooperation, communication, and other 21st-century skills.

PBL was chosen for this study because "its utilization in [English as Foreign Language] (EFL) classrooms may serve as one of the appropriate methods to incorporate culture into language contexts" (Sanatullova-Allison, 2015, p. 1). This chapter's main goal is to put PBL into practice as a cutting-edge method through which students can design authentic final products that require the investigation of cultural aspects of English-speaking countries in combination with various transversal subjects such as geography, history, literature, music, economics, and more. This project aims to boost foreign language students' motivation to learn more about English culture. The following subjects that were taught at the University of Almería, Spain in the 2021/22 academic year were included in this study: (1) Sociocultural Competence, as an element of programming in the teaching of a foreign language (part of the Primary Education degree program); (2) English teaching as a foreign language in infant education (part of the Infant Education degree program); (3) English III (part of the Hispanic Philology degree program); (4) English II (part of the Humanities degree program); and (5) English for Academic Purposes (part of the Hispanic Philology degree program).

In order to accomplish this general objective, several specific objectives were also set: (1) encouraging student autonomy in learning and the transversal application of knowledge; (2) promoting the use of information and communication technology (ICT); (3) examining the benefits and drawbacks of PBL in the university setting, specifically in the context of English language courses; and (4) evaluating PBL performance using a pre-task questionnaire that measured students' motivation levels and expectations in relation to this methodology at the tertiary level as well as a post-task survey that was used to monitor the consequent effects of PBL on students after its implementation in the classroom. The findings of this research are based on a comparative analysis of the results from the first survey and the information collected in the second questionnaire. In this study, special attention was placed on both PBL's impact on students' motivation levels and the changes in their opinions toward this didactic method as well as PBL's suitability and efficiency when implemented in various foreign language courses at the university level.

APPLICATIONS, ADVANTAGES, AND DISADVANTAGES OF TEACHING ENGLISH LANGUAGE AND CULTURE THROUGH PBL AT THE TERTIARY LEVEL

Language and culture coexist and support one another, which makes them closely related (Bryam et al., 2001). Hence, understanding culture is crucial as, otherwise, language learners may lack the skills needed to properly comprehend the thoughts, practices, and traditions of native speakers of the target language (Gandía, 2012). PBL is suitable for learning about both language and culture since "the students, actively and with a very clear personal intention, investigate their environment, form ideas and concepts around it, and form opinions of it and act upon it" (Bautista et al., 2017, p. 46). Therefore, through the use of PBL, foreign language teachers can instruct students in both language and culture. PBL is founded on constructivism and encourages creativity, independent learning, and intrinsic motivation while also allowing students to actively access and generate new information about a particular topic connected to the target language (Krajcik & Blumenfeld, 2005). Additionally, PBL is based on active and inquiry-based learning in which students must use their knowledge and abilities over an extended period to research a particular question, problem, or challenge. The goals and practices of PBL strongly emphasize the practice of higher-order thinking skills and the subskills identified in Bloom's (1956) taxonomy, including applying (doing and implementing), analyzing (differentiating, organizing, and attributing), evaluating (checking and criticizing), and creating (generating, planning, and producing).

The main goal of PBL is challenging students to create, plan, and complete a project that yields an authentic result, which is developed by exploring an issue, query, or challenge. The theoretical underpinnings of various curricula serve as the students' starting points, with them then building their understanding on their prior schemata or knowledge. PBL is a type of transversal teaching because of the clear connections it establishes between several academic disciplines, including history, geography, the arts, mathematics, economics, sociology, and literature, to name a few. The variety of methods appeals to students who have various interests and preferences, thereby increasing their motivation. The activities and the final output coalesce, allowing the learning outcomes to be met, with the assessment of the work and the product's elaboration also enhancing students' participation.

According to Prieto (2006), PBL is an efficient and adaptable approach that can raise the standard of academic learning in multiple ways. Firstly, it allows students to acquire various abilities, such as problem-solving, decision-making, and teamwork (de Miguel, 2015). For example, each student is a member of a group that collaborates to build knowledge, and as they debate options and solve problems that arise during the project, their logical reasoning, originality, inventiveness, and communication abilities are stimulated. Secondly, students also improve their fluency and practice oral communication through reasoning, negotiating, presenting information, discussing, and refuting their peers' comments in the foreign language, thus exemplifying the integration of language skills in PBL. These interpersonal and social skills improve peer connections, which benefits classes and other group activities by enhancing the environment and positively impacting the dynamics of the classroom. Thirdly, PBL encourages the growth of positive attitudes and values, such as critical thinking, because it involves obtaining information. When students learn to make precise and selective decisions as well as how to condense knowledge and present it to others, they gain the ability to more effectively distinguish between trustworthy and unreliable sources. Fourthly, students' writing comprehension and production improve through their cooperation when presenting the knowledge contained in the final output, and the reading required in the class also teaches students how to use new vocabulary and grammar correctly in the given contexts.

The drawbacks of PBL are common to other learner-centered methodologies. In a regular classroom, students must pay attention to the teacher, but they must speak up in a PBL setting. In previous approaches, students were given clear concepts and explanations, but PBL requires them to conduct research, gather data, assess its accuracy, and determine whether to include it in the final product based on its relevance to the topic. As a result, the first drawback is this occasionally troublesome shift between acting as an instructor and being a student because the PBL method is unsuccessful if learners are not engaged, are confused, or they do not have sufficient direction or guidance during the process. In this learning environment, the teacher is reduced to the position of a mentor, facilitator, and guide who gives instructions to aid students in finding and comprehending knowledge and processing it into a final product. Therefore, the educator must also scaffold the assignments while letting the students explore how real-world resources can be used. This allows classes to become learner-centered and facilitates students gaining agency, thus stepping out of their status as merely knowledge receptors and taking an active role in finishing the proposed tasks.

Regarding the challenges associated with the educator's role, it is important to highlight the sizeable time commitment required to conceptualize a project that fits a class's requirements, objectives, and content. Although PBL is time-consuming and laborious, its dynamic nature helps to offset some of these drawbacks. For example, the created resources are flexible and adaptable, and they can be extrapolated to other class activities and adapted for groups with various educational backgrounds. PBL's transversality necessitating greater cooperation between academic disciplines is also crucial as, otherwise, students may feel lost because the new knowledge they have gained cannot be built upon their existing knowledge.

In terms of the challenges or disadvantages experienced by learners, it is vital to note that students may be reluctant to participate in many group and class activities. Moreover, due to PBL's multi-layered structure, students may initially believe that it will increase their workload. Therefore, the division of the content into various sessions should reflect a distribution of goals and assignments that approach the topic from various angles. The present study evaluates the benefits and drawbacks of PBL at the tertiary educational level and concludes that the advantages outweigh the drawbacks of this important method of teaching language and culture.

SOCIOCULTURAL COMPETENCE AND USEFUL DIDACTIC RESOURCES FOR INTEGRATING CULTURE INTO ENGLISH LANGUAGE CLASSROOMS

Being communicatively competent in a foreign language implies not only mastering its grammar but also knowing the nuances of the culture within certain contexts. Therefore, sociocultural competence can be defined as the knowledge of the society and culture of the community or communities in which a language is spoken, including these communities' everyday lives, living conditions, values, body language, social conventions, and ritual behaviors. Sociocultural competence can be developed in English language classrooms through PBL by helping students develop transversal competence, fostering creativity, exchanging ideas, and searching for information, which are only some of the principles that should be included and fostered in this educational approach. These are detailed below.

- Transversality: PBL allows students to learn content from different subjects in their curricula.
- Creativity: All the activities require a high degree of creativity, which students can use to develop new learning methods.
- Exchanging ideas and collaborating: All activities are discussed in class, so consensus is essential for ensuring successful group projects.
- Searching for and compiling information: All activities require students to search for information, thus teaching them new methods of information mining, which leads to meaningful learning.
- Exposing students to different didactic resources: For tertiary students to learn and elaborate on their projects, they should be exposed to different didactic resources such as songs, stories, ICT materials, and other methods such as gamification, thus providing an eclectic nature to the didactic approach.

Songs are resources that make the learning of languages and cultures more pleasant and efficient as they introduce students to and familiarize them with the culture related to the foreign language they are learning. Songs reveal interesting information about cultures' contemporary social and cultural backgrounds. For example, pop music is an essential part of British culture as it indicates the multi-cultural nature of contemporary Britain. Ilari et al. (2013) claimed that "singing songs from different cultures may play important roles in the construction of our identities and in how we perceive and understand others and ultimately ourselves" (p. 202). Therefore, teachers should use songs in the classroom because they are common to all cultures. Moreover, songs are useful for developing students' grammar as well as for revising, consolidating, extending, and recycling vocabulary, and they can increase students' motivation and improve their pronunciation, among other advantages.

Similarly, culture is often embedded and transmitted through stories. Storytelling promotes diversity, a positive attitude towards individuals of all nationalities, races and religions, and intercultural understanding, which is crucial in today's world. By developing intercultural awareness, students become more open to different views and learn how to respect diversity since stories provide "ideal opportunities for presenting cultural information and encouraging cross-cultural comparison" (Ellis & Brewster, 2014, p. 46). Therefore, when students listen, read, and repeat stories, they are learning aspects of a culture that they can discuss later and compare to their own. Stories provide information about both the authors' cultures as well as the culture's folklore and fairytales (Povey, 2019). Storybooks, for example, are excellent methods of communicating cultural information and fostering international understanding since they reflect the environments and cultures of their authors and illustrators. There are several ways in which storytelling can enhance intercultural understanding and communication:

- It allows students to experience diverse cultures.
- It fosters students' ability to empathize with unfamiliar people, places, and situations.
- It offers insights into different traditions and values.
- It helps students understand how wisdom is common to all peoples and cultures.
- It offers insights into universal life experiences.

- It encourages students to consider new ideas, and it reveals the differences and commonalities in cultures around the world (Dvalidze, 2017, p. 52).

Apart from the resources discussed above, this paper considers the importance of gamification by analyzing the use of games in the classroom to facilitate students' learning of sociocultural information. Miller (2013) states that "the term of gamification was documented for the first time in 2008" (p. 196). It was initially related to the business world, but it has transformed into an adequate educational methodology. Gamification is defined as "the use of elements in the design of games within no ludic contexts" (Deterding et al. 2011, p. 9). This methodology plays an important role in the integration of culture into the English language classroom in addition to boosting students' motivation and participation, as clarified by Moriera et al. (2018):

In the current context of teaching, one of the key challenges is how to engage students and increase their involvement in achieving the objectives of the proposed learning activities. One of the suggested solutions is to combine learning strategies that involve active methods... with traditional methods.

Furthermore, Leung and Pluswik (2018) stated that this mixture of less and more active approaches can lead to further improvements in problem-solving skills, long-term retention of knowledge, and learner engagement. These ideas are also supported by Armie and Membrive (2022), who claim that teaching a second or foreign language

requires creativity as well as a large number of attractive resources and distinctive approaches that intermingle old methodologies and new techniques. The implementation of games in education might offer answers to some quandaries regarding class planning, management, teaching, and learning since games are desirable experiences that set motivational mechanisms into motion, engage learners by adding dynamism to lessons, and foster the consolidation of content through play. (p. 1)

As certain traditional teaching methods are often unengaging for some students, the use of active learning methodologies has been verified to be very useful as gamification is "one [technique] of active learning... which is a challenge activity defined by rules where players interact with each other or the gaming platform to provoke an emotional reaction and, finally, result in a quantifiable outcome" (Kapp, 2012, p. 118). This gamification process is often developed not only by challenge and competition but by pleasure, regardless of the outcome (Malaby, 2007). Thus:

Gaming often promotes authentic learning and ample problem-solving opportunities. When playing games, individuals are likely to become immersed in the experience and, as a result of this engagement, are more likely to remember information and develop an enduring understanding of [a] concept. (Gee, 2003)

Apart from contributing to improving students' 21st-century skills, gamification can trigger players' intrinsic motivation due to their typically immersing themselves in the game (Huotari & Hamari, 2017).

The particularity of the games is the autonomous nature of the activity as well as the engagement and pleasure of the activity that encourages the player. This is the nature of the games that gamification technology tries to capture, harness, and implement in contexts that generally have a more instrumental purpose. (Vesa et al., 2017)

The use of ICT in PBL is another very important aspect that has also been investigated. ICT plays a very important role due to its presence in students' life, so educators should take advantage of the students' familiarity with electronic devices and their interest in technology, thereby allowing learning experiences to be more productive and attractive. Likewise, the use of ICT in PBL facilitates the acquisition of technological competencies, can be used to develop social and communicative competencies, and facilitates cooperative-collaborative work, among other positives.

METHODOLOGY

The Participants

This study was carried out in an Andalusian public institution in Spain during the 2021/22 academic year. The sample comprised 110 undergraduate students (20 men and 90 women) who volunteered their participation. The students were enrolled in courses related to English language learning, including the Infant Education, Primary Education, Humanities, and Hispanic Philology degree programs. Among the 110 participants, 48 (43.6%) were international exchange students and 62 (56.4%) were local Spanish students.

Methods and Procedure

Two anonymous online surveys were employed to collect data from the participants about their opinions. The first consisted of a pre-task questionnaire that was distributed at the beginning of the 2021/22 academic year, and the second comprised a post-task questionnaire that was shared with this same group of students after the implementation of PBL in class. Both surveys, which were created using Google Forms, consisted of a maximum of 20 closed- and open-ended questions about the students' demographic information in addition to their 1) prior understanding and experiences with PBL; and 2) their perceptions of the positive and negative aspects of PBL in higher education.

Data Analysis

The students' answers were assessed both quantitatively and qualitatively through a frequency analysis of the closed-ended items—including the yes or no questions and the questions answered on a Likert scale—and the categorization and coding of the open-ended questions. In order to obtain a more accurate view of the students' experiences with PBL, the post-task questionnaire contained a wider range of options for students to indicate their degree of agreement with the items presented in the pre-task questionnaire. For example, if the pre-task survey provided the three options of "yes," "maybe," and "no" (see Table 2), the post-task survey included up to five options, including "strongly disagree," "disagree," "neutral," "agree," and "strongly agree" (see Table 3).

RESULTS

The majority of the participants stated that they had previously used PBL in an EFL classroom and that they were aware of what PBL was (see Table 1). In fact, certain students provided accurate definitions of PBL. For example, Student 26 stated that "it is a teaching method that allows the student to acquire knowledge and skills through the development of a project that responds to a specific problem in real life," and Student 34 defined it as "a teaching method in which students gain knowledge and skills by working for an extended period of time to respond to an authentic question."

Table 1. Students' prior experience with and knowledge of project-based learning

Item	Students (*N* = 110)	
	Yes (%)	No (%)
Previous experience with project-based learning (PBL) in English as a Foreign Language classrooms	69.1	30.9
Knowledge about PBL	70	30

In terms of the benefits of PBL in higher education, the students asserted that PBL enhanced their understanding of the course content and provided them with more relevant and helpful information than traditional teaching methods (see Table 2). Most of them also thought that this approach would facilitate their acquiring more independence in learning, demonstrating initiative, and being more organized in terms of planning their work more effectively. Furthermore, they considered PBL a potential strategy for increasing their engagement with collaborative projects and activities along with enhancing their communicative skills in English. Most of the students also stated that PBL could enable them to use resources and materials in ways different to their usual practices while also fostering the growth of their digital communication skills. Ultimately, most of the participants thought that this approach would improve their critical-thinking and creative skills as well as provide them the chance to be more entertained while studying, which would boost their motivation.

Table 2. The advantages of PBL as indicated in the pre-task survey

Item	Students (*N* = 110)		
	Proportion (%)	Maybe (%)	Yes (%)
Gaining a better understanding of practical and useful content	6.3	15.8	77.9
Being more autonomous, and having better organizational and planning skills	12.6	14.1	73.2
Being more actively involved, and developing communicative skills	5.3	14.2	80.5
Using different resources and materials, and improving ICT skills	21.1	22.6	56.3
Thinking critically, and being creative	15.8	16.8	67.4
Having fun, and increasing motivation levels	12.1	17.9	70

The respondents noted other potential advantages apart from those described above, such as cultural enrichment and the development of intercultural competence through working in groups with people from diverse nations and sociocultural backgrounds. Indeed, Student 21 said that "I think this will help me to know more about the culture of other countries while improving my intercultural and communicative skills," and Student 76 asserted that "this experience could allow people to learn more about other cultures."

In terms of PBL's limitations, the students detailed aspects lacking in the PBL approach, which are discussed in the "Applications, Advantages, and Disadvantages of Teaching English Language and Culture through PBL at the Tertiary Level" section. For instance, some students remarked on the problems related to working in groups as well as the greater workload that is required compared to traditional teaching and learning methods. For instance, Student 44 stated that "it can be complicated to work in [a] group when we don't know the other members. Moreover, it also can be difficult to work together when we have different schedules." Similarly, Student 59 opined that "if somebody in your group doesn't work well [with others], you may find yourself doing all the work." Student 71 remarked that "this methodology demands [much more] time than the traditional methodology."

Despite these comments, the results derived from the pre-task survey were mostly positive: The least popular item had a disagreement rate of only 21.1%, with over half of the sample—56.3% being the lowest value—being in agreement. Nevertheless, these students may not have experienced the implementation of PBL, so whether they maintained these beliefs was analyzed through the use of the post-task survey. The results of this post-task survey are included in Table 3 and are analyzed below.

Table 3. The advantages of PBL as indicated in the post-task survey

Item	Students (*N* = 110)				
	Strongly Disagree (%)	Disagree (%)	Neutral (%)	Agree (%)	Strongly Agree (%)
Better understanding of practical and useful content	–	0.5	14.5	54.5	30.5
Being more autonomous, and having better organizational and planning skills	0.5	1.7	14.1	51.8	31.9
Being more actively involved, and developing communicative skills	0.5	0.7	9.4	55.4	34.0
Using different resources and materials, and improving ICT skills	–	3.7	30.5	65.8	–
Thinking critically, and being creative	–	3.6	20.0	76.4	–
Having fun, and increasing motivation	–	3.6	23.6	72.8	–

The results of the post-task survey demonstrate an improvement in students' perceptions of PBL. Nevertheless, it is worth highlighting that their views sit in the middle of the spectrum since they avoid the extremes on either side, namely "strongly disagree" and "strongly agree." In fact, the most popular answer provided by the students was "agree," which was followed by "neutral." Compared to the pre-task questionnaire (see Table 2), the negative responses to the items—namely, the "strongly disagree" and "disagree" options—decreased by an average of 9.73% per item. The most dramatic change in the students' opinions occurred for the item "Using different resources and materials, and improving ICT skills" as disagreement shifted from 21.1% to 3.7%, with 0% strongly disagreeing and 3.7% disagreeing. The "Being more actively involved, and developing communicative skills" item had less variability since the difference between the pre-task and post-task surveys was only 4.1%, despite fewer students disagreeing with this statement in the pre-task survey. Furthermore, 5.3% of the students thought that PLB could lead to their being more actively involved and developing communicative skills, while only 1.2% of the sample still held this opinion in the post-task survey (0.5%=strongly disagree and 0.7%=disagree).

The neutral answers decreased for the first three items of "Better understanding of practical and useful content," "Being more autonomous, and having better organizational and planning skills," and "Being more actively involved, and developing communicative skills." Conversely, the neutral answers increased for the three items of "Using different resources and materials, and improving ICT skills," "Thinking critically, and being creative," and "Having fun, and increasing motivation." Nevertheless, the highest values (30.5% and 23.6%) did not represent even a third of the sample, which is a small percentage compared to the values obtained for those in agreement.

The percentage of the students who agreed with the positive values—including the "agree" and "strongly agree" options—increased by an average of 7.97% per item compared to the pre-task survey. The most dramatic change in the students' opinions was seen in the second item of "Being more autonomous, and having better organizational and planning skills," with an increase of the 10.5%. With a variation of 2.8%, the students' opinions on the "Having fun, and increasing motivation" item did not drastically vary between the pre-task and post-task surveys. This may be due to the fact that they already shared positive expectations on the fulfillment of this item in the pre-task questionnaire. In fact, in the pre-task survey, 70% of the students already believed that PBL could allow them to have fun and increase their motivation while learning. After implementing this approach, this increased to 72.8% (72.8%=agree and 0%=strongly agree). As above, there was also an absence of opinions on either extreme for this item. This was also observed in the "Using different resources and materials, and improving ICT skills," "Thinking critically, and being creative," and "Having fun, and increasing motivation" items as no students strongly disagreed nor strongly agreed with these items having improved. The fact that all of the students merely agreed or disagreed with the items may indicate that they were more skeptical about the fulfillment of these items in comparison to the "Better understanding of practical and useful content," "Being more autonomous, and having better organizational and planning skills," and "Being more actively involved, and developing communicative skills" items.

CONCLUSION

PBL is an innovative teaching approach that assists students in building both communicative skills and content knowledge simultaneously. This study proves that PBL is effective in the English language classroom at the university level and that learning via projects is more engaging, encouraging, and motivating for English language learners than traditional learning methods. Herein, 110 undergraduate students (both local and international) pursuing degrees in Infant Education, Primary Education, Humanities, and Spanish Philology were surveyed to assess the application of PBL in this context. This was achieved by having these students complete an anonymous pre-task questionnaire that explored their motivation levels and the benefits they expected from PBL being implemented in higher education. They were then asked to complete an anonymous post-task questionnaire after the implementation of PBL in the classroom, the results of which were contrasted against the results of the first survey.

The data from these surveys were gathered and analyzed both qualitatively and quantitatively. The first finding worth noting is that the students had previous experience with PBL in the EFL classroom. Secondly, most of the students agreed that this approach could enable them to be more autonomous while learning and that PBL would help them in effectively planning and organizing their work. They also noted that PBL would motivate them to build their communicative skills in EFL and enhance their digital communicative skills compared to conventional teaching methods. The participants also expected additional outcomes, such as cultural enrichment from working in groups with people who come from different nations and sociocultural backgrounds. The students also remarked on the disadvantages of using this approach, which included possible working-group management problems and the anticipated higher workloads compared to conventional approaches.

The results of the analysis of the post-task questionnaire demonstrated that the students agreement with each statement increased after the implementation of PBL in the classroom. There were dramatic increases in their agreement with the effects of PBL in terms of the "Being more autonomous, and having better organizational and planning skills" item. Nevertheless, they remained skeptical about the other improvements, such as the "Using different resources and materials, and improving ICT skills" item. Even though the degree of disagreement on this item decreased dramatically, most of the students remained neutral. Indeed, this was the item that the students disagreed with the most in the post-task survey. Therefore, future research should aim to improve this particular aspect. Moreover, further attention should also be paid to the items that, although they have a lesser degree of disagreement, have a greater percentage of neutrality, including the "Thinking critically, and being creative" and "Having fun, and increasing motivation" items. At less than 80%, these items also shared the lowest rates of agreement. In terms of the "Having fun, and increasing motivation" item, the students' agreement before the implementation (70%) remained noticeably similar to their agreement after the implementation (72.8%).

Despite further implementations needing to be made for particular aspects—"Using different resources and materials, and improving ICT skills," "Thinking critically, and being creative," and "Having fun, and increasing motivation"—this study has demonstrated that most of the students believed that the implementation of PBL improved their skills in the English language learning classroom. Moreover, it is worth highlighting the high percentage of the sample that agreed with the "More actively involved, and developing communicative skills" item in the post-task questionnaire as well as the dramatic increase in the students' agreement with the "Being more autonomous, and having better organizational and planning skills" items, which increased by 10.5%.

REFERENCES

Armie, M., & Membrive Pérez, V. (2022). *Escaping boredom in the classroom: Breakouts, breakout boxes and escape rooms*. Cambridge Scholars Publishing.

Bautista, J. M., Espigares, M. J., & Hernández, R. M. (2017). Aprendizaje Basado en Proyectos (ABP) ante el reto de una nueva enseñanza de las Ciencias. Rev. Bras. *Ensino Ciènc. Tecnol., 10*, 43–60.

Bell, S. (2010). Project-based learning for the 21st century: Skills for the future. *The Clearing House: A Journal of Educational Strategies, Issues and Ideas, 83*(2), 39–43. doi:10.1080/00098650903505415

Brown, H. D. (2007). *Principles of language learning and teaching*. Pearson Education.

Byram, M., Nichols, A., & Stevens, D. (2001). Introduction. In M. Byram, A. Nichols, & D. Stevens (Eds.), *Doing intercultural competence in practice* (pp. 1–8). Multilingual Matters Ltd.

De Miguel, M. (Ed.). (2005). *Metodologías de enseñanza para el desarrollo de competencias. Orientaciones para el profesorado universitario ante el Espacio Europeo de Educación Superior*. Alianza.

Deterding, S., Dixon, D., Khaled, R., & Nacke, L. (2011, September). From game design elements to gamefulness: defining "gamification". In *Proceedings of the 15th international academic MindTrek conference: Envisioning future media environments* (pp. 9-15). 10.1145/2181037.2181040

Dewey, J. (1938). *Experience and education*. Simon and Schuster.

Dvalidze, N. (2017). Teaching English language through storytelling. *Journal of Teacher Education, 6*(2), 53–58.

Ellis, G., & Brewster, J. (2014). *Tell it Again! The storytelling handbook for primary teachers* (3rd ed.). British Council.

Gandía, J. B. (2012). La rentabilidad del cómic en la enseñanza de la cultura en E/LE. *Foro de profesores en E/LE*, 8, 1-8.

Gee, J. (2003). *What video games have to teach us about learning and literacy*. Palgrave Macmillan. doi:10.1145/950566.950595

Huotari, K., & Hamari, J. (2017). A definition for gamification: Anchoring gamification in the service marketing literature. *Electronic Markets, 27*(1), 21–31. doi:10.100712525-015-0212-z

Ilari, B., Chen-Hafteck, L., & Crawford, L. (2013). Singing and cultural understanding: A music education perspective. *International Journal of Music Education, 31*(2), 202–216. doi:10.1177/0255761413487281

Kapp, K. (2012). *The Gamification of Learning and Instruction: game-based methods and strategies for training and education*. Pfeiffer.

Kavlu, A. (2017). Implementation of project-based learning (PBL) in EFL (English as a foreign language) classrooms in Fezalar Educational Institutions (Iraq). *International Journal of Social Sciences and Educational Studies, 4*(2), 67–79.

Kingston, S. (2018). Project-based learning & student achievement: What does the research tell us? *PBL Evidence Matters, 1*(1), 1–11.

Krajcik, J. S., & Blumenfeld, P. C. (2005). Project-based learning. In R. K. Sawyer (Ed.), *The Cambridge handbook of learning sciences* (pp. 317–334). Cambridge University Press. doi:10.1017/CBO9780511816833.020

Kramsch, C. (1998). *Language and culture*. Oxford University Press.

Kuang, J. F. (2007). Developing students' cultural awareness through foreign language teaching. *SinoUS English Teaching, 4*(12), 74–81.

Leung, E., & Pluswik, E. (2018). Effectiveness of Gamification Activities in a Project-based Learning Classroom. *Asee Annual Conference and Exposition*. 10.18260/1-2--30361

Malaby, T. M. (2007). Beyond play: A new approach to games. *Games and Culture*, *2*(2), 95–113. doi:10.1177/1555412007299434

Miller, C. (2013). The gamification of education. *Developments in Business Simulation and Experiential Learning: Proceedings of the Annual ABSEL conference,* 40.

Moreira, F., Ferreira, M. J., & Cardoso, A. (2018, November). ECLECTIC approach applied to data transformation curriculum course. In *Proceedings. 11th Annual International Conference of Education, Research and Innovation (ICERI 2018),* (pp. 6021-6029). 10.21125/iceri.2018.2411

Povey, E. (2019). Storytelling in young learner EFL education: From theory to practice. *International. Journal of Issues in Language and Teacher Education*, *4*, 1–14.

Prieto, L. (2006). Aprendizaje activo en el aula universitaria: El caso del aprendizaje basado en problemas. *Miscelánea Comillas. Revista de Ciencias Humanas y Sociales*, *64*(124), 173–196.

Sanatullova-Allison, E. (2018). Contextualized culture integration through project-based learning in EFL classrooms in Vietnam. *Journal of NELTA*, *20*(1-2), 5–15. doi:10.3126/nelta.v20i1-2.19771

Savignon, S. J., & Sysoyev, P. V. (2005). Cultures and comparisons: Strategies for learners. *Foreign Language Annals*, *38*(3), 357–365. doi:10.1111/j.1944-9720.2005.tb02222.x

Schulz, R. A. (2007). The challenge of assessing cultural understanding in the context of foreign language instruction. *Foreign Language Annals*, *40*(1), 9–26. doi:10.1111/j.1944-9720.2007.tb02851.x

Suzie, B., & Krauss, J. (2014). *Reinventing project-based learning: Your field guide to real-world projects in the digital age*. Corwin, Sage Publishing.

Thomas, J. (2000). *A review of research on project-based learning*. Retrieved from http://www. bobpearlman.org/BestPractices/PBL_Research.pdf

Vesa, M., Hamari, J., Harviainen, J. T., & Warmelink, H. (2017). Computer games and organization studies. *Organization Studies*, *38*(2), 273–284. doi:10.1177/0170840616663242

Chapter 10
Innovative Active Methodologies That Promote Learning in Postgraduate Students in the 21st Century:
Project-Based Learning and Flipped Classroom

Edgar Oliver Cardoso Espinosa
https://orcid.org/0000-0001-7588-9439
ESCA Santo Tomás, Instituto Politécnico Nacional, Mexico

María Elena Zepeda Hurtado
https://orcid.org/0000-0001-9764-5013
CECyT 11, Instituto Politécnico Nacional, Mexico

Jésica Alhelí Cortés Ruiz
https://orcid.org/0000-0002-5459-4874
CIECAS, Instituto Politécnico Nacional, Mexico

ABSTRACT

The objective of the chapter is to propose the incorporation of active methodologies for the development of the competencies of postgraduate students in order to promote quality professional training. This comparative study comes from research whose results emanate from the use of active and innovative teaching strategies, put into practice in the classroom, with the aim of defeating the problems that frequently arise such as academic lapsing, lack of interest and communication, work team up. Within the solution alternatives, strategies and materials were used to promote learning related to the needs and in the context of the students, to develop skills and knowledge for solving administrative situations. Therefore, a proposal for successful training practices in project-based learning and the flipped classroom is presented to promote quality academic training in postgraduate students.

DOI: 10.4018/978-1-6684-8861-4.ch010

INTRODUCTION

The appearance of the coronavirus has generated a series of social, economic, political and educational effects in individuals and countries that have fostered an environment of uncertainty that has gradually altered the dynamics of organizations. Given this context, it is important to train human talent capable of living in a world of uncertainty to promote the transformation of the rigid and centralized operating strategies of the institutions, which is why new methodological approaches are required for the development of competencies in postgraduate students that allow sustainable organizational development in the new normality (Chehaibar, 2020). Thus, the postgraduate program maintains in the 21st century the priority function of training high-level human capital oriented towards excellent performance in the organizations that work.

In this sense, the design and development of flexible, attractive, and dynamic training activities are recommended in order to promote communication and exchange of strategies through collaborative work that results in useful professional learning for students. Also, it should be noted that the implementation of active methodologies increases the perceived usefulness of the content, in addition to motivating and reducing the negative feelings generated by confinement (González et al., 2018). Thus, it is relevant to formulate strategies that make it possible to generate formative experiences from the digital resources and tools available to teachers that are effective in student learning (Molinero and Chávez 2019). In this way, active methodologies are the set of activities and strategies that teachers use to achieve greater student participation in the construction of their learning, highlighting Project-Based Learning, Gamification, and the Flipped Classroom (Roig-Vila & Álvarez, 2019).

In this sense, higher education, where the postgraduate level is located, it must be transformed to respond to this new context where learning enables the development of various capacities in the individual that allow them not only to solve the various problems of the real world but also the possibility of interacting collaboratively. The foregoing coincides with what was indicated by Sanz and López (2022) regarding the fact that the pandemic has promoted new opportunities to reflect on and potentiate academic training due to the fact that in countries a strategic factor that promotes economic development such as quality of life of citizens is education.

Therefore, during the forced transition from face-to-face to virtuality, educational actors learned and were trained by experience, so it is important to identify new ways of working to take advantage of the incorporated resources that allow a new teaching culture when the course is completed confinement (Lira & Uribe, 2022). Similarly, Dzandu et al. (2022); Guarino et al. (2022); Kang and Park (2022) establish that the continuous introduction of digital tools such as videoconferences, information management services in the cloud or knowledge management platforms have been highly accepted by both students and teachers.

Based on the above, it is required that the educational system in the post-COVID-19 era be oriented towards promoting the development of skills for life through a change in the ways of teaching and learning content based on methodologies that promote learning active, collaborative and by projects that are considered as tools for the acquisition, assimilation, and accommodation of new knowledge by the student body. Thus, the active methodologies considered in this chapter are the Flipped Classroom and Project-Based Learning.

The objective of the chapter is to propose the incorporation of active methodologies for the development of competencies of postgraduate students in order to promote quality professional training.

BACKGROUND

Skills for the 21st Century

The skills for the 21st century are considered as knowledge, skills and attitudes that must be developed or strengthened in school systems and through other learning spaces, which makes us think and rethink as suggested by Economic Commission for Latin America and the Caribbean (ECLAC, 2020), in contents that prepare students to understand reality and act in a supportive and responsible manner in order to respond to diversity and interact in an environment of uncertainty, beyond times of crisis.

The Organisation for Economic Co-operation and Development (OECD, 2015) is a framework for the education of the future that proposes a vision of what children and young people require in the future, where they talk about the common competences of the 21st century and what are the skills of critical thinking and resolution from problems. Critical thinking skills tend to be found in all seven learning areas (national language, mathematics, humanities, science, technology/home economics, arts, physical education, and health). These transferable competencies embedded in the curricula are more likely to be cognitive, social or emotional skills, or composite concepts.

Based on The Future of Jobs Report 2020 (World Economic Forum, 2020) the skills that will be most in demand by employers by 2025: 1) Analytical thinking and innovation; 2) Active learning and learning strategies; 3) Resolution of complex problems; 4) Thinking and critical analysis; 5) Creativity, originality and initiative; 6) Leadership and social influence; 7) Use, control and monitoring of technology; 8) Technology design and programming; 9) Resilience, stress tolerance and flexibility and, 10) Reasoning, problem solving and idea generation (ideation).

Another point of view on what are the graduation profiles of students for the 21st century is that of Fradel et al. (2016) who propose four dimensions for education: 1) Skills "how to use what we know" and which refers to creativity, critical thinking, communication and collaboration; 2) Knowledge "what we know and understand" among which interdisciplinarity, big data and digital literacy stand out; 3) Attitudes "behavior and engagement with the world" among which are awareness, curiosity, resistance, ethics and leadership, and 4) Meta-learning "Reflection and adaptation" that refers to meta-cognition and growth mindset.

Duran et al. (2021) propose that the student develop adaptability to achieve goals and quality standards required by the workplace; tolerance to be able to listen and be heard, propose, create and manage multimedia material; study, conduct, constitute, value and create information using different means of communication and adapt to different roles to exercise empathy.

Based on the above, there is a coincidence between the cited sources with respect to the fact that the competencies that are required to be developed in students cover the same structure of the concept: knowledge, skills and attitudes that allow them to adapt to the changing world of work in the face of revolutions. industrial, which implies the use of ICT, encourage self-learning to stay up-to-date and contribute to solving problems.

Therefore, the use of digital technologies plays a central role in developing the necessary skills for the 21st century with free and open materials and resources, changing the way information is obtained, handled and interpreted. Thus, United Nations Educational, Scientific and Cultural Organization (UNESCO, 2022) considers that digital technologies are a necessity, the use of digital innovation expands educational opportunities, inclusion, improves the relevance and quality of learning, creates learning pathways throughout life enhanced by ICT, strengthen education and learning management systems, and monitor learning processes.

Another approach to graduation profiles for the 21st century is the one that goes towards the development of soft skills or also known as soft skills or non-cognitive skills, which are defined according to Ortega (2016) cited by Fuentes et al. (2021) as the practices, attitudes, and capacities that a person has to relate and interact with the world. These allow among other processes; the understanding of emotions, achievement of objectives, decision making and the ability to face adverse situations. Proactivity, critical thinking, communication skills, time management, among others, are part of soft skills, which refer to aspects that are not only cognitive of the human being; they are skills that allow you to interact with your peers within the work environment and function in society (Espinoza, 2020).

ACTIVE TEACHING METHODOLOGIES

The didactic methodologies are based on theoretical foundations that allow the teacher to currently understand his role and that of the students in order to design and promote flexible learning experiences for comprehensive training, leaving aside the programmatic content and focusing on the student. Gonzalez et al. (2016) cited by Reynosa et al. (2019) in their research established theoretical precisions of learning to take into account when drawing up didactic strategies for learning, considering that learning is: a change in behavior, it is significant is the assimilation or integration to the individual himself of what he learns, supposes for the student the need to be able to integrate meanings, representations and emotions into their thought structure, making it their own, it is cooperative, it points out that moments of interaction are necessary to learn significantly.

To achieve this, one way is to relate the curricular contents and objectives with activities where the student is the protagonist and which at the same time implies attention to the needs of their environment in order to improve it, developing skills and knowledge as useful tools for the personal, academic and professional life, as well as participation in solving social problems in real settings and the use of technology. Semante and Gómez (2021) propose that didactic strategies through the use of technology constitute mediation tools between the subject who learns and the teaching content that the teacher uses to achieve certain learning.

To achieve this, the ideal graduation profiles, it is necessary to design and implement active and innovative teaching strategies aimed at developing these skills from in the classroom for the future that has already passed us. Therefore, the present investigation focused on documenting the educational experience in which the Project-Based Learning (PBL) and Flipped Classroom (FP) were used in groups of postgraduate level in the Instituto Politécnico Nacional (IPN), which, in addition to contributing to the development of skills and abilities, also formed part of a sudden process of the use of digital technology, of online education and the transition from online to face-to-face education. There are different didactic learning strategies that meet these characteristics, such as Project-Based Learning and Flipped Classroom among others.

PROJECT-BASED LEARNING (PBL)

PBL, according to García-Varcálcel and Bacilota (2017) is defined as a task-focused teaching and learning modality, a shared negotiation process between the participants, its main objective being to obtain a final product. This method promotes individual and autonomous learning within a work plan defined by objectives and procedures. Trujillo (2017) proposes six phases to develop the PBL: 1) Entry. The learning situation is presented to the students with the intention of provoking their involvement and motivation. 2) Challenge. It establishes what will be the question, problem or challenge that will be resolved throughout the project. 3) Search for information in different sources 4) Information management. From the reading, the information is processed to convert it into knowledge and face the challenge. 5) Production. A final product is generated that includes the lessons learned and solves the challenge. 6) Dissemination.

The PBL requires a pedagogical scaffolding where the student reconstructs pre-existing knowledge in a comprehensive and progressive way of knowledge construction, while the teacher introduces and links said construction processes of the student, thus guiding and guiding through activities of each one of the stages of the process.

FLIPPED CLASSROOM (FP)

The creation of the Flipped Classroom is attributed to Bergmann and Sams (2014). The FP optimizes the time in the classroom because the teacher orients his advice in a personalized way to the student body (Araya et al., 2021). FP fosters both autonomous learning Chen et al. (2021); Sandoval et al. (2021), collaborative work (Quinteros and Cárdenas, 2021) and strengthens self-regulation in the student body (Lin et al. 2021; Alegre and Silva, 2020).

The implementation of the FP requires the responsibility of the student body to review the contents of the course outside the classroom with the help of technology to later go to class with comments and doubts that allow them to both identify and apply what they have learned to contextualized situations in an environment where the teacher is a guide and reviewer of the progress of each student (Barral et al., 2018; González and Huerta, 2019).

The flipped classroom according to González and Abad (2020) is a method and a pedagogical instrument. Born in the nineties, it will not be considered as an innovative strategy. The fundamental basis of the method, which in English means "to go around the class", consists of reversing certain learning processes that used to take place in the classroom, transferring them outside of it, that is, carrying them out at home, and vice versa.

According to Gaviria et al. (2019) roles and moments are reversed, such as: practical activities—which are usually delegated to be carried out outside the classroom—are carried out during class, while master sessions—in which concepts and others are addressed—are analyzed the student in spaces outside of class. González and Huerta (2019). They mention that the success of this methodology depends on: the quality of the material available online, the appropriate organization of the face-to-face classes, where the teacher detects the difficulties that the previous learning of the students entails and uses the appropriate resources to correct the errors of understanding that can occur.

Based on the above, it is necessary to carry out a comparison between these two active methodologies:

Both approaches have benefits for deeper and more meaningful learning, fostering student creativity and autonomy when tackling complex real-world problems. Projects involve research, planning, execution and presentation, developing key 21st century skills such as problem solving and critical thinking. Additionally, by working as a team, students learn to collaborate and communicate effectively, essential skills for today's world of work. There is a clear need to adapt the infrastructure, human resources, logistics processes and everything that intervenes directly or indirectly in the educational process to be able to face what the future brings (Daza, 2021).

The FP can have the same scope; however, this depends on the methodologies, strategies, and activities that the teacher implements in the classroom, which go beyond the repetition of content. According to Fidalgo, Sein and García (2020). It is about transforming the educational model from a 1.0 model (students as recipients of content) to a 2.0 model, where students and teachers are producers and consumers of content. Likewise, teachers must provide new tools to classify, identify, organize and guarantee the use of content in an easy way. It is also necessary to consider that a course completely with a flipped classroom model implies making the most of the potential of ICT in education, especially with the use of virtual platforms that favor interaction outside the classroom, without losing the interactions typical of the face-to-face classroom (Hernández and Tecpan Flores, 2017).

As a summary, Table 1 identifies the similarities and differences between both methodologies.

Table 1. Similarities and differences between the FP and PBL

Similarities	Differences
● Students are responsible for their own learning and must take an active role. ● They promote participation and collaboration among students. ● They encourage the development of critical thinking skills, problem solving and decision making. ● Requires students to apply acquired knowledge to real-world situations	FLIPPED CLASSROOM ● It is flexible in terms of material review time, allowing students to acquire prior knowledge at their own pace. ● The student can review the learning materials as many times as necessary, which favors a more personalized learning and adapted to the individual needs of each student. ● The class time is optimized for the application and practice having previously acquired the theoretical concepts. ● There is more interaction between students. PROJECT-BASED LEARNING ● Allows students to address real world problems or questions. ● Projects are designed to reflect authentic situations, increasing motivation and relevance of learning for students. ● Integrates knowledge from different disciplines. ● It fosters skills such as critical thinking, problem solving, teamwork, and effective communication. ● Motivates the student to continue with other projects.

Source: Own elaboration.

Based on table 1, it is necessary to indicate that the FP and PBP are two innovative educational approaches that are transforming the way students learn and teachers teach. Both approaches focus on the student as the protagonist of their own learning, promoting active participation, autonomy and critical thinking.

MAIN FOCUS OF THE CHAPTER: METHODOLOGY

The methodological approach of the research was quantitative with a descriptive scope based on a comparative study in order to determine the effectiveness of each active methodology used at the postgraduate level in the academic training of the student body.

Regarding the research variables, they were: 1) Effectiveness of the active methodology: Activities and practical cases, time management, use of digital teaching resources, content linking and quality in academic training and, 2) Mathematical training: Mathematical skills developed by students.

The study participants were students who were studying a postgraduate degree in administration at the Instituto Politécnico Nacional in the first-semester subject called "Techniques and Models for Decision Making" corresponding to the master's Program in Business Administration. The sample was of a non-probabilistic type with the convenience criterion with a total of 78 students distributed in two groups: Flipped Classroom (G1) and Project-Based Learning (G2). The main characteristics are shown in Table 2.

Table 2. Study participants

Variable		G1 (n = 23)	G2 (n = 25)
Gender	Female	12	13
	Male	eleven	12
Age (years)	25 to 29	9	10
	30 to 34	eleven	13
	35 to 39	3	2
Work experience (years)	0 to 4	8	10
	5 to 9	14	14
	more than 9	1	1

Source: Own elaboration

Based on table 2, it is identified that the participation of the female sex is 52% and 48% corresponds to the male. For age, the range from 30 to 34 years stands out with 47% participation; while for work experience the majority interval is 5 to 9 years with 61%.

Regarding the didactic intervention of each methodology, it was carried out in five stages as shown in Table 3.

The information gathering instrument was a questionnaire with the following structure: 1) Socio-labor data; 2) Type of technological device and time of use per day; 3) Problems faced; 4) Assessment of the active methodology with items measured by a four-point frequency scale: Totally disagree (TD=1), disagree (D=2), neither disagree nor agree (DA=3), agree (A=4) and totally agree (TA=5), which were converted to a numerical value in order to obtain a quantitative measure.

For the content validity test of the questionnaire, it was carried out through the judgment of seven experts, of which four have more than 15 years of teaching experience in the subject of mathematics in the postgraduate course, two have 10 years in the construction and validation of instruments for the quantitative approach and one with 20 years in carrying out research projects on academic training at the postgraduate level. From this process, modifications were made with respect to the clarity and wording of each of the statements in the questionnaire. Subsequently, the reliability measurement was carried out, where a Cronbach's alpha coefficient of 0.886 was obtained, indicating a good degree of internal consistency and, therefore, reliable. Regarding the application of the instrument, it was carried out by email where the participants were indicated the objective and relevance of the study, as well as the confidentiality of the information obtained was guaranteed. In the same way, the time for collecting the information was after the end of the course in order to avoid possible biases for the study.

Once the collection of information was completed, the data analysis was carried out using descriptive statistics, mainly the mean (M) and standard deviation (SD) that allowed comparisons between the active methodologies to be carried out.

Table 3. Stages of the active methodology

First Stage: Theme	**Activities**
First session	The objective, competencies to be developed, contents and evaluation of the subject were presented to the students. Teacher explanation about the didactic methodology to be used. The digital teaching resources to be used (videos, articles, and presentations) were explained. Assembly with the student body for the election and delimitation of the subject, highlighting that it is related to their graduate degree in administration.
Second session	The topics to be developed: Quantitative model for strategic planning in the business sector.
Second Stage: Search and Selection of Information	**Activities**
Research in magazine articles and books	Teams were organized to develop the theoretical and contextual framework, establishing as criteria: topicality and significance for the administrative field. During this process, the teaching staff carried out a review of progress, indicating areas for improvement.
Third Stage: Socializing	**Activities**
Online sessions	Advances before the group: mention the study variables and estimated model. Each synchronous session is concluded with a socialization space on the relevance and usefulness in professional training where the teacher is the mediator.
Fourth Stage	**Activities**
Report	Preparation of the report that includes the statement of the problem, the formulation of the model, estimation, and interpretation of results. The conclusions and proposed solution were established.
Fifth Stage	**Activities**
Project presentation	Each team presented its final model indicating its scope and decision-making for the administrative field.
Course Evaluation	The students did it to express the comments of the course and their experience using the methodology. The research instrument was implemented.

Source: Own elaboration

RESULTS

During confinement, the type of technological device frequently used by postgraduate students was Laptop with 67% and the remaining 33% desktop computer, while the average daily time was more than three hours with 78% of the participants. With regard to internet connectivity, it was found that 89% of the students are fixed type.

Meanwhile, the situations faced by the student body during the pandemic were 92% logistical (organization and management of time, class, work and family schedules, physical space to work remotely); 65% of an affective type (emotional and health experiences) where the students acknowledged experiencing feelings of sadness, frustration, anxiety, tiredness, and demotivation, as well as of a technological type in 36% (internet connectivity).

While the evaluations of the postgraduate students regarding the effectiveness of the active methodology used in the course are shown in Table 4.

Table 4. Descriptive statistics on the active methodology used

During Confinement, the Active Methodology Has Been Effective Because:	G1		G2	
	M	SD	M	SD
Promotes learning from the practical activities carried out in the online session	4.22	0.92	4.15	0.86
Contributes to participation and motivation for learning by combining the previous activities with practical cases	4.45	0.33	4.33	0.46
Encourages the resolution of practical situations through collaborative work	4.69	0.47	4.48	0.51
Enables teacher accompaniment during the course	4.15	0.91	4.37	0.85
Foster positive relationships with team members	4.23	0.95	4.49	0.51
Plan and organize the time for my academic training	4.62	0.58	4.31	0.69
Each course content is supported by digital teaching resources (videos, articles, presentations, spreadsheets, and text)	4.95	0.12	4.96	0.11
Promote experiential and meaningful learning for my professional training	4.79	0.21	4.53	0.55
It allows me to track my academic progress	4.26	0.87	4.42	0.64
Link the contents in an updated way to use them in the workplace	4.87	0.26	4.81	0.24
Promotes quality in the training received	4.94	0.17	4.85	0.33

Source: Own elaboration

Table 4 shows the results in relation to the effectiveness generated by the two active methodologies used at the postgraduate level during confinement, so it was found that the highest average evaluations by the participants were: The use of digital teaching resources; a quality in academic training; connection with the workplace and encourages experiential and meaningful learning.

These findings imply the recognition by postgraduate students of the relevance and constant use of technology in academic training, highlighting the use of videos, articles, presentations, spreadsheets and text, mainly, which in combination with the participation and teachers' guide have motivated quality learning that has applications in the workplace. Therefore, both active methodologies are effective because they favor both time management, positive relationships, motivation to learn together with academic monitoring during the course, which allows a successful school trajectory of the course. In the same way, both methodologies encourage collaborative work, which enables the exchange of concepts and strategies for resolving the situations to be resolved, which complements the academic training of each participant.

Regarding the mathematical skills developed by the use of active methodologies at the postgraduate level, they are shown in Table 5.

Table 5. Descriptive statistics on mathematics training

During the Lockdown, the Ability To	G1		G2	
	M	SD	M	SD
Identify the variables	3.81		3.82	
Use mathematical terminology	3.95		3.92	
reason mathematically	3.77		3.69	
formulate a strategy	3.66		3.56	
argue mathematically	3.53		3.35	
Use tables and graphs in spreadsheets	3.89		3.85	
interpret the results	3.22		3.18	

Source: Own elaboration

From Table 3, the highest scores on average by the participants are the "use of mathematical terminology"; "the use of tables and graphs in spreadsheets"; "identify the variables of the situation" and "mathematical reasoning". These findings allow us to deduce that the two active methodologies used promote a mathematical training based on the use of situations that involve the use of technological tools that encourage communication and exchange of ideas among the students that make it possible to formulate a strategy based on mathematical arguments that conclude in the interpretation of the results.

SOLUTIONS AND RECOMMENDATIONS

In this sense, the recommendations for its implementation are the following:

PROJECT-BASED LEARNING (PBL)

- Start. Recover with the students what a project is, why they work and how. Present as an example different problem studied and different cases worked by projects, as well as the products.
- Brainstorm so that they express projects in which they have participated that were successful and why, in the same way those that were not successful and what could be the causes.
- Challenge, problem, or challenge. As a team, analyze the various problems that can be addressed in various fields: social, economic, and political, choosing one and the possible product to be developed to disseminate the results.

- In an assembly, the proposals of the teams are exposed, by way of brainstorming other elements are integrated to select a topic, problem, or challenge for each group.
- Establish a short time to search for information, in different sources.
- Return to the challenge, problem, or challenge, to make some reconsiderations, delimit it in time and space, as well as define the final product more clearly.
- Continue with the documentary research process and the elaboration in teams of partial evidence that give proof of the information collected. Infographics, reports, digital presentations.
- Continuous group feedback to teams on evidence
- Flexibility to integrate, substitute or eliminate partial evidence and the criteria or elements that the final evidence must have.
- Allow each team member to develop or strengthen the differentiated skills that each one possesses.
- Use technologies that are available to students, that are current, varied, and interesting.
- The presentation of the final product fulfills its function of disseminating results through different channels, making it attractive and group.
- The leitmotiv of the project will always be the competencies, objectives, or purposes of a learning unit through which the project is coordinated.

FLIPPED CLASSROOM (FP)

- Have a platform or technological tools where the student is provided with the materials for prior review: texts, videos, audios.
- The material must be carefully selected, in several ways: it is related to the topic to be addressed in the classroom, consider the characteristics of the student (age, vocabulary, length, time allocated for review), that the content is brief and attractive.
- Suggest the preparation of evidence that will be presented at the beginning of the session for feedback.
- Start the session with an exercise that implies a challenge for the student, where they put into practice the previously reviewed information as a way to corroborate the handling of concepts or procedures, clarify doubts.
- Plan the activities and strategies that will be carried out in the classroom, which match the material previously reviewed: preparation of graphic schemes, analysis of a case, solution to a problem, analogies.

FUTURE RESEARCH DIRECTIONS

The future studies to be carried out based on the research carried out is to consider the perspective of the teaching staff with respect to the active methodologies and the evaluation instruments used.

Also, carry out studies on another active methodology that is gamification in order to determine its effectiveness and usefulness in the academic training of postgraduate students. Likewise, carry out mixed-type research in order to expand the findings at the postgraduate level considering other subjects, educational programs, and even higher education institutions at a national and international level.

In addition, it is important to carry out studies on the learning experiences lived by the student body, teachers, and educational leaders in order to design and implement attention and intervention strategies with respect to their psychological, emotional, and physical well-being.

CONCLUSION

The consequences of the COVID-19 pandemic were diverse for individuals, organizations, and countries, so in the case of the education sector, it was faced with complying with the attention and educational continuity of students through virtual scenarios based on the use of tools. technological. The foregoing implied both planning and implementation of methodologies aimed at promoting the constant participation of the student body in order to achieve the objectives of each course and academic program. In this way, the relevance of the research is to provide results regarding the effectiveness of both the PBL and the flipped classroom during the period of confinement at the postgraduate level. Therefore, the general objective of the study was achieved since both methodologies generate a high level of student satisfaction in three categories: the continuous use of digital resources, experiential learning and motivation for academic training.

In the same way, what was stated by Cabero and Llorente (2020) is confirmed regarding the fact that the learning generated through the flipped classroom was possible through the use of technological tools that allowed a synchronous and asynchronous relationship between teacher-student. Thus, the information obtained from the study is relevant because it allows the identification of the characteristics of academic training at the postgraduate level, as well as access and management in terms of technological tools and digital teaching resources that influence school performance.

Therefore, it is important to consider that these two active methodologies require a training process on the part of the teacher and an attitude of commitment and flexibility to put the student as the center with pre-existing knowledge, interests, technological resources and adapt to circumstances without lose sight of the educational objective that it intends to reach and the skills or competencies to be developed in the student body.

Having integrated both the PBL and the FP to achieve the development of mathematical skills is a success as active methodologies, because they consider previous knowledge as a starting point; On the other hand, it is necessary to consider the context of the students to develop the projects, as already mentioned, the available technological resources, since sometimes, in addition to being a support, it becomes an obstacle when it is not available. The foregoing implies continuous monitoring by the teaching staff to make the adjustments to which there is space, time, materials, distribution of roles in the teams, among others.

For the implementation of these two active methodologies, it is necessary to consider the following challenges: Teachers require training on use and scope; the students need a digital culture to review the teaching materials; the educational institution has the infrastructure and connectivity that allows its operation and establish research processes aimed at considering its effectiveness in each course and school level.

ACKNOWLEDGMENT

This research was supported by the Instituto Politécnico Nacional, México [SIP202301890] and [SIP20221267].

REFERENCES

Alegre, M., & Silva, C. (2020). Teaching practices in Higher Education: Didactic analysis of institutional conferences at the IES of Puerto Tirol—Chaco, Argentina. *Magazine of Studies and Experiences in Education*, *19*(39), 287–304. doi:10.21793/rexe.20201939alegre17

Araya, S., Rodríguez, A., Badilla, N., & Marchena, K. (2021). The flipped classroom as a teaching resource in the Costa Rican context: Case study on its implementation in a secondary school. *Education Magazine*, *46*(1), 103–119. doi:10.15517/revedu.v46i1.44333

Barral, A., Ardi-Pastores, V., & Simmons, R. (2018). Student Learning in an Accelerated Introductory Biology Course is Significantly Enhanced by a Flipped Learning Environment. *CBE Life Sciences Education*, *17*(3), 1–9. doi:10.1187/cbe.17-07-0129 PMID:30040530

Bergmann, J., & Sams, A. (2014). *Our story: How to create an "upside down class"?* SM Editions.

Cabero, J., & Llorente, C. (2020). Covid-19: radical transformation of digitization in university institutions. *Virtual Campus, 9*(2), 25-34. (www.revistacampusvirtuales.es)

Chehaibar, L. (2020). Curriculum flexibility. Tensions in times of pandemic. In H. Casanova (Coord.), Education and pandemic. An academic vision (pp. 83-91). UNAM.

Chen, L., Lin, T., & Tang, S. (2021). A qualitative exploration of nursing undergraduates' perceptions towards scaffolding in the flipped classroom of the Fundamental Nursing Practice Course: A qualitative study. *BMC Family Practice*, *22*(1), 245–255. doi:10.118612875-021-01597-4 PMID:34903171

Daza, C. (2021). Teaching skills in the 21st century facing the virtuality of education on the occasion of COVID-19. *Redipe Bulletin*, *10*(5), 177–188.

Durán, C., Páez, D., & Nolasco, C. (2021). Profile, challenges and challenges of the university student in the 21st century. *REDIPE Bulletin Magazine, 10*(5), 189-198. https://revista.redipe.org/index.php/1/article/view/1296

Dzandu, M., Pathak, B., & Cesare, S. (2022). Acceptability of the COVID-19 contact-tracing app—Does culture matter? *Government Information Quarterly*, *39*(4), 101750. Advance online publication. doi:10.1016/j.giq.2022.101750 PMID:35909915

Economic Commission for Latin America and the Caribbean. (2020). *Education in times of the COVID-19 pandemic*. https://www.cepal.org/es/comunicados/la-cepal-la-unesco-publican-documento-que-analiza-desafios-la-educacion-que-ha-brido-la

Fadel, C., Bialink, M., & Trilling, B. (2016). *Education in four dimensions*. Editorial Granika.

Fidalgo-Blanco, Á., Sein-Echaluce, ML, & García-Peñalvo, F. (2020). *Real advantages in the application of the Aula Invertida-Flipped Classroom method.* GRAIL Group.

Fuentes, G., Moreno, L., Rincón, D., & Silva, M. (2020). Assessment of soft skills in higher education. *University Formation, 14*(4), 49–60. doi:10.4067/S0718-50062021000400049

García-Varcálcel, A., & Gómez-Pablos, V. (2017). Project-based learning (ABP): evaluation from the perspective of primary school students. *Educational Research Journal, 1*(35), 113. https://digitum.um.es/digitum/bitstream/10201/52389/1/246811-963851-1-PB.pdf

Gaviria, D., Arango, J., Valencia, A., & Bran, L. (2019). Perception of the flipped classroom strategy in university settings. *Mexican Journal of Educational Research, 24*(81), 1–15. https://www.redalyc.org/journal/140/14062583011/html/

Gaviria Bustamante, K., Carriazo Diaz, C., & Perez Reyes, M. (2020). Educational planning as a fundamental tool for quality education. *Utopia and Latin American Praxis, 25*(3), 87-95. https://www.redalyc.org/articulo.oa?id=27963600007

González, D., Jeong, J., Gallego, A., & Cañada, F. (2018). Influence of the flipped methodology on the emotions felt by students of the Primary Education Degree in science classes depending on the baccalaureate studied. *Chemical Educator (Boise, Idaho), 29*(1), 77–88. doi:10.22201/fq.18708404e.2018.1.63698

González, M., & Abad, E. (2020). The flipped classroom: A challenge for university teaching. *VEC Virtuality. Education in Science, 11*(20), 75–91. https://revistas.unc.edu.ar/index.php/vesc/article/view/27449

González, M., & Huerta, P. (2019). Flipped classroom experience to promote prosumer students of the higher level. *RIED. Iberoamerican Journal of Distance Education, 22*(2), 245–263. doi:10.5944/ried.22.2.23065

González, M., & Huerta, P. (2019). Flipped classroom experience to promote prosumer students of the higher level. *RIED Ibero-American Journal of Distance Education, 22*(2), 1–15. doi:10.5944/ried.22.2.23065

Guarino, I., Aceto, G., Ciuonzo, D., Montieri, A., Persico, V., & Pescapé, A. (2022). Contextual counters and multimodal Deep Learning for activity-level traffic classification of mobile communication apps during COVID-19 pandemic. *Computer Networks, 21*(9), 109452. Advance online publication. doi:10.1016/j.comnet.2022.109452 PMID:36447639

Hamadi, M., El-Den, J., Azam, S. & Sriratanaviriyakul, N. (2022). Integrating social media as cooperative learning tool in higher education classrooms: An empirical study. *Journal of King Saud University—Computer and Information Sciences, 34*(6), 3722-3731. doi:10.1016/j.jksuci.2020.12.007

Hernández-Silva, C., & Tecpan Flores, S. (2017). Flipped classroom mediated by the use of virtual platforms: A case study in the training of physics teachers. *Studies Pedagogical (Valdivia), 43*(3), 193–204.

Kang, D., & Park, M. (2022). Interaction and online courses for satisfactory university learning during the COVID-19 pandemic. *International Journal of Management Education, 20*(3), 100678. Advance online publication. doi:10.1016/j.ijme.2022.100678

Lin, H., Hwang, G., Chang, S., & Hsu, Y. (2021). Facilitating critical thinking in decision making-based professional training: An online interactive peer-review approach in flipped learning context. *Computers & Education, 173*, 104266. Advance online publication. doi:10.1016/j.compedu.2021.104266

Lopez, Lira, L., & Uribe Lopez, A. (2022). Emerging pedagogies developed in higher education from the confinement by covid-19. *Opening, 14*(1), 114–131. doi:10.32870/Ap.v14n1.2149

Espinoza Mina, M. A., & Gallegos Barzola, D. (2020). Soft skills in education and business: Systematic Mapping. *UISRAEL Scientific Magazine, 7*(2), 39-56. doi:10.35290/rcui.v7n2.2020.245

Molinero Bárcenas, M., & Chávez Morales, U. (2019). Technological tools in the teaching-learning process in higher education students. *RIDE. Ibero-American Journal for Educational Research and Development, 10*(19). Advance online publication. doi:10.23913/ride.v10i19.494

Organisation for Economic Co-operation and Development. (2015). *Future of Education and Skills 2030.* https://www.oecd.org/education/2030-project/

Quinteros, C., & Cárdenas, N. (2021). Flipped classroom and role play: Implementation in the agricultural technical baccalaureate. *Interdisciplinary Koinonia, 6*(3), 106–127.

Reynosa Navarro, E., Serrano Polo, E. A., Ortega-Parra, A. J., Navarro Silva, O., Cruz-Montero, J.M., & Salazar Montoya, E. (2019). Teaching strategies for scientific research: relevance in the training of researchers. *University and Society, 12*(1), 259-266. http://scielo.sld.cu/scielo.php?script=sci_arttext&pid=S2218-36202020000100259

Roig-Vila, R., & Álvarez Herrera, J. (2019). Repercussion on Twitter of the active methodologies ABP, Flipped Classroom and Gamification. *RIED. Iberoamerican Journal of Distance Education, 22*(2). https://www.redalyc.org/articulo.oa?id=331460297005

Sandoval, V., Marín, B., & Barrios, T. (2021). The flipped classroom as a didactic strategy for the generation of competences: A systematic review. *RIED Ibero-American Journal of Distance Education, 24*(2), 285–308. doi:10.5944/ried.24.2.29027

Sanz, R., & Lopez, E. (2022). Educational learning after the COVID-19 pandemic. What role should the school play in the new world scenario? *Complutense Journal of Education, 33*(2), 215–223. doi:10.5209/rced.73928

Semante, D., & Gomez, V. (2021). Active didactic strategies to improve academic performance in the subject of Social Studies. *Episteme Koinonia, 4*(8). Advance online publication. doi:10.35381/ekv4i8.1386

Trujillo, F. (2017). Project-based learning: lines of progress for an innovation. *Language and Literature Didactics Texts, 78,* 42-48. https://www.researchgate.net/publication/320961737_Aprendizaje_basado_en_proyectos_Lineas_de_avance_para_una_innovacion_centenaria_Textos_Didactica_de_la_Lengua_y_de_la_Literatura_num_78_pp_42-48_octubre_2017

United Nations Educational, Scientific and Cultural Organization. (2022). *What you need to know about digital learning and education transformation.* https://www.unesco.org/es/digital-education/need-kno

Word Economic Forum. (2020) *Report on the future of employment 2020.* https://www-weforum-org.translate.goog/reports/the-future-of-jobs-report-2020/?_x_tr_sl=en&_x_tr_tl=es&_x_tr_hl=es&_x_tr_pto=sc

ADDITIONAL READING

Cabanillas, E. (2023). The impact of interdisciplinary Project Based Learning on young learners' speaking results. *Porta Linguarum*, *39*(39), 129–145. doi:10.30827/portalin.vi39.22864

Cardenas, E., & Deroncele-Acosta, A. (2023). Integration and dialogue in the era of virtual education: B-learning, E-learning and M-learning in virtual learning environments. *Philosophical Magazine*, *40*(103), 442–454. doi:10.5281/zenodo.7600964

Cervantes, M., Cruz, J., Sánchez, D., & Hernández, U. Z. (2023). Flipped classroom model: Validation of the instrument to assess the perception and satisfaction of university students. *Journal of Social Sciences*, *29*(2), 229–241. https://produccioncientificaluz.org/index.php/rcs/index

Crespí, P., García-Ramos, J., & Queiruga-Dios, M. (2022). Project-Based Learning (PBL) and Its Impact on the Development of Interpersonal Competences in Higher Education. *Journal of New Approaches in Educational Research*, *11*(2), 259–276. doi:10.7821/naer.2022.7.993

Deng, H. (2023). *Rethinking the* innovation, effectiveness, *and perceptions of flipped* classroom. *Infancia y Aprendizaje*, *46*(2), 322–351. doi:10.1080/02103702.2023.2170109

Diningrat, S., Setvosari, P., Ulfa, S., & Widiati, U. (2023). The Effect of an Extended Flipped Classroom Model for Fully Online Learning and its interaction with Working Memory Capacity on Students' Reading Comprehension. *Journal of New Approaches in Educational Research*, *12*(1), 77–99. doi:10.7821/naer.2023.1.1073

García-Monge, A., Pellicer-Iborra, C., Jiménez-Ruíz, M., & Gómez-González, F. (2023). The space in the work by projects: The case of the projects of the Rubik Model. *Interuniversity Electronic Journal of Teacher Training*, *26*(2), 37–56. doi:10.6018/reifop.559421

Gomez-Puig, M., & Stoyanova, A. (2022). Learn participant: L'ús d'estrategies d'aprenentatge active in higher education. *REIRE Revista d'Innovació I. Research in Education*, *15*(2), 1–4. doi:10.1344/reire.39770

Mamani, M., Martínez, G., Mamani, J., & Montero, A. (2023). Logical-mathematical thinking: Review of the STEAM assessment model to develop mathematical skills. *The Philosophical Journal*, *40*(103), 83–98. doi:10.5281/zenodo.7558099

Ochoa, E., Victores, M., & Dominguez, D. (2022). Information technologies and the flipped classroom: a review of their integration and implementation. *Scientific Series of the University of Informatics Sciences, 15*(2), 159-177. http://publicaciones.uci.cu

Proud, S. (2022). If I built them, will they come? A review of the evidence of barriers to active learning in university education. *REIRE Revista d'Innovació I. Research in Education*, *15*(2), 1–14. doi:10.1344/reire.38120

Sánchez-Rivas, E., Ramos-Núñez, M., Linde-Valenzuela, T., & Sánchez-Rodríguez, J. (2023). Perception of university students regarding project-based learning with technology. *Interuniversity Electronic Journal of Teacher Training*, *26*(1), 71–84. doi:10.6018/reifop.543281

Santillan-Aguirre, J. (2022). Flipped Classroom: Approach or Methodology? *Pole of Knowledge*, *7*(67), 1–22.

Sanz-Arcega, E. (2022). Economics papers: Potentialities for teaching Public Economics topics from a flipped classroom experience with non-economists. *E-Pública Electronic Magazine on the Teaching of Public Economics*, *30*, 1–16.

Velasco Moreno, M. (2023). On project-based learning approach and future foreign language teachers. *Human Review. International Humanities Review/Revista Internacional De Humanidades, 17*(1), 1-8. doi:10.37467/revhuman.v12.4719

KEY TERMS AND DEFINITIONS

Assessment: A cyclical, ongoing process of establishing clear, measurable expected outcomes of student learning.

Competency: A human potentiality for action requiring knowledge and abilities that can be learned and involves cognitive and non-cognitive elements.

Formative Assessment: It is concerned with the promotion of learning during a period of instruction.

Information Communication Technology: It refers to the use of technological tools such as the internet, computers, etc., to communicate information wirelessly to users and to enable them to access, store, share, and use information.

Program: Instrument curriculum where teaching and learning activities, allowing scholars to guide their practice with regard to the objectives to be achieved, are organized to develop competences in students based on a set of content, resources, and facilities.

Statistical Data Analysis: Data analysis by statistical methods, which ranges from the collection of data, the measures using data, to the interpretation of data.

Statistics: Science which analyses and draws conclusions from data. It can be applied in. most areas of human knowledge.

Chapter 11
The Future of Distance Education:
An International Comparative Study

Patricia Fidalgo
Emirates College for Advanced Education, UAE

Joan Thormann
Lesley University, USA

Oleksandr Kulyk
Oles Honchar Dnipro National University, Ukraine

José Alberto Lencastre
Universidade do Minho, Portugal

Maria João Figueiras
https://orcid.org/0000-0003-3941-0229
Zayed University, UAE

ABSTRACT

This chapter explores students' opinions and experiences with distance education (DE) from a multinational perspective, focusing on Portugal, Ukraine, and the United Arab Emirates. The research questions address students' previous experiences with online courses, their opinions about DE, reasons for enrolling or not enrolling in DE courses, perceived preparedness for DE, receptiveness to DE courses after the pandemic, preferences for DE formats, and differences based on age, gender, and country. By examining these aspects, the study aims to provide research-informed support for future DE models and contribute to understanding students' preferences, needs, and opinions regarding DE. The findings can potentially inform IHEs' digitalization strategies, recruitment efforts, learning approaches, and the development of purposefully designed DE and blended learning programs. The multinational approach strengthens the analysis and enhances understanding of the impact of DE on students' learning experiences in different countries.

DOI: 10.4018/978-1-6684-8861-4.ch011

INTRODUCTION

With the pandemic, it became necessary for many Institutions of Higher Education (IHE) to use the Distance Education (DE) format to keep schools open (UNESCO, 2020; UNESCO & IESALC, 2020). Before the pandemic, IHEs mainly used a face-to-face teaching model, although an increasing number of IHEs have been offering online classes (Seaman et al., 2018; Kentnor, 2015). Substantive changes in the educational environment were required from IHEs to address the challenges of the sudden crisis.

Purposefully designed DE programs could not be developed due to the need to transition from face-to-face to DE rapidly. IHEs need to plan for post-pandemic scenarios in which "digitalization expands and complements student-teacher and other relationships" (Schleicher, 2020, p. 4).

Marinoni et al. (2020) found that DE brought many challenges to IHEs during the pandemic, such as a negative impact on the quality of educational activities and a decrease in the equity of learning opportunities. According to Zhaohui (2020), organizational agility allows many IHEs to respond quickly to challenges. Many IHEs focused primarily on transitioning their course content to online platforms rather than effective online pedagogy (Crawford et al., 2020), which led to questioning if IHEs are prepared to switch to online classes (Houlden & Veletsianos, 2020). During the pandemic, many IHEs provided training for instructors on how to create and deliver online courses. Students' opinions and engagement with DE depend on the quality of DE classes; thus, instructors need further training in online pedagogy.

This empirical study aims to support the theoretical framework of DE by providing a multinational perspective on students' opinions and experiences with DE learning. The study intends to offer research-informed support for future DE models that IHEs can use and contribute to understanding students' preferences, needs, and opinions regarding DE. The study's results have the potential to assist IHEs in planning their digitalization strategies, recruitment efforts, and learning approaches while also informing the development of purposefully designed DE and blended learning (BL) programs. To address these objectives, the authors of this chapter adapted and modified research questions from a pre-pandemic study to reflect the current pandemic context. The research questions include exploring students' previous experiences with online courses, their opinions about DE, reasons for enrolling or not enrolling in DE courses, their perceived preparedness for DE, receptiveness to DE courses after the pandemic, preferences for DE formats, and examining similarities and differences in responses based on age, gender, and country.

Including three countries, Portugal, Ukraine, and the United Arab Emirates, will allow for a comprehensive analysis of the impact of DE on students' learning experiences. A multinational approach strengthens the results and understanding of the future educational landscape for IHEs in different countries.

BACKGROUND

Because of the pandemic, all educational stakeholders were affected and found it necessary to address pedagogical and technological issues. Assessing students' opinions and needs regarding DE has gained a renewed interest in the last two years. The DE format can be used to reach a large number of students (Perraton, 2008) due to its scalable nature (Fidalgo, 2012). According to Perraton (2008) "If the right model could be found, the techniques of [...] distance learning might be adapted to the needs of some of the millions who drop out of school, or who dropped out in a previous generation" (p. 6).

The theoretical framework for DE addresses definitions, different types, opportunities and challenges. Keegan (2002) defined DE as a learning experience where students and instructors are located in other places and times. A degree and credential can be earned by students who do not physically attend a brick-and-mortar institution (Gunawardena et al., 2006). Several formats are used within DE, such as blended, synchronous, asynchronous, open scheduled online courses, and massive online open courses (MOOC). The definition of the Blended Learning (BL) format used in this chapter is the one described by Roblyer (2016) and Garrison and Kanuka (2004) as a rotation model, where a percentage of the course is delivered remotely online — either synchronously or asynchronously — and part face-to-face. DE models refer to classes taught 100% remotely, synchronously, asynchronously, or both (Watts, 2016). All study material is provided online in open scheduled online courses, and students work asynchronously, at their own pace, although assignments have deadlines (Campus Explorer, 2019). Cormier et al. (2010) described a MOOC as a free online course where course resources are accessible to all participants. Many participants can attend MOOCs as they are available for anyone to enroll. Researchers have well-documented opportunities and challenges of DE. Among the opportunities are flexibility and time-saving. Motivation and interaction are some of the challenges mentioned (De Paepe, L., Zhu, C., & Depryck, K., 2018; Mahlangu, 2018; Vasilevska et al., 2017).

Research by González et al. (2020) compared students' learning strategies before and after the lockdown and concluded that COVID-19 confinement positively impacted students' performance. According to this study, students performed better on all types of assessments after the beginning of the lockdown. In contrast, a study by Aristovnik et al. (2020) about students' opinions found that the "lack of computer skills and the perception of a relatively higher workload prevented students from perceiving a higher performance while adapting to the 'new normal'; namely, education from a distance" (p. 22).

Despite the many opportunities DE afforded IHEs in several countries, DE was not a widely used format until the pandemic. Among those countries were Portugal, Ukraine and the United Arab Emirates (UAE).

THE STUDY

This study investigates students' preferences, needs, and opinions regarding DE and offers research-based data that can support IHEs in expanding their learning formats. Providing a multinational perspective assists in broadening the understanding of the use of DE. This chapter's authors used the research questions in a previous pre-pandemic study (Fidalgo et al., 2020). Some of the research questions were changed to reflect the pandemic context:

Research Questions

- What were students' experiences with DE and their opinions of the DE format?
- What reasons do students have for enrolling/not enrolling in DE courses, what preparation do they need to take DE courses, and how confident are they using the computer and the Internet?
- After the pandemic, how receptive are the students to taking DE courses, and which formats do they prefer?
- What are the similarities and differences between students' responses to the survey according to their age, gender, and country?

Data Collection and Analysis

Data collection was performed during the first semester of the 20/21 academic year. The participants in this study were graduate and undergraduate students from five universities in Portugal, four in Ukraine, and two in the UAE. Several universities in the three countries were invited to participate in this study. This resulted in a convenience sample that included those who agreed to involve their students.

A total of 980 participants responded to the online survey. Participants were both male and female. For the UAE (n=176), the mean age was 21.5 (SD = 5.1), 91% were female, for Portugal (n=460), the mean age was 27.4 (SD = 11.2), 74% female, and for Ukraine (n=344), the mean age was 19.6 (SD = 3.8), 75% female. Portugal, Ukraine and UAE were selected because of their similarity regarding DE in higher education. None of them had a broad DE offering before the pandemic and had to move to a DE model that was not purposefully designed.

An online survey of twelve questions was e-mailed to students in the three countries. The survey was translated into Portuguese, Ukrainian and Arabic. The working language at UAE universities is English. Thus, the survey was sent both in Arabic and English. Participants could choose to respond in either English or Arabic, their native language. The survey aimed to assess students' opinion towards DE, level of confidence using a computer and the Internet, and interest in taking DE courses in the future.

A descriptive data analysis was performed by country, looking at the frequencies of the items related to reasons for enrolling/not enrolling in DE courses, requirements, and formats of DE courses. A composite score was developed by adding all the items of the enroll/not enroll, and requirements for DE, respectively. To investigate the pattern of association between the choice of an item and the country, a Chi-square analysis was performed. Also, a Pearson correlation was performed to explore the association between the student's opinion of DE courses and their levels of confidence in using the computer and the Internet. A one-way ANOVA was used to investigate differences between countries on the dependent variables ("What is your opinion towards distance education?", "Are you interested in enrolling in distance education courses when the pandemic is over?" and "How confident do you feel using a computer and the Internet?"). A multivariate analysis of variance (MANOVA) was performed to analyze the main effects and interactions on the same dependent variables according to age group and country. Unanalyzed qualitative data gathered through the comments made by participants in the open-ended question were used to illustrate results and support the discussion of the research questions.

Limitations

One limitation of this study is that a convenience sample was used to gather the data. Another limitation involves collecting data during the pandemic. This may have restricted the number of participants and their responses.

Portugal, Ukraine, and UAE DE during COVID-19 Pandemic

In the UAE, the first case of COVID-19 was confirmed at the end of January 2020. The first cases were reported in Portugal and Ukraine at the beginning of March 2020. All three governments quickly introduced measures to prevent the spread of the coronavirus.

In mid-March, all educational institutions in the three countries were mandated to transition to a DE format (Ministry of Education and Science of Ukraine, 2020; Governo da República Portuguesa, 2020;

Ministry of Education of the United Arab Emirates, 2020c). Teachers and students participated in all online learning activities, and most on-campus facilities were closed. Numerous challenges occurred with this transition. For example, in Ukraine, according to a survey conducted at the beginning of May 2020, the following issues were mentioned: weak Internet connection, the lack of skills using software, the lack of video communication, and instability of the educational institutions' servers (State Service of Education Quality of Ukraine, 2020). Many IHEs offered professional development with online sessions to help instructors cope with and strengthen distance learning practices. For instance, in Portugal, the Centre IDEA-UMinho, which promotes innovation, and the development of teaching and learning (Centro IDEA-UMinho, 2020), provided weekly webinars and online sessions for teachers to share good DE practices and advanced use of technology. To ensure the success of DE in the UAE, the Ministry of Education established a system with the Telecommunications Regulatory Authority (TRA) and two Internet providers (Du and Etisalat) to offer mobile Internet at no cost for those who did not have access at home (Ministry of Education of the United Arab Emirates, 2020c).

Most IHEs could decide which technologies to use for distance learning. Many universities held online classes synchronously, using Microsoft Teams, WebEx, Zoom, etc. Many teachers used e-mail and messaging (e.g., Viber, Skype, Messenger, Telegram) as tools for asynchronous communication (State Service of Education Quality of Ukraine, 2020). DE was the format used in most IHEs of the three countries for the remainder of the 2019-20 academic year and beyond.

After a long lockdown period, the Portuguese and Ukraine governments decided that the students should return to school in September 2020 for the new academic year. However, since October 2020, all Ukrainian IHEs have worked in DE format because of the growth of the pandemic. In Ukraine, the added war scenario was a variable that contributed to keeping the DE format longer. In Portugal, each IHE was left to choose whether to keep online classes in specific contexts. Despite initially opening all IHEs in the UAE, the government decided it was safer to continue online education. The decision was revised periodically depending on the evolution of COVID-19 cases (Ministry of Education of the United Arab Emirates, 2020a).

Post-pandemic scenarios are being planned to include different learning formats (Toquero, 2020). In all three countries (National Agency for Higher Education Quality Assurance, 2020; Armando et al., 2021; Ministry of Education of the United Arab Emirates, 2020b), the need for and relevance of introducing and developing BL is under consideration for IHEs.

This research is an empirical study that aims to support the theoretical framework of DE. It provides a multinational perspective of students' opinions about the DE learning experience. It aims to give research-informed support to DE models that IHEs can use in the future. The results of this study may contribute to understanding students' opinions, needs and preferences regarding DE. This may also assist IHEs in planning their future digitalization, expanded recruitment, and learning strategies. The results can also inform the inclusion of purposely designed DE and BL programs.

RESULTS

This study reflects opinions about DE during the pandemic. Concerning whether students had taken any DE courses before the pandemic, the highest percentage was reported by UAE students (98%), followed by Ukrainian students (68%) and Portuguese students (67%). Of those who indicated they took online courses, the type of courses was not shared, and 27% did not report how many courses. For those who did respond to the number of courses taken, the mean number reported was 4.5 (SD = 4.1) and ranged from 1 to 36. For descriptive purposes, the means and standard deviations of the study variables (when applicable) are presented in Table 1.

Table 1. Means and standard deviations of the study variables

Variables	Range	UAE Mean (sd)	Portugal Mean (sd)	Ukraine Mean (sd)
What is your opinion towards Distance Education?	(1 – 5)	3.38 (1.40)	3.29 (1.07)	3.54 (1.10)
Are you interested in enrolling in Distance Education courses when the pandemic is over?	(1 – 5)	3.27 (1.59)	3.05 (1.37)	3.26 (1.32)
How confident do you feel using a computer and the Internet?	(1 – 5)	4.38 (0.94)	4.22 (0.82)	4.15 (0.99)
Number of advantages of enrolling in a Distance Education course	(0 – 9)	3.01 (2.37)	2.35 (1.70)	3.45 (2.03)
Number of disadvantages of enrolling in a Distance Education course	(0 – 10)	3.88 (2.33)	2.98 (1.82)	2.76 (1.92)
Number of requirements to enroll in a Distance Education course	(0 – 10)	3.56 (2.39)	2.60 (1.70)	2.58 (1.62)

STUDENTS' OPINIONS ABOUT DE (RESEARCH QUESTION 1)

The mean values concerning students' opinions about DE are presented in Figure 1, showing that the average response was neutral or unable to judge DE.

Figure 1. Students' opinions about distance education

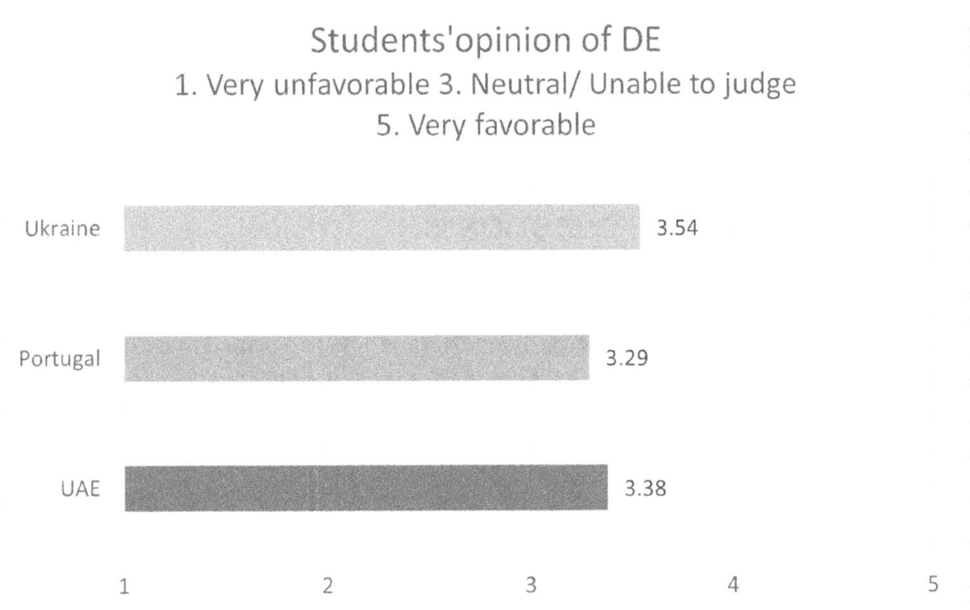

To explore whether there were significant main effects of gender and country and interaction effects, using age as a covariate, a univariate general linear model (GLM) was performed. Age as a covariate was significant [F (1,97) = 73,62, p = .000, ηp 2 = .070]. There was a significant main effect of the country [F (2,973) = 15,05, p = .000, ηp 2 = .030] (Figure 2). There were significant differences concerning opinions about DE between Portugal and Ukraine and between the UAE and Portugal. The Ukrainian students reported more favorable attitudes than the Portuguese students; the UAE students reported more favorable attitudes than the Portuguese. There were no significant differences between the UAE and Ukraine. There was an interaction effect between country and gender [F (2,973) = 7,22, p = .000, ηp 2 = .015]. In the UAE, males indicated a more positive attitude towards DE, while females reported more positive views about DE in Ukraine than males. There was no difference regarding gender between the Portuguese students.

Figure 2. Students' opinions about distance education according to age group

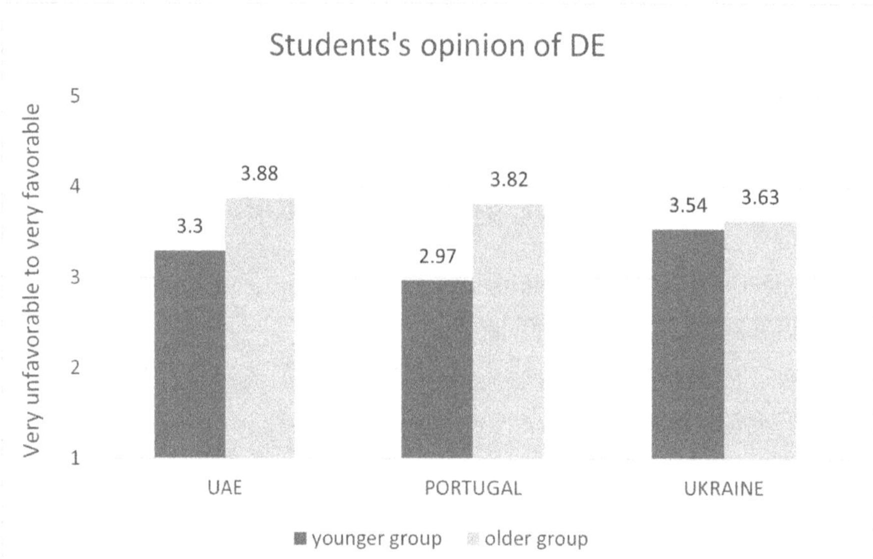

STUDENTS' REASONS FOR ENROLLING/NOT ENROLLING IN DE COURSES (RESEARCH QUESTION 2)

Figure 3 indicates the percentage of the chosen items measuring reasons for enrolling in DE courses, according to each country. A chi-square test of independence was performed to examine the association between each of the reasons to enroll and each of the three countries. The association between these variables was significant for the following items "I don't have to travel to school" ($\chi2$ (2, 980) = 47.9, p = .000), "I can manage my own class time" ($\chi2$ (2, 980) = 6.88, p = .032), "I can save time by studying from a location I choose" ($\chi2$ (2, 980) = 87.9, p = .000), "I have more options when selecting courses and/or a college to attend" ($\chi2$ (2, 980) = 43.2, p = .000), and "I can enroll in a program that I really want" ($\chi2$ (2, 980) = 101.78, p = .000). The latter was more likely to be considered as an advantage by Ukrainian students.

The following reasons to enroll in DE courses were more likely to be chosen by UAE students: "It is less expensive" ($\chi2$ (2, 980) = 6,66, p = .036), and "I can work at my own pace" ($\chi2$ (2, 980) = 23.38, p = .000).

The Portuguese students were more likely to consider "I had to take online courses because of the pandemic" ($\chi2$ (2, 980) = 15.07, p = .001) than the other students.

Figure 3. Reasons for enrolling in distance education courses by country

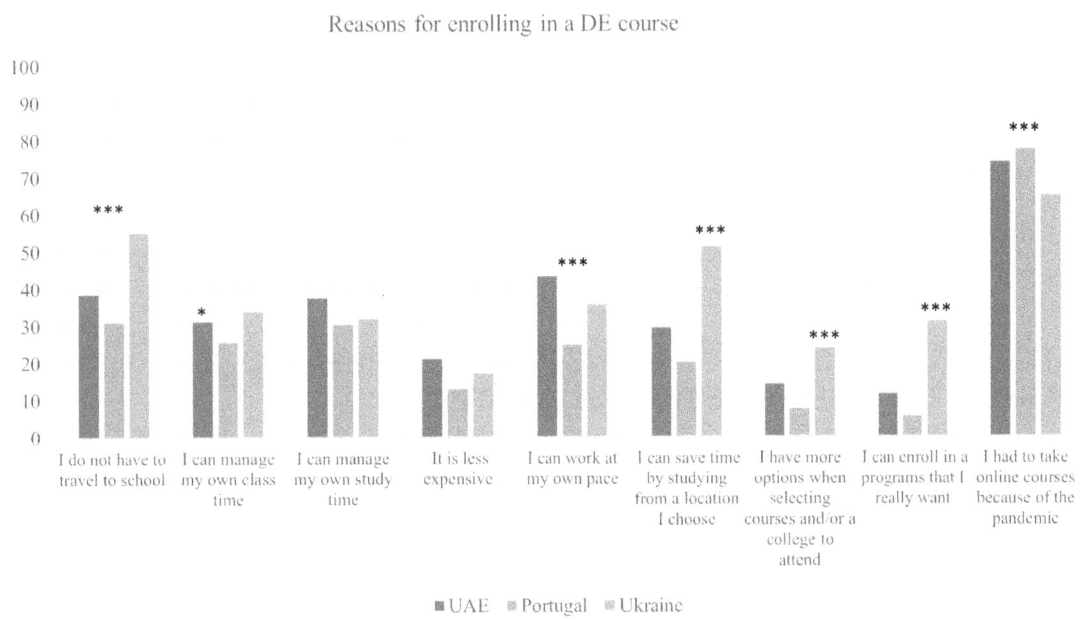

The percentage of the chosen items measuring reasons to not enroll in DE courses according to each country is reported in Figure 4. A chi-square test of independence was performed to examine the association between each of the reasons not to enroll and the three countries. The association between these variables was significant for the following items: "I do not know enough about distance education courses" ($\chi2$ (2, 980) = 13.55, p = .001); and "I do not have constant, reliable access to technology" ($\chi2$ (2, 980) = 30.53, p = .000). Ukrainian students were more likely to consider lack of technology access as a reason to not enroll than the other students. For the item "I would have difficulty getting immediate feedback" ($\chi2$ (2, 980) = 18.27, p = .000), UAE students were more likely to consider this reason than the others. For the items "I would feel isolated" ($\chi2$ (2, 980) = 7.07, p = .029), "I prefer face-to-face classes" ($\chi2$ (2, 980) = 20.34, p = .000), and "I would have difficulty interacting with peers" ($\chi2$ (2, 980) = 21.17, p = .000), Portuguese students were more likely to consider these reasons not to enroll than the other students.

Figure 4. Reasons for not enrolling in distance education courses by country

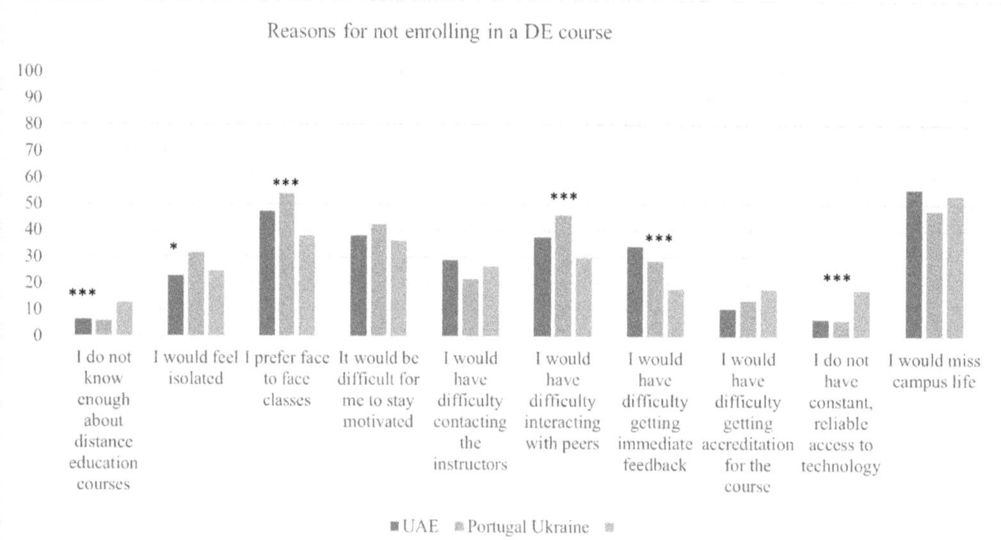

WHAT DO STUDENTS THINK THEY NEED TO BE PREPARED TO TAKE DE COURSES? (RESEARCH QUESTION 2)

The percentage of the chosen items measuring the requirements of enrolling in DE courses according to each country is reported in Figure 5. A chi-square test of independence was performed to examine the association between each requirement and the three countries. The association between these variables was significant for the following items: "I would need better computer equipment" ($\chi2$ (2, 980) = 6.87, p = .032), and "I would need better language skills (mainly English)" ($\chi2$ (2, 980) = 27.91, p = .000).

The previous requirements were more likely to be selected by Ukrainian students. The following requirements were more likely to be considered by the UAE students than the other students. "I would need a better internet connection at home" ($\chi2$ (2, 980) = 28.26, p = .000), "I would need better writing skills" ($\chi2$ (2, 980) = 48.62, p = .000), "I would need to develop a study plan" ($\chi2$ (2, 980) = 19.95, p = 000), "I would need a dedicated study space" ($\chi2$ (2, 980) = 85.24, p = 000), and "I would need better time management skills" ($\chi2$ (2, 980) = 25.50, p = 000). There was no significant association between the requirements to enroll in a DE course and being a Portuguese student.

To explore whether there were significant main effects of age group, country, and interaction effects, a MANOVA was performed for the number of reasons to enroll/not enroll and requirements (composite score) needed for taking DE courses. The latter were the dependent variables, and the country and age group were the independent variables. There was a significant main effect for age group regarding the number of reasons to not enroll in DE courses [F(1, 974) = 13.2, p =.000, ηp 2 = .013] and the preparation required [F(1, 974) = 4.6, p =.033, ηp 2 = .005]. The younger participants reported more reasons and preparation not to enroll than the older students. There was a significant main effect for country on reasons to enroll [F (2, 974) = 6.3, p =.002, ηp 2 = .013] and preparation [F (2, 974) = 8.8, p =.008, ηp 2 = .013]. Portuguese students reported a significantly lower number of reasons to enroll than the other students.

Figure 5. Requirements needed to enroll in distance education courses by country

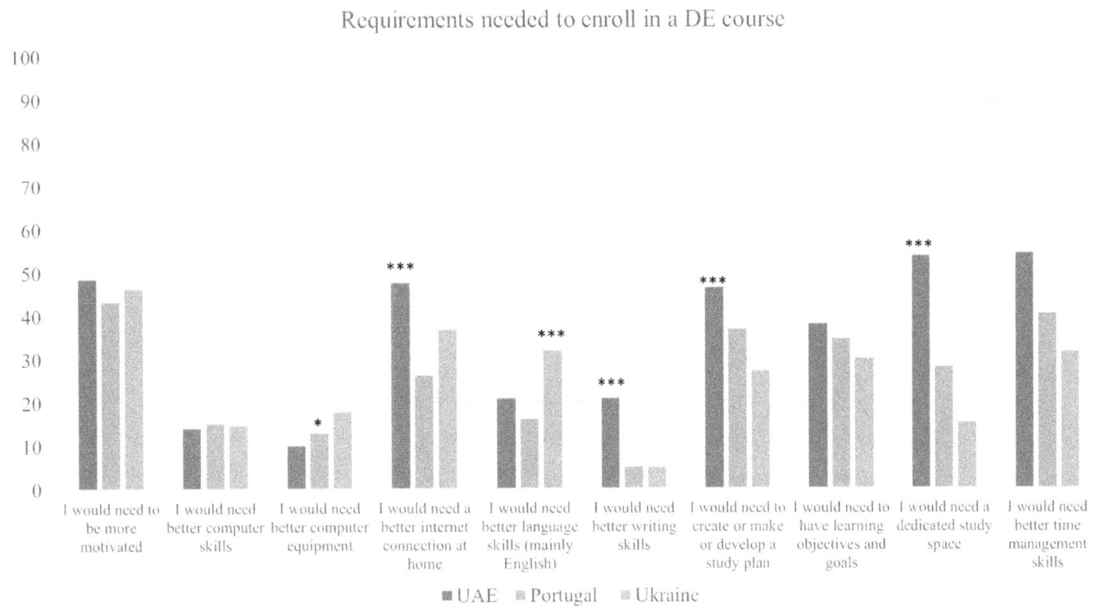

The UAE students reported a significantly higher number of requirements than the other students. A significant interaction was found between age group and country on number of reasons to enroll in DE courses [F (2, 974) = 4.9, p =.008, ηp 2 = .010] (Figure 6). Through planned contrasts the younger group in Portugal reported a significant smaller number of reasons to enroll than in Ukraine (t(586.3)=-10.4; p = .000) and in the UAE (t(209.8)=-4.75;p=.000), which also differs significantly from Ukraine (t(256.1)=-2.03; p=.044). There were no significant differences by country in the older group.

HOW CONFIDENT ARE STUDENTS IN USING THE COMPUTER AND THE INTERNET? (RESEARCH QUESTION 2)

To explore whether there were significant main effects of age group and country, as well as interaction effects, a MANOVA on the dependent variable "How confident do you feel using a computer and the Internet?" was performed. There was a significant main effect of age group [Pillai's Trace = .026, F = 8.59, df = (3,972), p = .000]. The older group reported higher confidence levels than the younger group [F (1, 974) = 8.59, p =.003, ηp 2 = .009]. There was a significant main effect of the country. The UAE students reported higher levels of confidence than the Portuguese (t(268)=-3.03; p = .003) and Ukrainian students (t(306.4)=2.5;p=.011). Also, a significant positive correlation was found between students' opinions of DE courses and the level of confidence in using the Internet and the computer in each country (r= .38; p =.000) for the UAE, (r=.26; p=.000) for Portugal and (r = .42; p=.000) for Ukraine. More positive opinions about DE courses are associated with higher computer and Internet use confidence levels.

Figure 6. Students' opinion about distance education by gender

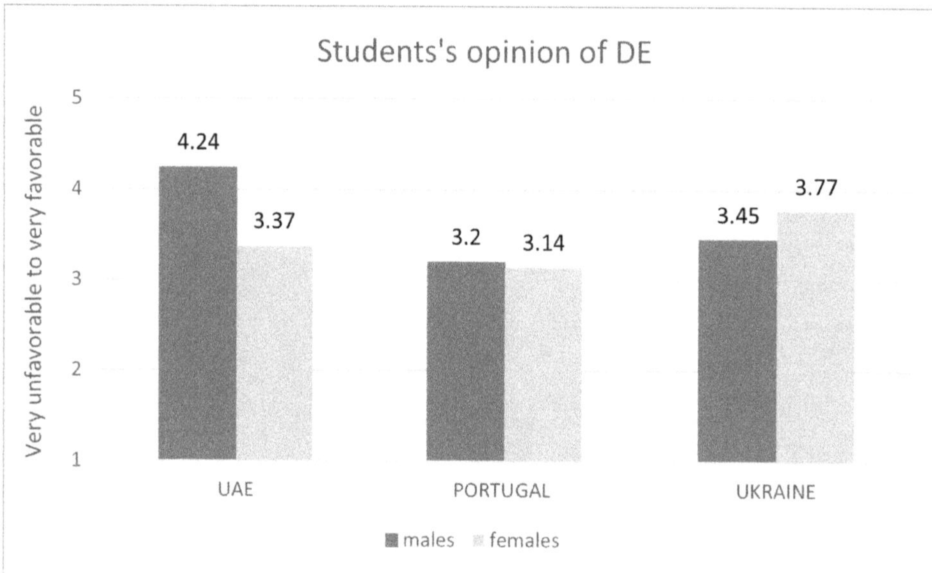

AFTER THE PANDEMIC, HOW RECEPTIVE ARE THE STUDENTS TO TAKING DE COURSES? (RESEARCH QUESTION 3)

The means for the variable "Are you interested in enrolling in distance education courses when the pandemic is over?" did not show significant differences between the countries. To explore whether there were significant main effects of age group and country, as well as interaction effects, a MANOVA was performed. There was a significant main effect of age group [Pillai's Trace = .026, F = 8.59, df = (3,972), p = .000]. The older group reported higher interest in enrolling in a DE course [F (1, 974) = 22.26, p =.000, ηp 2 = .022] than the younger group. A significant interaction was found between age group and country in the interest in enrolling in a DE course after the pandemic [F(2, 974) = 6.05, p =.002, ηp 2 = .012] (Figure 7). Through planned contrasts, significant differences were found in the younger group. Portuguese participants indicated less interest in taking DE courses after the pandemic than the Ukrainian participants (t(607.9) = -6.7; p=.000), and the UAE students (t(249.4)=-3.0;p=.003).

AFTER THE PANDEMIC, WHAT DE FORMATS ARE STUDENTS WILLING TO TAKE? (RESEARCH QUESTION 3)

A frequency analysis assessed students' preferences regarding the DE course formats they would be willing to take after the pandemic. Portuguese and Ukrainian students stated they prefer BL courses, while UAE students chose synchronous courses closely followed by BL courses (Figure 8).

Figure 7. Interest in enrolling in distance education courses in the future

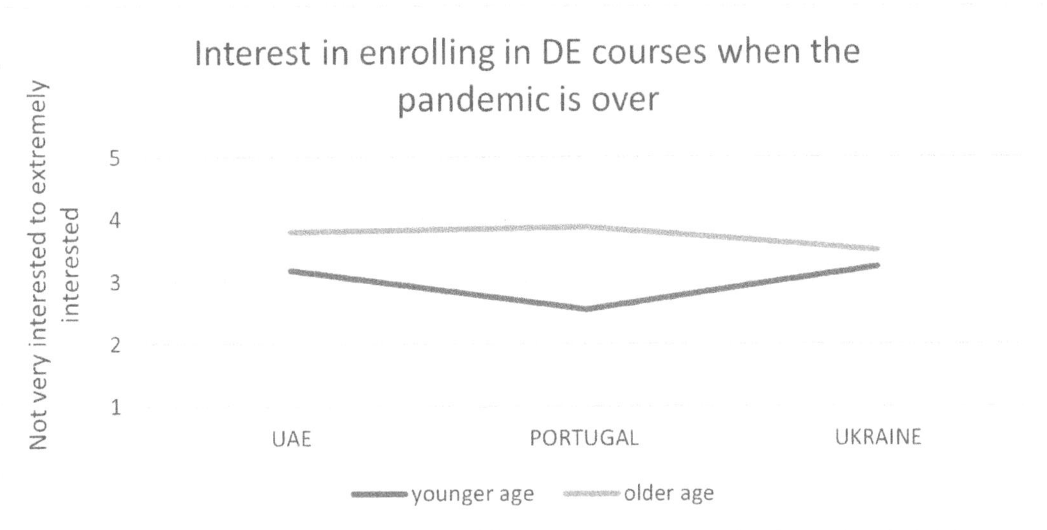

Figure 8. Types of distance education courses students would be interested in taking in the future

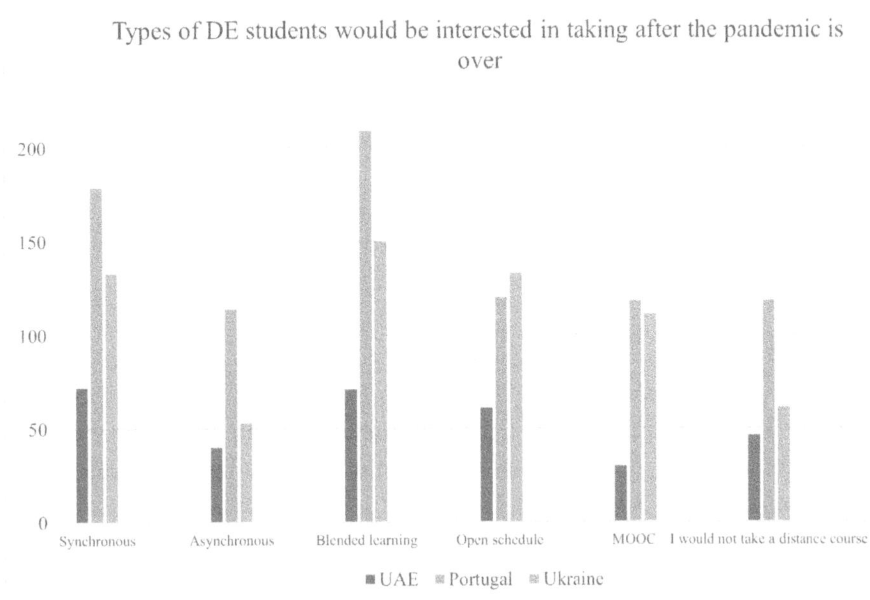

A chi-square test of independence was performed to examine the association between each of the types of DE and the three countries. The association between these variables was significant for the following items: "Asynchronous" ($\chi2$ (2, 980) = 10.72, p = .005) and "I would not take a distance course" ($\chi2$ (2, 980) = 8.2, p = .017). The Portuguese students were more likely to choose these items. For the items "Open schedule" ($\chi2$ (2, 980) = 14.97, p = .001) and "MOOC" ($\chi2$ (2, 980) = 14.14, p = .001), the Ukrainian students were more likely to choose these items.

DISCUSSION

As a result of the pandemic, most of the participants in this study had already taken DE courses before responding to the online survey, which was expected to affect their opinions. The frequency analysis of students' opinions about DE revealed that the majority of responses were neutral or unable to judge, indicating that students' experience with DE did not strongly sway their opinions of the online format.

However, the frequency analysis did not account for the positive or negative opinions expressed by many students in the open-ended comment item, which provided more nuanced feedback beyond the neutral or undecided responses. Two UAE students wrote, "Some educators are very helpful, however, there were some who made things much harder for us" and "I feel less anxious when participating online because I get nervous whenever I'm being the center of attention or put on the spotlight". Other studies performed during the pandemic reported that students had positive and negative opinions about DE (Bawa'aneh, 2021; Toquero, 2020).

Overall, this study found that older students had a more favorable view of DE and were more inclined to take online courses in the future. This could be attributed to their better understanding of the benefits and opportunities presented by DE and the increased flexibility it offers working students and those with family obligations. Previous studies also found older students more receptive to enrolling in DE courses (Kara et al., 2019; Amponsah et al., 2018). In Portugal, the participants in the study had a higher mean age than the other two countries. Among the older participants in Portugal, there were no gender differences. In contrast, in the UAE and Ukraine, there were varying opinions between men and women in the younger age groups, with men exhibiting a more positive attitude towards DE. The reasons for this gender disparity in the younger age groups are unclear.

However, Marley's meta-analysis (2007) on gender differences in DE concluded that several factors might influence gender opinion. Attitude towards and use of technology, learning and communication styles are some of those factors. Marley's study reported that men spend more time using the computer, have a more independent learning style, and communicate to report rather than make connections. These three factors may influence the DE experience and account for the more favourable opinion by men about DE.

While students from Ukraine and Portugal had comparable levels of previous experience with DE courses, as well as similar levels of computer and internet proficiency, Ukrainian students demonstrated a more favorable perception of DE in contrast to their Portuguese counterparts. This disparity may be explained by Ukraine's educational system's modernization efforts, which have included DE over the past decade, as noted by Voloshinov et al. (2020). However, the war in Ukraine has most likely disrupted these investments.

Due to the pandemic, the study participants had at least a semester and a half of DE courses. Although this group of students had more DE experience than the group in previous research by the authors (Fidalgo et al., 2020), their opinion is similar (on average neutral or unable to judge). The experience with DE provided in the pandemic context may have reinforced students' previous opinions about DE. This presents a challenge for IHEs considering expanding their online offerings and needing strategies to improve students' experiences.

Similar results to this study have been found by other researchers investigating students' opinions about DE during the pandemic (Rad et al., 2021; Lall & Singh, 2020; Hebebci et al., 2020; Yılmaz İnce et al., 2020;). Although many students reported a positive opinion about DE in these studies, several issues were mentioned as playing a negative role in their experience. Communication, interaction, language and technical problems were some of those issues (Lall & Singh, 2020). Participants in the Hebebci et al. (2020) study reported "their dissatisfaction as a result of not understanding the subject, finding the teacher inadequate, inadequate time, and lack of infrastructure." (p. 278)

Most students said the pandemic was the main reason to enroll in DE courses. Another reason the students chose was "Not having to travel to school". A Ukrainian student reported: "As it turned out, distance learning is very convenient, especially for those students who were spending a lot of time commuting to university", while another student from the UAE said "(…) this is an easy way to save lots of money and also time." The other most selected reasons were related to time management, as reflected in some of the students' comments. "Distance learning allows greater flexibility in time management. For those who work, it is very positive," wrote a Portuguese student, and another wrote, "(…) distance learning forces us to create a routine and follow it carefully. Knowing how to manage time is crucial (…)". Several of the challenges and positive aspects of DE mentioned by the students align with the findings of Allo (2020) and Toquero (2020) studies.

Students' choices of reasons not to enroll in DE courses were very similar for all three countries and are also confirmed by the studies from Allo (2020) and Toquero (2020). Most of the students stated that they would miss campus life and that they prefer face-to-face classes. A Ukrainian student wrote "It is very convenient in today's reality, but if there is an opportunity to attend face-to-face courses, I will choose that option", and a Portuguese student wrote, "I don't learn as well with online classes as I do in face-to-face classes, since it is easier to be distracted". In addition, students from the UAE and Ukraine expressed concerns about maintaining motivation in DE courses, while Portuguese students noted that interacting with peers may be challenging in an online format. Two Portuguese students wrote, "I don't think it's beneficial for us, because we isolate ourselves more and we don't have the ability to create "bonds" with our classmates and even teachers" and "In my opinion, distance learning is very complicated and demotivating. . . . I hope it doesn't last for long".

The students from all three countries exhibited similar preferences for the requirements they deemed necessary before enrolling in DE courses. The most frequently mentioned requirements were motivation and time management, followed by having a dedicated learning space, clearly defined learning goals, and a reliable internet connection. A Ukrainian student wrote, "The [Internet] connection is sometimes awful. Sometimes a lot of information is lost", while a UAE student wrote that he/she could easily be "distracted by a family member". The pandemic forced students and instructors into intensive digital learning environments. This may also have expanded digital inequalities (OECD, 2020). The current research results showed that the student's access to technology was a concern.

The frequency analysis of students' post-pandemic preferences for DE courses indicated that the BL format was the most popular choice among Portuguese and Ukrainian students, while those from the UAE were divided between synchronous and BL formats. The BL format combines the benefits of traditional face-to-face and online classes, and students appear to appreciate this hybrid approach.

Moreover, BL provides a possible pathway to a fully online format. Several authors have discussed that the post-pandemic educational scenario will likely include BL approaches (Pham & Ho, 2020; Sarkar & Biswas, 2021; Zhao & Watterston, 2021; Code et al., 2020; Eveleigh, 2020). However, several conditions need to be met to ensure the success of BL. These conditions include good ICT infrastructure and support, access to technological devices, software and online platforms, and professional development for instructors.

Although the students in the present study reported high confidence in using computers and the Internet, it is worth noting that the assumption of digital proficiency among "digital natives" is misguided, as Hargittai (2010) argued. Previous research has demonstrated that students' self-perceptions of their technological competencies do not always align with their actual abilities, as highlighted in studies by Fidalgo et al. (2020) and Hargittai (2010). Moreover, significant disparities exist in technology access between countries and even within different regions of the same country. Enhanced technological infrastructure and improved access to technology within each country could promote greater digital literacy.

FUTURE RESEARCH DIRECTIONS

Future studies might explore learning models adopted by IHEs after the pandemic. Research similar to this book chapter focusing on faculty opinions about DE and comparing it to students' views would be beneficial. An assessment of the types of counseling services that IHEs provide for DE students and student satisfaction would enhance universities' adoption of differentiated learning models.

CONCLUSION

According to OECD (2020), there is a need to reflect on long-term educational opportunities because of the pandemic. Exploring different educational models will help countries better understand how digital tools may assist student learning. The usual school timetable and workload will need to change, and "new solutions may be explored to provide students with opportunities to have more agency by being given more autonomy." (OECD, 2020, p.3)

Increased demand for flexible and accessible education is expected to continue, and DE will likely play a crucial role in meeting this demand. This research and other studies provide recommendations for developing quality DE programs that will increase student satisfaction. Following are some of those recommendations:

- Invest in technology and infrastructure: Ensure students access reliable and user-friendly technology.
- Ensure that course materials are well-organized, easy to access, and available in multiple formats.
- Create opportunities for students to engage with each other and with faculty, such as virtual study groups or online discussions.
- Offer flexible scheduling to accommodate students with varying needs.
- Provide support services such as tutoring, advising, and counseling.
- Maintain quality standards by regularly reviewing and updating course content, assessment methods, and instructional design.

IHEs can enhance their marketing strategies for promoting DE courses. One strategy is to advertise the advantages of DE, such as convenience, accessibility, flexibility, and cost savings, to prospective students.

Another strategy for marketing DE courses is to showcase successful graduates and the career outcomes of previous students, including job placement and salary increases. It is also important to use digital marketing strategies, such as e-mail campaigns, social media, and targeted online ads, to reach a wider audience. IHEs can also offer prospective students interactive experiences, such as live virtual classes and online discussion forums, to simulate the classroom experience and foster a sense of community among students.

Governments can support universities in offering DE in various ways. Funding universities to upgrade their technological infrastructure and purchase necessary equipment is one way to support DE. Another way for governments to help ensure quality programs is to establish and enforce regulations and accreditation standards for DE programs to meet the needs of students. For example, the Portuguese government instituted a law in 2019 to regulate DE (Governo da República Portuguesa, 2019). A way to address digital skills was instituted in Ukraine, where all students at a secondary level are required to take a minimum of one computer course (Ministry of Education and Science of Ukraine, 2017). However, this may have been disrupted by war. To help IHEs stay at the forefront of technological advancements and provide innovative and effective learning experiences for students, governments can fund research and development about DE.

DE is rapidly evolving and expanding, with technological advancements making it increasingly accessible and effective. As more learners turn to online education, the future of education will likely include DE. However, challenges such as ensuring quality education, addressing the digital divide, designing effective assessments, and promoting student engagement and motivation will need to be addressed. As long as these challenges are met, DE will continue to play a significant role in providing accessible and flexible learning opportunities for people all over the world.

Based on current trends and emerging technologies, the future of DE will likely include personalization as learners expect a tailored learning experience that meets their individual needs and preferences. This is being achieved through the use of adaptive learning technologies and data analytics. Also, Artificial Intelligence (AI) and Virtual Reality (VR) are being integrated into DE. These provide intelligent tutoring, automated assessment, personalized feedback to learners, and immersive and engaging learning experiences that can be used to teach complex concepts and skills. Finally, as learners seek to gain recognized qualifications and demonstrate their skills and knowledge, credentialing and micro-credentialing are becoming increasingly crucial in DE.

In conclusion, the future of DE will require a concerted effort from educators, institutions, and policymakers to address the challenges and ensure that learners have access to high-quality education characterized by a continued focus on personalization, the integration of emerging technologies, and the use of innovative pedagogies to deliver effective and engaging learning experiences, regardless of their location or circumstances.

REFERENCES

Allo, M. D. G. (2020). Is the online learning good in the midst of Covid-19 pandemic? The case of EFL learners. *Journal Sinestesia*, *10*(1), 1–10. https://sinestesia.pustaka.my.id/journal/article/view/24

Amponsah, S., Torto, B. A., & Badu-Nyarko, S. K. (2018). Ghanaian mature students' motivation to pursue degree programmes through distance education. *International Review of Education*, *64*(5), 585–606. doi:10.100711159-018-9728-8

Aristovnik, A., Kerži˘, D., Ravšelj, D., Tomaževi˘, N., & Umek, L. (2020). Impacts of the COVID-19 Pandemic on Life of Higher Education Students: A Global Perspective. *Sustainability, 12*, 1-34. https://doi.org/https://doi.org/10.3390/su12208438

Armando, N., Almeida, R., Fernandes, J. M., Silva, J. S., & Boavida, F. (2021). End-to-end experimentation of a 5G vertical within the scope of blended learning. *Discover Internet of Things, 1*(9), 9. Advance online publication. doi:10.100743926-021-00008-5

Bawa'aneh, M. S. (2021). Distance learning during COVID-19 pandemic in UAE public schools: Student satisfaction, attitudes and challenges. *Contemporary Educational Technology, 13*(3), ep304. Advance online publication. doi:10.30935/cedtech/10872

Campus Explorer. (2019). *Different Types of Distance Learning*. Retrieved December 12, 2022, from https://www.campusexplorer.com/college-advice-tips/7021E31E/Different-Types-of-Distance-Learning/

Centro IDEA-UMinho. (2020). *Quais são os objetivos do Centro IDEA-UMinho?* Retrieved December 12, 2022, from https://idea.uminho.pt/pt/quem-somos/Paginas/missao.aspx

Code, J., Ralph, R., & Forde, K. (2020). Pandemic designs for the future : perspectives of technology education teachers during. *Information and Learning Sciences, 121*(5/6), 419-431. https://doi.org/https://doi.org/10.1108/ILS-04-2020-0112

Cormier, D., McAuley, A., Siemens, G., & Stewart, B. (2010). *What is a MOOC?* [Video file]. YouTube. https://www.youtube.com/watch?v=eW3gMGqcZQc

Crawford, J., Butler-Henderson, K., Rudolph, J., Malkawi, B., Glowatz, M., Burton, R., Magni, P. A., & Lam, S. (2020). COVID-19: 20 countries' higher education intra-period digital pedagogy responses. *Journal of Applied Learning & Teaching, 3*(1), 9–28. https://doi.org/https://doi.org/10.37074/jalt.2020.3.1.7

De Paepe, L., Zhu, C., & Depryck, K. (2018). Online Dutch L2 learning in adult education: Educators' and providers' viewpoints on needs, advantages and disadvantages. *Open Learning, 33*(1), 18–33. doi: 10.1080/02680513.2017.1414586

Eveleigh, B. (2020). *Embracing virtual and blended learning post COVID-19*. Pamoja. https://pamoja-education.com/blog/embracing-virtual-and-blended-learning-post-covid-19

Fidalgo, P. (2012). *Learning networks and moodle use in online courses: a social network analysis study* [Unpublished doctoral dissertation]. Universidade Nova de Lisboa, Lisbon.

Fidalgo, P., Thormann, J., Kulyk, O., & Lencastre, J. A. J. A. (2020). Students' perceptions on distance education: A multinational study. *International Journal of Educational Technology in Higher Education, 17*(1), 18. doi:10.118641239-020-00194-2

Garrison, D. R., & Kanuka, H. (2004). Blended learning: Uncovering its transformative potential in higher education. *The Internet and Higher Education, 7*(2), 95–105. doi:10.1016/j.iheduc.2004.02.001

Gonzalez, T., De la Rubia, M. A., Hincz, K. P., Comas-Lopez, M., Subirats, L., Fort, S., & Sacha, G. M. (2020). Influence of COVID-19 confinement on students' performance in higher education. *PLoS ONE, 15*(10), 1-23. https://doi.org/https://doi.org/10.1371/journal.pone.0239490

Governo da República Portuguesa. (2019). *Portaria n.o 359/2019. Diário da República n.o 193/2019, Série I de 2019-10-08*. Retrieved December 12, 2022, from https://dre.pt/home/-/dre/125085420/details/maximized

Governo da República Portuguesa. (2020). *Comunicado do Conselho de Ministros de 19 de março de 2020*. Comunicados do Conselho de Ministros.

Gunawardena, C., McIsaac, M., & Jonassen, D. (2006). Distance education. In D. H. Jonassen (Ed.), *Handbook of Research for Educational Communications and Technology* (pp. 355–396). Association for Educational Communications and Technology.

Hargittai, E. (2010). Digital Na(t)ives? Variation in internet skills and uses among members of the "net Generation.". *Sociological Inquiry*, *80*(1), 92–113. doi:10.1111/j.1475-682X.2009.00317.x

Hebebci, M. T., Bertiz, Y., & Alan, S. (2020). Investigation of Views of Students and Teachers on Distance Education Practices during the Coronavirus (COVID-19) Pandemic. *International Journal of Technology in Education and Science*, *4*(4), 267–282. doi:10.46328/ijtes.v4i4.113

Houlden, S., & Veletsianos, G. (2020). *Coronavirus pushes universities to switch to online classes — but are they ready?* Retrieved December 11, 2022, from https://theconversation.com/coronavirus-pushes-universities-to-switch-to-online-classes-but-are-they-ready-132728

Kara, M., Erdoğdu, F., Kokoç, M., & Cagiltay, K. (2019). Challenges Faced by Adult Learners in Online Distance Education: A Literature Review. *Open Praxis*, *11*(1), 5. doi:10.5944/openpraxis.11.1.929

Keegan, D. (2002). *The Future of Learning: From eLearning to mLearning*. https://files.eric.ed.gov/fulltext/ED472435.pdf

Kentnor, H. E. (2015). Distance education and the evolution of online learning in the United States. In D. J. Flinders & C. M. Moroye (Eds.), *Curriculum and Teaching Dialogue* (pp. 21–34). American Association for Teaching & Curriculum.

Lall, S., & Singh, N. (2020). COVID-19: Unmasking the new face of education. *International Journal of Research in Pharmaceutical Sciences*, *11*(1), 48–53. doi:10.26452/ijrps.v11iSPL1.2122

Mahlangu, V. P. (2018). The Good, the Bad, and the Ugly of Distance Learning in Higher Education. In M. Sinecen (Ed.), *Trends in E-learning* (pp. 17–29). IntechOpen. doi:10.5772/intechopen.75702

Marinoni, G., Van't Land, H., & Jensen, T. (2020). *The impact of Covid-19 on higher education around the world*. Retrieved November 17, 2022, from https://www.iau-aiu.net/IAU-Global-Survey-on-the-Impact-of-COVID-19-on-Higher-Education-around-the

Marley, J. L. (2007). Gender Differences and Distance Education: Major Research Findings and Implications for LIS Education. *Journal of Education for Library and Information Science*, *48*(1), 13–20. https://doi.org/http://www.jstor.org/stable/40324317

Ministry of Education and Science of Ukraine. (2017). *Програма курсу "Інформатика" для 5–9 класів загальноосвітніх навчальних закладів* [Program of the Course "Informatics" for Grades 5–9 of Secondary Schools]. Retrieved November 12, 2022, from https://mon.gov.ua/storage/app/media/zagalna serednya/programy-5-9-klas/onovlennya-12-2017/programainformatika-5-9-traven-2015.pdf

Ministry of Education and Science of Ukraine. (2020). *Letter No. 1/9-154*. Retrieved November 12, 2022, from https://mon.gov.ua/storage/app/media/news/Новини/2020/03/11/1_9-154.pdf

Ministry of Education of the United Arab Emirates. (2020a). *Confirmed options for re-opening schools in the academic year 2020-21*. Retrieved November 13, 2022, from https://u.ae/en/information-and-services/education/confirmed-options-for-reopening-schools-in-the-academic-year-2020-21

Ministry of Education of the United Arab Emirates. (2020b). *Partial re-opening of higher education institutes in the academic year 2020-21*. Retrieved November 13, 2022, from https://u.ae/en/information-and-services/education/higher-education/partial-re-opening-of-higher-education-institutes-in-the-academic-year-2020-21

Ministry of Education of the United Arab Emirates. (2020c). *Distance learning in times of COVID-19*. Retrieved November 13, 2022, from https://u.ae/en/information-and-services/education/distance-learning-in-times-of-covid-19

OECD. (2020). *Education responses to COVID-19: Embracing digital learning and online collaboration*. Retrieved November 23, 2022, from https://www.oecd.org/coronavirus/policy-responses/education-responses-to-covid-19-embracing-digital-learning-and-online-collaboration-d75eb0e8/

Perraton, H. (2008). *Open and distance learning in the developing world*. Routledge.

Pham, H., & Ho, T. (2020). Toward a 'new normal' with e-learning in Vietnamese higher education during the post COVID-19 pandemic. *Higher Education Research & Development*, *39*(7), 1327–1331. doi:10.1080/07294360.2020.1823945

Rad, F. A., Otaki, F., Baqain, Z., Zary, N., & Al-Halabi, M. (2021). Rapid transition to distance learning due to COVID-19: Perceptions of postgraduate dental learners and instructors. *PLoS One*, *16*(2), 1–22. doi:10.1371/journal.pone.0246584 PMID:33556131

Roblyer, M. D. (2016). *Integrating Educational Technology into Teaching*. Pearson.

Sarkar, A., & Biswas, P. (2021). Blended Learning as an Approach in Post Pandemic Era in India. In D. K. Kundu (Ed.), *E-Learning-Teaching Strategies and Teachers' Stress in Post Covid-19* (pp. 161–168). Kanishka Publishers.

Schleicher, A. (2020). *The Impact of Covid-19 on Education Insights From Education at a Glance 2020*. OECD Publishing. https://www.oecd.org/education/the-impact-of-covid-19-on-education-insights-education-at-a-glance-2020.pdf

Seaman, J. E., Allen, I. E., & Seaman, J. (2018). *Grade Increase: Tracking Distance Education in the United States*. Babson Survey Research Group. https://files.eric.ed.gov/fulltext/ED580852.pdf

State Service of Education Quality of Ukraine. (2020). *Report on the Results of the Survey on the State of Use of Distance Learning Technologies in Higher Education Institutions of Ukraine*. State Service of Education Quality of Ukraine. https://sqe.gov.ua/images/materials/опитування/зво/Опитування_дистанційне навчання_ЗВО.pdf

Toquero, C. M. (2020). Challenges and Opportunities for Higher Education amid the COVID-19 Pandemic: The Philippine Context. *Pedagogical Research, 5*(4), 1-5. https://doi.org/https://doi.org/10.29333/pr/7947

UNESCO. (2020). *Distance learning strategies. What do we know about effectiveness?* https://unesdoc.unesco.org/ark:/48223/pf0000373350

UNESCO & IESALC. (2020). *COVID-19 and higher education: Today and tomorrow. Impact analysis, policy responses and recommendations*. https://bit.ly/34TOSvu

Vasilevska, D., Rivza, B., Pivac, T., Alekneviciene, V., & Parlińska, A. (2017). Analysis of the Demand for Distance Education at Eastern and Central European Higher Education Institutions. *Journal of Teacher Education for Sustainability*, *19*(1), 106–116. doi:10.1515/jtes-2017-0007

Voloshinov, S., Kruglyk, V., Osadchyi, V., Osadcha, K., & Symonenko, S. (2020). Realities and prospects of distance learning at higher education institutions of Ukraine. *Ukrainian Journal of Educational Studies and Information Technology*, *8*(1), 1–16. doi:10.32919/uesit.2020.01.01

Watts, L. (2016). Synchronous and asynchronous communication in distance learning: A review of the literature. *Quarterly Review of Distance Education*, *17*(1), 23–32. doi:10.1007/BF02763377

Williamson, B., Eynon, R., & Potter, J. (2020). Pandemic politics, pedagogies and practices: Digital technologies and distance education during the coronavirus emergency. *Learning, Media and Technology*, *45*(2), 107–114. doi:10.1080/17439884.2020.1761641

Yılmaz İnce, E., Kabul, A., & Diler, İ. (2020). Distance Education in Higher Education in the COVID-19 Pandemic Process: A Case of Isparta Applied Sciences University. *International Journal of Technology in Education and Science*, *4*(4), 343–351. doi:10.46328/ijtes.v4i4.112

Zhaohui, W. (2020). *How a top Chinese university is responding to coronavirus.* World Economic Forum. Retrieved December 10, 2022, from https://www.weforum.org/agenda/2020/03/coronavirus-china-the-challenges-of-online-learning-for-universities/

KEY TERMS AND DEFINITIONS

Asynchronous: Learning that happens without real-time interaction, allowing participants to access and engage with instructional materials at their own pace and convenience. Interaction between participants may also occur in this format.

Blended Learning: An approach to education that combines in-person instruction with online distance activities.

Distance Education: An educational format involving physically separated students and instructors, typically facilitated through Internet services and online platforms.

Digital Skills: Competencies and abilities required to effectively navigate, utilize, and communicate using digital technologies and tools.

Educational Models: Frameworks or structures that guide the design and delivery of educational programs, including methods, approaches, and strategies employed in teaching and learning.

Massive Online Open Courses (MOOC): Free online courses open to many participants globally, with access to educational resources and interactive activities.

Synchronous: Learning that occurs in real-time, where participants communicate simultaneously.

Time Management Skills: The ability to plan, prioritize, and allocate time to accomplish tasks, meet deadlines, and balance various responsibilities.

Chapter 12
Inclúyete Virtual Reality:
A Teaching Innovation Proposal to Work on Mental Health Stigma in University Students

Noelia Navarro Gómez
Universidad de Almería, Spain

Ana María Martínez-Martínez
Universidad de Almería, Spain

Anabel Corral-Granados
https://orcid.org/0000-0001-8870-8844
Universidad de Almería, Spain

María del Pilar Díaz-López
https://orcid.org/0000-0002-0888-925X
Universidad de Almería, Spain

Remedios López-Liria
https://orcid.org/0000-0002-3416-2791
Universidad de Almería, Spain

ABSTRACT

Stigma towards mental health disorders is a powerful obstacle to the recovery and social inclusion. This is why social awareness and empathy must be addressed as soon as possible. Among the possibilities for intervention, new technologies, and more specifically virtual reality (VR), are a very attractive resource for young people who spend a large part of their free time playing video games. The aim of this chapter is to present a proposal that can be applied to university students to address stigma in mental health whose central axis is a virtual reality programme, Inclúyete VR. This programme fosters empathy towards those who suffer from a mental health disorder, allowing them to "experience" how a person feels when they have hallucinations and what the possibilities are for a psychosocial approach to the disorder. The session should be led by a mental health professional familiar with the software. The total duration of the VR programme is approximately 10 minutes, which is very appropriate to initiate a discussion-reflection about mental health disorders.

DOI: 10.4018/978-1-6684-8861-4.ch012

INTRODUCTION

The stigma inherent in mental health disorders is a powerful burden that hinders recovery and seriously interferes with the social integration of those who suffer from it. We can define stigma as "a set of attitudes, usually of a negative nature, that a social group holds towards another minority group based on a series of characteristics that allow them to be differentiated, thus being a universal phenomenon with biological, psychological, and social bases that simplify information" (López et al., 2008, p. 46). It would be a chronic and acute stressor that would act as an invisible burden in addition to the obstacles caused by the existing clinical symptomatology (Major, et al., 2003; Petersen, et al., 2014). Thus, among other disadvantages, stigma would prevent people from asking for help in the presence of psychopathologies (Mellor, 2014; Pinto--Foltz et al., (2012). As noted above, the effects of stigma are multiple and varied, such as social isolation or loneliness, which affects self-esteem by generating feelings of shame, inferiority, and self-rejection that end up being internalized (Quinn et al., 2015). In terms of the process of stigma acquisition and development, the available research literature shows controversial results. For example, some studies indicate that young people lack clear ideas about what a mental disorder is, its treatment, or what characteristics would be associated with it (Spitzer and Cameron, 1995). Other studies indicate that young people manifest initially low levels of stigma (Schülze et al., 2003) and they have basic knowledge about what mental illness is and they do not have negative attitudes towards people who have a mental disorder (Watson et al., 2010). On the other hand, we also find in the literature several studies showing that under-18s do manifest certain types of prejudices, and beliefs about dangerousness and unpredictability, as well as attributed weakness of character, which would lead to an increase in social distance (Schachter et al., 2008).

BACKGROUND

Stigmatizing beliefs would appear at these ages to become evaluative standards later on (Bulanda, et al., 2014; Georgakakou-Koutsonikou, & Williams, 2017). The presence of these attitudes in the school environment is not only relevant to the adult stereotype but also, to a large extent, because if they are activated, they will provoke potential rejection among peers, making it even more difficult for young people with emotional problems to adapt. The coronavirus pandemic has highlighted how fragile mental health can be and the need for investment in health resources. Differentiating between age groups, adolescence is a crucial period for intervention as it is at this time that mental disorders begin to make their debut, with the lack of information and preventive strategies being the main causes behind the high numbers of psychopathologies that have been found in recent years to mention some figures, it is estimated that, mental health problems will be the main cause of disability among young people in 2030 and that approximately one out of four people will suffer from a psychological disorder during their lifetime. Around 20% of children and adolescents suffer from mental health problems, with 50% of these starting at the age of 14 and 75% before the age of 18 (Deolmi & Pisani, 2020). In addition, suicide is the second or third leading cause of death in young people aged 15-19, with high prevalence rates for anxiety disorders, depressive disorders, disruptive behavior, substance use, etc. These are just some of the available data, which show a clear upward trend in the incidence and prevalence of mental disorders, fostered by the socio-historical conditions of our context. Taking these figures into account, it is necessary to raise awareness in society about mental disorders, an awareness that should start from childhood and adolescence and where classrooms can play a crucial role. On the other hand, the expansion of new information and communication technologies (ICTs) has profoundly changed the way we relate to the world and offers new alternatives for intervention. For example, there is already available a Serious game (Stigma-stop), which presents characters suffering from a mental disorder (specifically, depression, panic disorder with agoraphobia, bipolar disorder, and schizophrenia). Its usefulness in secondary school students, university students or psychology professionals has been shown (Cangas et al., 2017; Mullor, et al., 2019). What had not been resorted to until now was the use of Virtual Reality, a tool that has brought about an important revolution that is presented to us as one of the most attractive for young people, while opening the door to new educational possibilities and experiential and meaningful learning. Currently, the use of Virtual Reality in the pedagogical field is booming (Dos Santos & Dos Santos, 2019; Sural, 2018), being present in various curricula (Huttar & Brintzenhofeszoc, 2019), being praised for the effectiveness of its role as an effective pedagogical resource that supports student learning (Jamil et al., 2019), in addition to its profoundly innovative character (Slater et al., 2019). Numerous reasons make VR highly attractive for use with young people and in educational contexts, including its ubiquity and ergonomics, being extrapolated to any context, while adapting to specific teaching and learning processes (Fombona & Pascual, 2017), increasing student motivation and participation in educational dynamics (Marín & Muñoz, 2018). The reception by students is usually quite positive, encouraging dynamism and interaction, as well as active participation, and connection with peers, teachers, as well as with educational content and material resources (Radu, 2014). Their success lies in their versatility, the possibility of application in group or collective format, the ease with which they can be applied, and how economical they are once designed. The present work arises from the fusion of these thematic axes, i.e. awareness of mental health disorders and the fight against stigma, and the new information technologies and virtual reality. The aim is to present a proposal for raising awareness of mental health stigma in the secondary school classroom, more specifically about psychotic disorders, based on Inclúyete VR, a Virtual Reality program. This virtual program allows empathizing with a person suffering from a mental disorder while illustrating the treatment possibilities offered in the inclusive initiative "Inclúyete," managed by the University of Almería, FAISEM, and Almería, Tierra sin Estigma, where university students, mental health users and people interested in the experience, jointly carry out courses on art and culture, sport and personal wellbeing.

METHODOLOGY

The Participants

The proposal is susceptible to be applied at any age, in this case, a relevant proposal is made to be applied to university students. It will ensure that the study is as homogeneous as possible in terms of variables such as sex.

Instruments

To evaluate the effectiveness of the intervention, it is proposed to apply the following tools, both in the baseline evaluation and after the intervention, to determine the effectiveness of the intervention and to obtain an assessment from the students:

- Attributional Questionnaire (AQ27; Corrigan et al., 2003). A questionnaire designed to measure stigmatizing attitudes and beliefs towards people with severe mental disorders. A reduced version of 14 items was used for the present study (Saavedra et al., 2020). This questionnaire is composed of four factors, all of them with adequate psychometric properties (Fear of danger=.88; Lack of solidarity=.837; Coercion=.864 and Avoidance=.758). Thus, the dangerousness/fear factor refers to the belief that people with schizophrenia pose a threat to others or to being afraid of those with a disorder (e.g., "Henry would terrify me" or "Do you feel Henry is dangerous?". The lack of solidarity factor refers to the willingness to care for people with a severe mental disorder (e.g. "How likely is it that you would help Enrique?;" "How sure are you that you would help Enrique?);" the coercion factor refers to the desirability of forcing people with schizophrenia to be treated (e.g. "I think it is in the best interest of the community to treat people with schizophrenia," "I think it would be best for the community where Henry lives if he were placed in a psychiatric hospital;" "To what extent do you think an asylum, where Henry can be away from his neighbors, is best for him?") and the avoidance factor of willingness to live or work near someone with a severe mental disorder ("If I were an employer, I would interview him for a job;" "I would share my car with Henry every day").
- Inclúyete-VR assessment questionnaire. This is a brief instrument composed ad hoc for the present study, consisting of 8 items. Participants must respond by giving scores from 1-10 to different aspects related to interest, the usefulness of the program, whether they consider that it helps users of mental health services, or whether they would recommend its use, among others.

Inclúyete VR is a virtual reality program designed with the UNITY tool and funded by a grant from the Spanish Ministry of Science, Innovation, and Universities. Inclúyete VR aims to offer information in an immersive way about mental disorders and different psychosocial resources that are used today in the recovery of these problems. The idea is that this information is easy to experience and, therefore, to remember and understand. It is also intended to encourage empathy, through greater closeness and understanding, for people suffering from mental health problems. On the first screen, a scenario is shown with "monsters" that the player must annihilate. The monsters quickly multiply, and it is impossible to kill them all. At this point, "a big monster" appears, and the only possible option is to escape. The meaning of this scene is to show a person invaded by hallucinations, delusions, or recurring thoughts (represented by the monsters). Initially, they are not very disruptive and can be controlled, but over time, these sensations become more persistent. Subsequently, a character appears and offers alcohol to the player. The screen then fades to black and ambulance sirens are heard. In many cases, the use of substances is common, which aggravates the difficulties and makes the need for professional help more and more evident. In the second part of the program, the person wakes up in a "Crisis Centre," where other people tell of similar situations. At this point in the program, different activities appear in which the player can interact, which are being used in the comprehensive and rehabilitative approach to psychosis in the Inclúyete VR program (canine therapy, sports activities, painting, computers, radio, and gardening). After completing each of these activities, a message appears explaining the usefulness of the different tasks, and the person is rewarded with "a star," so that when they have all of them, they can leave the room, like an "Escape Room," which is also very popular among young people. Thus, in one of the rooms, two characters appear who comment on their experience with psychosis and offer help to the player. In this way, the crucial role of social networks and self-help groups in recovery is highlighted. In the same room, a dog also appears, which can be petted and represents the benefits of the canine therapy. After this action, the first "star" is given and the message appears: "Animals can also improve our wellbeing," showing the importance that animals can play as allies in overcoming this type of difficulties. Further on in the game, we find a gymnasium with a basket and a lot of basketballs that the player can try to make a basket. Once the player has made a basket, another star appears together with the message "Physical activity is an ally of mental health" appears. Indeed, the sport has multiple benefits not only on a physical level but also on a psychological level. In another room, a blank canvas appears, and a paintbrush with a palette of colors that the player can pick up to make a drawing on the canvas. After doing this activity, this message appears: "Through art, we can share our emotions." In another room, there is a radio studio with a mixing desk that allows the player to interact with it to mix sounds and talk on the radio. After doing so, another of the "stars" appears, and the message: "All people have a lot to say." It is important to listen to people with mental health, give them a voice, and allow them to express how they feel. Another room contains three pots with tomato plants and a watering can, which once turned over the pots makes the tomatoes grow. Once it does, the star appears, and the message: "Routines help us to organize ourselves." Finally, another room contains computer parts that must be assembled where they fit on the motherboard so that once this task is done, the last of the stars is obtained and the following message appears: "Work helps us to value ourselves." Once the person has achieved the 6 stars, a door opens, and a sign appears with the sentence by Booker T. Washington: "I have learned that success is not so much measured by the position one has reached in life, as by the obstacles one has had to overcome in the effort to reach it." Below are some images captured from the virtual program.

Figure 1. Images from Inclúyete VR

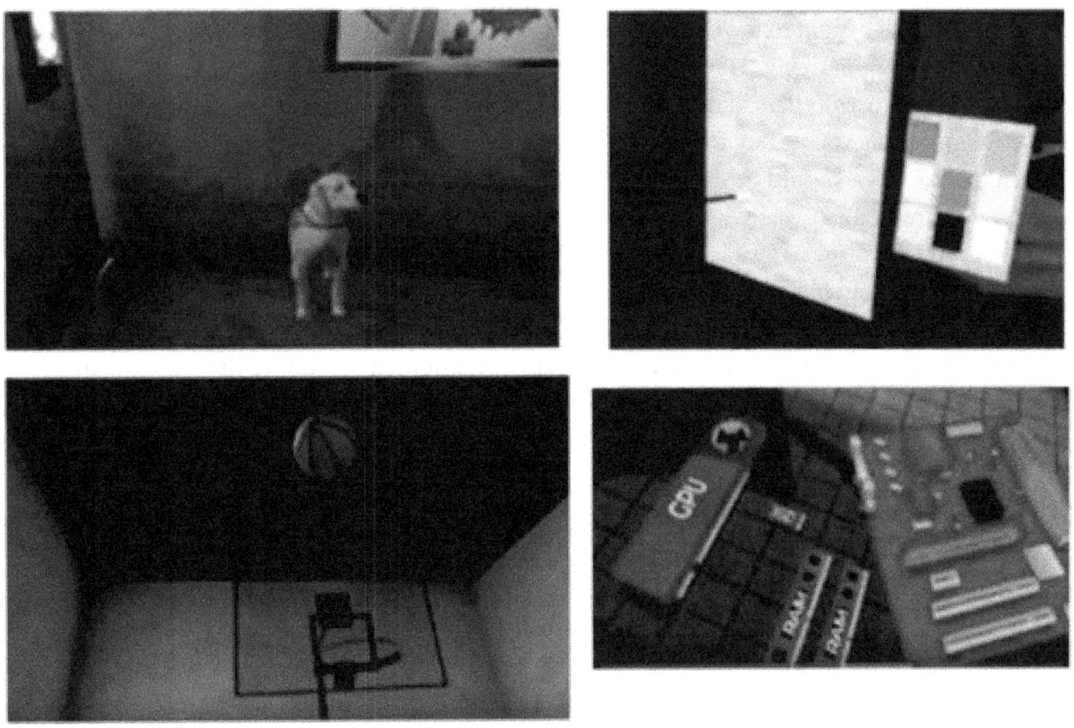

Procedure

The proposal consists of a session of approximately 1 hour and a half, which can be integrated into the classroom context. The session should be led by a mental health professional familiar with the Inclúyete VR program. Mentors can actively participate in the activity if they feel it is appropriate. The session will begin by completing the evaluation questionnaire, (AQ-14), and then the VR program will be implemented using Oculus Rift goggles. This can be done virtually, or by choosing a volunteer and connecting the goggles to a screen so that the rest of the participants can watch the development of the game. The program provides an initial immersive contact that can open a debate to debunk myths and reinforce the ideas presented. As an example, here are some of the questions that can be asked: How did you feel when the monsters appeared? Were the feelings always the same? Why do people, in many cases, resort to drug use? What consequences do you think substance use can have on psychological problems? If you had a mental health problem, what do you think would help you the most to overcome it? Do you know anyone (or maybe yourself), who has experienced a mental health problem? What happened? What helped them to get out of that situation? What do you think can be done to help a person suffering from mental health problems?

The aim is to reflect on the variability of symptomatology. During the course of a mental disorder, there are times when the persons may feel "powerful" and others when they may be distressed, desperate, depressed and try to eliminate these feelings. On the other hand, drug use can sometimes be a trigger that, mixed with other factors, can trigger a mental disorder. In this way, drugs make the person even more "disconnected" from others, thinking only about obtaining substances and ceasing to experience interest in activities that previously interested them (together with the different economic, judicial, and family consequences that drug use generally entails). It is also fundamental to emphasize that mental health problems are relatively common (1 in 4) and that they can happen to all of us, in certain circumstances. It is important to give weight to the context, to life's problems, avoiding resorting to biological explanations that are more stigmatizing. Having a support network that accompanies others without judgment, with the understanding of others, and, when necessary, with the help of professionals, it is necessary to overcome them. In this sense, the importance of psychologists should be emphasized, and the existing mental health resources should be explained. It is also necessary to explain the warning signs and the most common symptoms of mental disorders to be able to detect them and seek help as early as possible. In addition, this discussion can be accompanied by a first-person testimony from a user of mental health services to explain his or her experience. This last component is highly effective in previous pilot studies, as "putting a face" to the disorders breaks down prejudices, myths, and barriers and powerfully fosters empathy. Subsequently, and to conclude the intervention, the AQ-14 will be administered again, along with a short questionnaire in which the students' opinions about the VR software will be collected.

FUTURE RESEARCH DIRECTIONS

To analyze the existence of statistically significant differences in the pre-test and post-test, it is proposed to apply a Student's t-test for related samples. These analyses should be complemented with a statistical measure to determine the magnitude of the change brought about by the virtual reality experience through Cohen's d. All statistical analyses will be performed using SPSS 23.0 statistical software.

CONCLUSION

This paper aimed to present a brief anti-stigma intervention proposal for university students, whose central component is virtual reality software, Inclúyete-VR. designed by a team of mental health service users and professionals, it provides a first-person experience of psychosis and the possibilities of recovery.

These resources represent an innovative panorama in the way of conceiving teaching-learning, tools that can be applied at different educational levels of formal and informal education (Rodríguez, et al., 2018), promoting new ways of understanding teaching methodology (Viñals & Cuenca, 2016). In this way, empathy towards those suffering from some kind of mental health problem is fostered, increasing their knowledge of mental illness, and raising awareness of the importance of mental health in emotional well-being. Thus, it has been used in the treatment of psychopathologies as topographically different as phobias, post-traumatic stress, addictions, or eating disorders (see Gregg and Tarrier, 2007). Among its many advantages, its versatility makes it applicable to multiple fields (e.g., medicine, architecture, or aviation).

The possibility of 360° visualization is another factor that adds to its attractiveness and has a favorable influence on motivation by making it possible to analyze facts and details carefully. The proposed preventive and intervention strategy would therefore fulfill a triple function: on the one hand, to present information regarding the symptomatology of the disorders (in the video game, for example, the "monsters" that do not disappear would come to represent the thoughts and hallucinations, their persistent and intrusive character), on the other hand, it educates about the different options to achieve an integral rehabilitation, by presenting different treatment options (canine therapy, painting, gardening, sport ...), and finally, by complementing with debate-reflection, it encourages empathy, understanding of life circumstances and the need for a better understanding of the symptoms of the disorders, and the need for a better understanding of the different options available. It is essential to highlight the importance of giving a more active role to the users of Mental Health services in the design of intervention strategies, which favors the normalization of psychological disorders and social integration. Despite the advances that are taking place in the way mental disorders are conceived and approached, myths and stereotypes are still present in the general population (Angermeyer et al., 2016; Baba et al., 2017; Barry, McGinty, Pescosolido and Goldman, 2014). In the case of the target population in question, youth is a period of change in which the beliefs that will shape the adult personality and lifestyle begin to crystallize, while some mental disorders begin to "debut" and are particularly vulnerable to stigma (Clement et al., 2015; Lansu, 2018), so intervention at this time is of great importance. Increasing age hurts attitudes and social distance expressed toward mental disorders, and there is also variability according to the type of emotional problems exhibited (Dolphin and Hennessy, 2017). Compared to gender, girls tend to show less rejection towards their peers with emotional problems (Silke et al., 2017; Swords, Heary, and Hennessy, 2011).

Numerous experiences have been developed to reduce stigma and promote mental health in the school context. Despite the enormous disparity between them in terms of methodology, degree of structure, extension, and other relevant parameters, in general terms, most of the studies note the initial positive effect they have on stigmatizing attitudes, highlighting their cost-effectiveness (Bentham et al., 2013; Bulanda et al., 2014; Yang et al., 2018). In our context, efficacy studies have been published with programs such as "What's Up" (Andrés-Rodríguez et al., 2017), "EspaiJove" (Casañas et al., 2018) or "Stigma-Stop" (Cangas et al., 2019; Cangas et al., 2017), including in some cases technologies such as virtual reality as in "Inclúyete VR" (Cangas and Galván, 2020). Inclúyete-VR already has a pilot validation study applied to 12 psychologists, showing its usefulness in reducing stigma, while they expressed their interest and usefulness (Cangas and Galván, 2020). Likewise, the participants who tested the video game in the experimental group rated it highly, with all the items exceeding a high mark.

In this sense, we postulate that Inclúyete-VR will have positive results in the scores of the questionnaire that is applied in terms of stigma reduction, awareness of the problem, and reduction of social distance. On the contrary, we hope to encourage an attitude of help, not of segregation or ridicule or paternalism (which, despite the good intentions it hides, only conveys the message of fragility, underestimating their possibilities). The aim is to promote active accompaniment, without value judgments, based on understanding.

We also hypothesize that it will be well received by the participants. In addition to all this, the new technologies provide amplified learning environments that allow us to implement more flexible, active, dynamic, and playful methodologies, by the diverse characteristics of the students, which are an incentive for young people.

As discussed in the conceptualization of the concept, students seem to have confused ideas about the etiology, pointing mostly to failures in the nervous system (Angermeyer and Matschinger, 2005; Stone and Merlo, 2011). It is also interesting to note that in the age comparison, older students show more agreement with statements such as "mentally ill people are driven by their emotions," "normal people think things through," and "many people become mentally ill to avoid the difficult problems of everyday life" or "one of the main causes of mental illness is a lack of moral strength or willpower" or with phrases emphasizing that it is easy to distinguish those with disorders from those without disorders. Since education and awareness-raising are the best strategies to combat this type of difficulty, we consider the need to design interventions that can be applied at this age. Given this issue, it is essential to intervene as early as possible, bearing in mind that many psychological disorders make their debut during adolescence and youth, particularly those related to the depressive-anxiety sphere (Del Barrio, 2010). Failure to address mental disorders and their effects could have pernicious effects that extend into adulthood and could undermine the chances of leading a fulfilling life. To cite other effects of stigma, the internalization of stereotypes and prejudices, and the impact on the self-esteem and personal and social image of the person and their environment (López et al., 2008), could lead to internalized self-stigma or stigma. Moreover, stigma intimately affects the formation of our identity and that of others (Vilardaga, 2012). It is nothing more than a heuristic process that allows us to define ourselves and others by economizing on our cognitive resources. That is a simplified and automatic construction of others to predict and manipulate our social environment.

Research in this sense is essential, as it will open up new lines of research into the mechanisms underlying the processes of deformation and development of stigmatizing attitudes, the specific moment in evolutionary development at which it occurs, and what type of procedures are most useful in tackling this problem. In this sense, the present work provides further evidence to support the use of innovative interventions to fight stigma. Given that most of the existing anti-stigma programs are based on educational talks, videos, and presentations of what mental disorders are, their origin, and how to deal with them, this study is proof of the benefits of VR as an educational tool.

REFERENCES

Angermeyer, M. C., & Matschinger, H. (2005). Labelling -Stereotype- Discrimination. An Investigation of the Stigma Process. *Social Psychiatry and Psychiatric Epidemiology*, *40*, 391–395. doi:10.100700127-005-0903-4 PMID:15902409

Aznar-Díaz, I., Romero-Rodríguez, J., & Rodríguez-García, A. M. (2018). La tecnología móvil de realidad virtual en educación: Una revisión del estado de la literatura científica en España. *EDMETIC*, *7*(1), 256–274. doi:10.21071/edmetic.v7i1.10139

Aznar-Lou, I., Serrano-Blanco, A., & Luciano, J. V. (2017). Effectiveness of the "what's up!" intervention to reduce stigma and psychometric properties of the Youth Program Questionnaire (YPQ): Results from a cluster non-randomized controlled trial conducted in Catalan high schools. *Frontiers in Psychology*, *8*, 1608. doi:10.3389/fpsyg.2017.01608 PMID:28959228

Baba, Y., Nemoto, T., Tsujino, N., Yamaguchi, T., Katagiri, N., & Mizuno, M. (2017). Stigma toward psychosis and its formulation process: Prejudice and discrimination against early stages of schizophrenia. *Comprehensive Psychiatry*, *73*, 181–186. doi:10.1016/j.comppsych.2016.11.005 PMID:28040575

Barry, C. L., McGinty, E. E., Pescosolido, B. A., & Goldman, H. H. (2014). Stigma, discrimination, treatment effectiveness, and policy: Public views about drug addiction and mental illness. *Psychiatric Services (Washington, D.C.)*, *65*(10), 1269–1272. doi:10.1176/appi.ps.201400140 PMID:25270497

Bentham, C., Daunt, A., Taylor, S., & Simmons, M. (2013). Mental health workshops delivered by medical students in Cambridge secondary schools: An evaluation of learning. *Psychiatria Danubina*, *25*, 224–230. PMID:23995181

Bulanda, J. J., Bruhn, C., Byro-Johnson, T., & Zentmyer, M. (2014). Addressing mental health stigma among young adolescents: Evaluation of a youth-led approach. *Health & Social Work*, *39*(2), 73–80. doi:10.1093/hsw/hlu008 PMID:24946423

Cangas, A. J., & Galván, B. (2020). Estudio preliminar de la utilidad de un programa de realidad virtual contra el estigma en salud mental. *International Journal of Developmental and Educational Psychology*, *1*(1), 45–52. doi:10.17060/ijodaep.2020.n1.v1.1760

Cangas, A. J., Navarro, N., Aguilar-Parra, J. M., Trigueros, R., Gallego, J., Zárate, R., & Gregg, M. (2019). Analysis of the Usefulness of a Serious Game to Raise Awareness about Mental Health Problems in a Sample of High School and University Students: Relationship with Familiarity and Time Spent Playing Video Games. *Journal of Clinical Medicine*, *8*(10), 1504. doi:10.3390/jcm8101504 PMID:31547017

Cangas, A. J., Navarro, N., Parra, J., Ojeda, J. J., Cangas, D., Piedra, J. A., & Gallego, J. (2017). Stigma-Stop: A serious game against the stigma toward mental health in educational settings. *Frontiers in Psychology*, *8*, 1385. doi:10.3389/fpsyg.2017.01385 PMID:28878702

Casañas, R., Arfuch, V. M., Castellví, P., Gil, J. J., Torres, M., Pujol, A., & Caussa, A. (2018). Espai-Jove. net"-a school-based intervention programme to promote mental health and eradicate stigma in the adolescent population: Study protocol for a cluster randomized controlled trial. *BMC Public Health*, *18*(1), 939. doi:10.118612889-018-5855-1 PMID:30064404

Clement, S., Schauman, O., Graham, T., Maggioni, F., Evans-Lacko, S., Bezborodovs, N., & Thornicroft, G. (2015). What is the impact of mental health-related stigma on help-seeking? A systematic review of quantitative and qualitative studies. *Psychological Medicine*, *45*(1), 11–27. doi:10.1017/S0033291714000129 PMID:24569086

Corrigan, P., Markowitz, F. E., Watson, A., Rowan, D., & Kubiak, M. A. (2003). An attribution model of public discrimination towards persons with mental illness. *Journal of Health and Social Behavior*, *44*(2), 162–179. doi:10.2307/1519806 PMID:12866388

Del Barrio, V. (2010). Salud mental infanto-juvenil: Requisitos para su evaluación, diagnóstico precoz y prevención. Infocoponline. *Revista de Psicología*, 13–19.

Dolphin, L., & Hennessy, E. (2017). Labelling effects and adolescent responses to peers with depression: An experimental investigation. *BMC Psychiatry*, *17*(1), 228. doi:10.118612888-017-1389-9 PMID:28646873

Dos Santos, M. J., & Dos Santos, C. A. (2019). Virtual reality in the classroom: Geography teaching practice. *Geosaberes*, *10*(22), 72–80. doi:10.26895/geosaberes.v10i22.814

Fombona, J., & Pascual, M. Á. (2017). La producción científica sobre Realidad Aumentada, un análisis de la situación educativa desde la perspectiva SCOPUS. *EDMETIC*, *6*(1), 39–61. doi:10.21071/edmetic.v6i1.5807

Georgakakou-Koutsonikou, N., & Williams, J. M. (2017). Children and young people's conceptualizations of depression: A systematic review and narrative meta-synthesis. *Child: Care, Health and Development*, *43*(2), 161–181. doi:10.1111/cch.12439 PMID:28090667

Gregg, L., & Tarrier, N. (2007). Virtual reality in mental health: A review of the literature. *Social Psychiatry and Psychiatric Epidemiology*, *42*(5), 343–354. doi:10.100700127-007-0173-4 PMID:17431528

Huttar, C. M., & Brintzenhofeszoc, K. (2019). Virtual Reality and Computer Simulation in Social Work Education: A Systematic Review. *Journal of Social Work Education*, •••, 1–11. doi:10.1080/10437797.2019.1648221

Lansu, T. A. (2018). Burn! How implicit and explicit attitudes predict early adolescents "hot sauce" aggression toward classroom peers. *Journal of Experimental Child Psychology*, *167*, 423–432. doi:10.1016/j.jecp.2017.11.006 PMID:29241999

López, M., Laviana, M., Fernández, L., López, A., Rodríguez, A. M., & Aparicio, A. (2008). La lucha contra el estigma y la discriminación en salud mental: Una estrategia compleja basada en la información disponible. *Revista de la Asociación Española de Neuropsiquiatría*, *101*(1), 43–83. doi:10.4321/S0211-57352008000100004

Lorenzo-Álvarez, R., Rudolphi-Solero, T., Ruiz-Gómez, M. J., & Sendra-Forte, F. (2019). Medical Student Education for Abdominal Radiographs in a 3D Virtual Classroom Versus Traditional Classroom: A Randomized Controlled Trial. *AJR. American Journal of Roentgenology*, *213*(3), 644–650. doi:10.2214/AJR.19.21131 PMID:31287725

Major, B., McCoy, S., Kaiser, C., & Quinton, W. (2003). Prejudice and self-esteem: A transactional model. *European Review of Social Psychology, 14*(1), 77–104. doi:10.1080/10463280340000027

Marín, V., & Muñoz, V. P. (2018). Trabajar el cuerpo humano con realidad aumentada en educación infantil. *Revista Tecnología. Ciência & Educação (Bauru), 9*, 148–158.

Mellor, C. (2014). School-based interventions targeting stigma of mental illness: Systematic review. *Psychiatric Bulletin, 38*(4), 164–171. doi:10.1192/pb.bp.112.041723 PMID:25237538

Moreno, N., Leiva, J., Galván, M., López, E., & García, F. (2017). Realidad aumentada y realidad virtual para la enseñanza-aprendizaje del inglés desde un enfoque comunicativo e intercultural. In J. Ruiz Palmero, J. Sánchez-Rodríguez, & E. Sánchez-Rivas (Eds.), *Innovación docente y uso de las TIC en educación* (pp. 1–11). UMA Editorial.

Muijen, M. (2012). The state of psychiatry in Europe: Facing the challenges, developing consensus. *International Review of Psychiatry (Abingdon, England), 24*(4), 274–277. doi:10.3109/09540261.2012.697264 PMID:22950764

Muijen, M. (2012). The state of psychiatry in Europe: Facing the challenges, developing consensus. *International Review of Psychiatry (Abingdon, England), 24*(4), 274–2777. doi:10.3109/09540261.2012.697264 PMID:22950764

Peterson, D., Pere, L., & Sheehan, N. (2004). *Respect cost nothing: a survey of discrimination faced by people with experience of mental illness in Aotearona*. Mental Health Foundation of New Zealand.

Pinto-Foltz, M. D., Hickman, R., Logsdon, M. C., & Burant, C. H. (2012). Psychometric evaluation of the revised attribution questionnaire (r-AQ) to measure metal illness stigma in adolescents. *Journal of Nursing Measurement, 20*(1), 47–58. doi:10.1891/1061-3749.20.1.47 PMID:22679709

Quinn, D. M., Williams, M. K., & Weisz, B. M. (2015). From discrimination to internalized mental illness stigma: The mediating roles of anticipated discrimination and anticipated stigma. *Psychiatric Rehabilitation Journal, 38*(2), 103–108. doi:10.1037/prj0000136 PMID:25844910

Radu, I. (2014). Augmented reality in education: A meta-review and cross-media analysis. *Personal and Ubiquitous Computing, 18*(6), 1533–1543. doi:10.100700779-013-0747-y

Rodríguez, A. M., Cáceres, M. P., & Alonso, S. (2018). La competencia digital del futuro docente: Análisis bibliométrico de la productividad científica indexada en Scopus. *International Journal of Educational Research and Innovation. IJERI, 10*, 317–333. doi:10.37954.e.v6i1.164

Rüsch, N., Angermeyer, M., & Corrigan, P. (2005). Mental illness stigma: Concepts, consequences and initiatives to reduce stigma. *European Psychiatry, 20*(8), 529–539. doi:10.1016/j.eurpsy.2005.04.004 PMID:16171984

Schachter, H. M., Girardi, A., Ly, M., Lacroix, D., Lumb, A. B., van Berkom, J., & Gill, R. (2008). Effects of school based interventions on mental health stigmatization: A systematic review. *Child and Adolescent Psychiatry and Mental Health, 2*(1), 18–32. doi:10.1186/1753-2000-2-18 PMID:18644150

Schülze, B., Richter-Werling, M., Matschinger, H., & Angermeyer, M. (2003). Crazy? So what! Effects of a school project on students' attitudes towards people with schizophrenia. *Acta Psychiatrica Scandinavica, 107*(2), 142–150. doi:10.1034/j.1600-0447.2003.02444.x PMID:12534440

Silke, C., Swords, L., & Heary, C. (2017). The predictive effect of empathy and social norms on adolescents' implicit and explicit stigma responses. *Psychiatry Research, 257*, 118–125. doi:10.1016/j.psychres.2017.07.033 PMID:28753461

Slater, P., Hasson, F., Guillen, P., Gallen, A., & Parlour, R. (2019). Virtual simulation training: Imaged experience of dementia. *International Journal of Older People Nursing, 14*(3), 1–11. doi:10.1111/opn.12243 PMID:31070851

Spitzer, A., & Cameron, C. (1995). School-age children's perceptions of mental illness. *Western Journal of Nursing Research, 17*(4), 398–415. doi:10.1177/019394599501700405 PMID:7660662

Swords, L., Heary, C., & Hennessy, E. (2011). Factors associated with acceptance of peers with mental health problems in childhood and adolescence. *Journal of Child Psychology and Psychiatry, and Allied Disciplines, 52*(9), 933–941. doi:10.1111/j.1469-7610.2010.02351.x PMID:21223258

Vilardaga, R. (2012). Estigma y prejuicios. In M. Páez-Blarrina & O. Gutiérrez-Martínez (Eds.), *Múltiples Aplicaciones de la Terapia de Aceptación y Compromiso* (pp. 251–267). Pirámide.

Villalustre, L., & Del Moral, M. E. (2017). Juegos perceptivos con realidad aumentada para trabajar contenido científico. *Educação. Formação & Tecnologias, 10*(1), 36–46.

Viñals, A., & Cuenca, J. (2016). El rol del docente en la era digital. *Revista Interuniversitaria de Formación del Profesorado, 30*(2), 103–114.

Watson, A. C., Ottati, V. C., Morabito, M. S., Draine, J., Kerr, A. N., & Angell, B. (2010). Outcomes of pólice contacts with persons with mental illness: The impact of CIT. *Administration and Policy in Mental Health, 37*(4), 302–317. doi:10.100710488-009-0236-9 PMID:19705277

Yang, J., Cervera, R. L., Tye, S. J., Ekker, S. C., & Pierret, C. (2018). Adolescent mental health education InSciEd Out: A case study of an alternative middle school population. *Journal of Translational Medicine, 16*(1), 84. doi:10.118612967-018-1459-x PMID:29615090

Yang, X. Z., Lin, L., Cheng, P. Y., Yang, X., & Ren, Y. (2019). Which EEG feedback works better for creativity performance in immersive virtual reality: The reminder or encouraging feedback? *Computers in Human Behavior, 99*, 345–351. doi:10.1016/j.chb.2019.06.002

KEY TERMS AND DEFINITIONS

Empathy: The ability to share someone else's feelings or experiences by imagining what it would be like to be in that person's situation.

Mental Disorder: Is characterized by a clinically significant disturbance in an individual's cognition, emotional regulation, or behaviour. It is usually associated with distress or impairment in important areas of functioning. There are many different types of mental disorders. Mental disorders may also be referred to as mental health conditions.

Mental Health: Is a state of mental well-being that enables people to cope with the stresses of life, realize their abilities, learn well and work well, and contribute to their community. It is an integral component of health and well-being that underpins our individual and collective abilities to make decisions, build relationships and shape the world we live in. Mental health is a basic human right. And it is crucial to personal, community and socio-economic development.

Self-Esteem: Is how we value and perceive ourselves. It's based on our opinions and beliefs about ourselves, which can feel difficult to change. We might also think of this as self-confidence.

Social Inclusion: Is the process by which efforts are made to ensure equal opportunities—that everyone, regardless of their background, can achieve their full potential in life. Such efforts include policies and actions that promote equal access to (public) services as well as enable citizen's participation in the decision-making processes that affect their lives.

Stigma: Is when someone views you in a negative way because you have a distinguishing characteristic or personal trait that's thought to be, or actually is, a disadvantage (a negative stereotype). Unfortunately, negative attitudes and beliefs toward people who have a mental health condition are common. Stigma can lead to discrimination.

Virtual Reality Program: Is a computer-generated environment with scenes and objects that appear to be real, making the user feel they are immersed in their surroundings.

Chapter 13

Creative Writing of Disinformation Narratives:

Teaching Innovation Using Artificial Intelligence

Anastasio García Roca
https://orcid.org/0000-0002-2277-3034
University of Almería, Spain

Ana Manzano León
https://orcid.org/0000-0001-6966-0355
University of Almería, Spain

José M. Rodríguez Ferrer
https://orcid.org/0000-0003-1811-0040
University of Almería, Spain

Rubén Camacho Sánchez
University of Almería, Spain

ABSTRACT

The ubiquity of new portable devices with internet access and recent events such as the global pandemic caused by COVID-19 and the Russian invasion of Ukraine has highlighted the importance of media and information literacy. In parallel, developing tools based on artificial intelligence will allow the developing of sophisticated disinformative content (deepfakes). This media context demands new and refined informational competences to identify and verify information. This research presents a proposal to develop media and information literacy using AI tools to create audiovisual content. Strategies such as flipped classrooms, thematic debates, and playful activities are used. A selection of evaluation instruments is presented, which, through a pre-and post-experimental design, will allow the evaluation of this proposal and the learning developed by the students.

DOI: 10.4018/978-1-6684-8861-4.ch013

INTRODUCTION

Disinformation Narratives

The digital era has brought about a myriad of transformative changes across different facets of society, leaving a profound impact on the economic, social, and cultural domains (Del-Fresno, 2019). Undeniably, our daily routines and behaviors have been significantly influenced by the widespread presence of electronic devices with Internet connectivity (Amo & Baldrich, 2021). The ability to access a vast repository of knowledge with just a few clicks has become a remarkable feature of our lives, offering instant and up-to-date information. This phenomenon has bestowed numerous advantages, closely linked to the development and education of individuals, empowering them in unprecedented ways.

However, amid the advancements of the information society, there exists a crucial distinction between mere information and genuine knowledge—a separation between the technical progress represented by the former and the transformative potential of the latter in shaping human beings (López-Borrull, Vives-Gràcia, & Badell, 2018). Both concepts, the information society and the knowledge society, hold immense promise for societal, cultural, economic, and democratic progress. It is undeniable that the Internet stands as the key facilitator in navigating this transition. Nevertheless, it is of utmost importance to recognize that while information serves as a powerful tool for acquiring knowledge, it does not encompass the entirety of what knowledge truly entails (UNESCO, 2005, p. 19).

In essence, the digital age has unlocked unprecedented opportunities for individuals to access and utilize vast amounts of information. This accessibility holds the potential to enhance learning, foster innovation, and promote social development. However, it is vital to acknowledge that true knowledge transcends beyond a mere accumulation of facts and figures. Knowledge involves critical thinking, interpretation, and the ability to apply information effectively in various contexts. Achieving a knowledge society necessitates nurturing these cognitive skills, promoting deep understanding, and fostering a culture of continuous learning.

While the digital revolution has indeed democratized access to information, it is imperative to bridge the gap between information and knowledge. This involves equipping individuals with the necessary skills to analyze, synthesize, and evaluate the information they encounter. Moreover, it calls for promoting a sense of digital literacy that encompasses not only technical competencies but also ethical considerations, such as media literacy and responsible online behavior.

The continuous creative circulation of information and the establishment of new, more flexible and decentralized interaction structures have posed a challenge for constructing the knowledge society model that defines the digital era (Chartier, 2018). In today's ever-changing, dynamic, plural, and highly technologized world, (hyper)information needs to be processed by users to be transformed into knowledge (UNESCO, 2005). In this sense, the epistemic function takes priority in the mechanisms of textual production and reception (Caro-Valverde, 2018; Millán, 2008).

Recent historical events, such as the global pandemic caused by Covid-19 or the Russian invasion in Ukraine, have underscored the vital importance of Media and Information Literacy: these are events in which information and communication are overwhelming and constantly changing and updating (Golovchenco, Mareike, & Adler-Nissen, 2018). In addition to traditional media that follow a top-down logic, social networks are taken as dissemination channels and sources of information. In these communication channels, hoaxes and fake news about political and governmental issues are transmitted (Salaverría et al., 2020; Destiny & Omar, 2020; Jenkins et al., 2013). Furthermore, many media transmit messages that require scientific literacy and geopolitical and historical knowledge to comprehend fully.

All this has emphasized the importance of literacy (in its broadest sense) in priority issues such as individual and community health (Brørs, Norman, & Norekvål, 2020). From this perspective, concepts such as health literacy (Sentell et al., 2020; Paakkari, and Okan, 2020); ehealth literacy (Brørs et al., 2020; Chong et al., 2020); social media literacy (Puri et al., 2020); scientific literacy (Vraga et al., 2020); information literacy (Naaem and Bhatti, 2020); digital literacy (Tejedor et al., 2020; Sánchez-Cruzado et al., 2021), etc., have gained greater relevance. Being functionally literate, understood as the ability to act effectively in one's community (UNESCO, 2005), implies acquiring new knowledge, skills, and attitudes linked to the digital era. Thus, 21st-century society and digital transformation require new competencies linked to (multi)literacy (New London Group, 1996), a concept that is constantly being updated due to its close connection with the sociocultural context in which it is used (Knobel and Lankshear, 2006; Cassany and Castellà, 2010).

Weiss et al. (2020, pp. 3-7) list the main factors that favor the dissemination and virtualization of this type of content:

- *as a result of information overload and "the principle of least effort."* Fake news, and other misinformative content, use a much more didactic, clear, and direct discourse, with headlines that exploit the curiosity gap and induce (mis)interpretations with a register close to that used by the recipients. In this sense, the dissemination channel is decisive: they are transmitted between users (*peer to peer*) through social networks and therefore these contents are not received almost unconsciously.
- *as a result of poisoned public discourse, logical fallacies, and overconfidence.* Misinformative narratives tend to generate emotions (generally negative) in the receiver/reader to provoke his/her participation in the dissemination of (mis)information.
- *as context-independent in a "Post-truth" society.* Narratives are more relevant than data because they have a much more obvious impact on the readership. Data are replaced by personal stories.
- *as propaganda/disinformation.* Generally, misinformative contents hide economic interests or political returns.
- *as a rumor, misinformation, and conspiracy theories.* Fake news and other misinformative content is presented as truth silenced by political or economic elites.
- *as parody, satire, and political kayfabe.*

News verification agencies and educational efforts have not proven sufficient in combating the spread of disinformation messages. In response, several European Union countries have resorted to censorship measures, restricting access to Russian media outlets like Sputnik and Russia Today. However, the challenge persists as social networks continue to play a significant role in shaping the information landscape (Newman et al., 2021), and they have become a fertile ground for the dissemination of fake news (Allcott and Gentzkow, 2017).

Interestingly, the average user believes they possess a reasonable level of confidence in identifying fake news (Eurometer, 2018). Unfortunately, disinformation tactics are evolving, becoming more intricate and sophisticated with the use of bots and Deepfakes (Tolosa et al., 2020), making it increasingly challenging to distinguish misinformative content from genuine information.

In this context, literacy, in its broadest sense, emerges as the most effective firewall against misinformation. UNESCO, in 2008, adopted the comprehensive term "Media and Information Literacy," encompassing interdependent and convergent competencies necessary to navigate communications and content through various channels such as libraries, media, and internet platforms (UNESCO, 2022). Among others, they integrate:

Table 1. Media and information literacy ecology. Extracted from Wilson et al. (2021)

Media literacy	Computer literacy
Information literacy	Internet Literacy
Freedom of expression literacy	Digital literacy
Freedom of information literacy	Film literacy
Library literacy	Game literacy
News literacy	

Media literacy refers to the ability to analyze, interpret and produce messages in different media, such as television, radio, newspapers, and social networks, among others (Livingstone, 2008). This literacy includes understanding the mechanisms of media production and distribution, as well as the ability to detect disinformation and manipulation of information. Information literacy refers to the ability to search for, evaluate and use information effectively. This includes the understanding of different sources of information, the ability to use search tools, and the ability to evaluate the relevance and reliability of information.

CREATIVE WRITING AND ARTIFICIAL INTELLIGENCE

The democratization of new technologies has significantly impacted our understanding of concepts such as creator, user-reader, and creativity. In this rapidly evolving landscape, as pointed out by Casacuberta (2003), the Internet has transformed the role of the creator into that of the public. The network is now comprised of a vast and intricate web of decentralized information nodes, where numerous individuals actively contribute to the creation and dissemination of content.

These independent users engage in various forms of expression, including blogs, social networks, personal websites, and more. Through these platforms, they share their unique perspectives, ideas, and information in the form of fragmented pieces. This shift has given rise to a media environment characterized by a much more horizontal distribution, in stark contrast to the traditional top-down approach of broadcasting.

As the internet empowers individuals to become creators, the distinction between creators and consumers has blurred. Users are no longer passive recipients of information but active participants in shaping the digital landscape. They contribute to the creation of collective intelligence, where the knowledge and insights of many interweave to form a rich tapestry of ideas.

Moreover, the concept of creativity itself has transformed. In the traditional sense, creativity was often associated with professionals in specific fields who possessed specialized skills and resources. However, in the age of democratized technologies, creativity has become accessible to anyone with an internet connection. Individuals can explore and experiment with various mediums, leveraging digital tools to bring their ideas to life and share them with the world.

The democratization of new technologies has opened up unprecedented opportunities for expression, collaboration, and innovation. It has broken down barriers to entry and given voice to diverse perspectives that were previously marginalized or overlooked. As a result, the landscape of content creation has become more inclusive, dynamic, and representative of the vast array of human experiences. To this we must add the ideals of the culture of convergence (Jenkins, 2008) and the culture of participation (Jenkins et al., 2015), (dis)information society, knowledge society. These, in turn, are related to countless concepts such as *transgender (*Scolari, 2015), social reader, digital reader, forensic reader, text pirate (Jenkins, 2008), *prosumer*, *coauthor* or *wreader*, *voter*, *hyperreader*, etc (García-Roca, 2021a, 2021b). This hyper-information and conceptual oversaturation is evidence of the vertiginous and profound cultural and social changes introduced by the new information and communication technologies.

From this perspective, digital reading, as summarized by García-Roca and Amo (2023), is multimodal (it combines a multitude of languages such as written text, static and moving images, audio, etc.), hypertextual (some texts refer to others following a rhizomatic structure) (Landow, 2009), personalized (the contents offered by the most popular search engines adapt to sociodemographic and socio-demographic variables, and are personalized) (Landow, 2009).), *hypertextual* (some texts refer to others following a rhizomatic structure) (Landow, 2009), personalized (the contents offered by the most popular search engines are adapted to the sociodemographic variables and interests of users, with the dangers implied by the bubble filter[1]), social (with *peer to peer* or *bottom up* contents, unlike traditional media that follow a *top-down* logic), ubiquitous (it is found in our pockets, computers, virtual assistants at home, cars, etc.), immediate and updated (the news are in our pockets, computers, virtual assistants at home, cars, etc.), immediate and up to date (the news are always in our pockets, computers, virtual assistants at home, cars, etc.), and are always in the hands of the user (the news are always in our pockets, computers, cars, etc., immediate and up-to-date (news and information have too short an obsolescence and are quickly buried by algorithms), horizontal and open (editorial intermediaries are eliminated) and comprehensive and immersive (with the global ideal of *metaverse*).

In this sense, the PISA 2022 report (OECD, 2022) measured the evolution of knowledge and skills in the context of the ubiquity of Information and Communication Technologies, as well as the ability to respond to the demands of a changing world in which innovation and creativity are backbone elements for adaptation. The development of creative and critical thinking through reading and writing skills must be considered as factors of great relevance (García-Roca, Manzano-León, & Rodríguez-Ferrer, 2023).

The creative possibilities of writers and artists have multiplied with new modes of expression and experimentation. Digital tools have facilitated the dissemination of written texts, video recordings, audio editing, graphic design, etc. In this sense, new tools based on creative Artificial Intelligence (hereafter AI) such as *Dall-e 2*, *Stable Diffusion,* and *Midjourney* stand out. These are image-generative *AIs* that rely on other images shared on the network. Other AIs (*Vall-e*, developed by *Microsoft*) are capable of imitating voices with a sample time of only three seconds. *Meta*, meanwhile, is developing another *AI* (*Make-A-Video*) that automates the creation of short videos from written text. The emergence of AI-generated compositions brings about unique results, as each composition is never repeated. However, this new frontier also brings legal uncertainty regarding copyright. With a simple operation, users can input a description of an image, and the AI will swiftly design it within seconds. Technological advancements in this field are progressing at an astonishing pace.

Nevertheless, these tools also introduce certain risks, as they enable the creation of manipulated images, videos, and audio that are easily accessible to all users. The ease with which false and misleading audiovisual content can be produced is a cause for concern. Deepfakes, for instance, are becoming increasingly sophisticated, with a rapid improvement in accuracy and realism. It is becoming progressively more challenging to differentiate between a fake video, image, or audio and a genuine one. It is important to note that spreading false information is much simpler than verifying and authenticating it.

The proliferation of manipulated content has significant implications for society. It can undermine trust, distort reality, and contribute to the spread of misinformation. Detecting and addressing the authenticity of media content has become an urgent priority. Efforts are being made to develop technologies and techniques that can identify deepfakes and counter their harmful effects. Additionally, legal frameworks are being explored to regulate the use of AI-generated content and ensure accountability for its creation and dissemination. It is crucial for users to be vigilant and critical consumers of media, understanding the potential for manipulation and the need for fact-checking. Promoting media literacy and digital literacy skills is vital to navigating this complex landscape effectively. By enhancing our ability to discern between genuine and manipulated content, we can mitigate the risks associated with AI-generated compositions and foster a more informed and resilient society.

In summary, while AI-generated compositions offer unique and rapidly produced results, they also pose legal uncertainties regarding copyright. The accessibility and simplicity of these tools introduce risks, including the creation of manipulated content such as deepfakes. The difficulty in distinguishing between real and fake media content calls for increased efforts in detection and regulation. Moreover, fostering media literacy and critical thinking skills is essential in navigating this evolving landscape and combatting the spread of misinformation.

TEACHING INNOVATION IN TEACHING TRAINING

This proposal will be developed in the subjects in which the signatories teach, but especially in the subject of *Didactics of Language and Literature* corresponding to the degree of *Primary Education Teacher* at the University of Almería. The main objective of this subject is to provide students with knowledge, skills, and attitudes that favor the promotion and a positive approach to the learning of the Spanish language and its literature for children and young people. According to the title, the training of Primary Education teachers must guarantee the acquisition of professional competencies, among which the most relevant are those that contribute to promoting an Education for Democracy. In addition, and accordance with the Spanish Framework of Qualifications for Higher Education (MECES), it is guaranteed that students:

- Act as mediators so that all the activity carried out is meaningful and stimulates the development potential of each student in cooperative group work.
- Have the ability to design and organize disciplinary and interdisciplinary work and collaboration with the world outside the school.

- Be professionals capable of analyzing the context in which their activity is developed and planning it, of responding to a changing society.
- Have training supported by a methodology appropriate to the paradigm of the research teacher.

Following the regulations of the University of Almeria, the basic competencies were defined in the R.D. 1393/2007, of October 29, and are aimed at the acquisition by the student of general training, in one or more disciplines, oriented to the preparation for the exercise of professional activities. They will be adapted progressively and according to what is established by the current R.D. 822/2021, of September 28, which establishes the organization of university education and the procedure for quality assurance.

- Ability to make judgments. Students should have the ability to gather and interpret relevant data (usually within their area of study) to make judgments that include a reflection on relevant social, scientific, or ethical issues.
- Ability to communicate and social skills. Students should be able to convey information, ideas, problems, and solutions to both specialized and non-specialized audiences.

The transversal competencies of the University of Almeria, approved by the Governing Council on June 17, 2008, are the following:

- Basic knowledge of the profession. Knowledge, skills, and attitudes that enable the understanding of new theories, interpretations, methods, and techniques within the different disciplinary fields, leading to optimally meeting professional requirements.
- Ability in the use of ICT. Use Information and Communication Techniques (*ICTs*) as a tool for expression and communication, for access to information sources, as a means of archiving data and documents, for presentation tasks, for learning, research, and cooperative work.
- Oral and written communication in one's language. Understand to express with clarity and opportunity ideas, knowledge, problems, and solutions to a wider audience, specialized or non-specialized (and feelings through words, adapting to the characteristics of the situation and the audience to achieve understanding and adhesion).
- Social competence and global citizenship. Respect fundamental rights and equality between men and women, human rights, the values of a culture of peace and democracy, environmental principles, and development cooperation that promote an ethical commitment in a global, intercultural, free, and just society.

Therefore, this proposal seeks to address the challenge of improving the identification of fake news, manipulative content, (un)informative data, and false images, audio, and videos generated by artificial intelligence. The objective is to develop an exceptionalist attitude towards potentially dangerous content, emphasizing the need for heightened scrutiny and critical evaluation.

To achieve this, a multi-faceted approach is suggested. Firstly, it is crucial to invest in research and development efforts to enhance the detection and verification mechanisms for AI-generated content. This includes advancing technologies that can analyze and assess the authenticity of media, and identifying patterns and anomalies that may indicate manipulation.

Furthermore, collaboration between technology companies, researchers, and policymakers is essential. By fostering partnerships, it becomes possible to share expertise, resources, and insights to develop effective strategies for combating manipulated content. This collaboration can also lead to the establishment of industry standards and best practices that address the challenges posed by AI-generated content.

Another important aspect of this proposal is the promotion of media literacy and digital literacy education. By equipping individuals with the skills to critically evaluate information and recognize potential manipulation, they can make informed decisions and contribute to a more resilient information ecosystem. This education should target not only the general public but also journalists, educators, and other key stakeholders who play a critical role in disseminating accurate information. As well, raising awareness about the existence and implications of AI-generated content is vital. This includes educating the public about the risks associated with deepfakes, manipulated images, and other forms of AI-generated media. By understanding the potential dangers, individuals can develop a more discerning approach to consuming and sharing content, thus mitigating the spread of misinformation.

In addition to technological advancements and education, legal and regulatory frameworks must be developed or adapted to address the challenges posed by AI-generated content. This includes establishing guidelines for the responsible use of AI technologies, ensuring accountability for the creation and dissemination of manipulated content, and defining the boundaries of intellectual property rights in this context.

OBJECTIVES

General

This proposal is based on the research results obtained by Amo and García-Roca (2023), which concluded that a training program aimed at developing this competency framework in the field of lifelong learning should be intensified as a priority line of action. Training critical and creative users implies:

- To develop the ability to question ideas, hypotheses, or legitimized practices, analyzing different points of view on a problem and reflecting on the position adopted, taking into account the different possible alternatives.
- To foster the ability to develop novel and original products, as well as to use innovative and personal ideas, establishing connections with other fields of knowledge.

In relation to the above, this proposal aims to develop students' creative and critical thinking, as well as to offer useful tools for its implementation in teaching in the Primary Education cycles.

Specific

The above general objectives are crystallized in the following operational objectives:

- To know the main misinformative manifestations existing in the network.
- To identify the manipulative communicative intent of fake news headlines.

- To recognize the basic discursive structure of *cyber-bait-based* headlines that explore the reader's curiosity gap.
- To write a misinformative news item with which to manipulate a sector of the population about a topic of current social relevance.
- To create false evidence based on IIAA about fictional events.

METHODOLOGY

The Participants

The study to be carried out will be quasi-experimental, in which groups A and B of *Didactics of Language and Literature* will be used as the experimental group in which this proposal will be implemented, and groups A of the subjects *Socio-educational Programs for Children, Youth, Adults and the Elderly* (Degree in Social Education) and *Family, Society and Social Networks* (Degree in Psychology) as control groups in which the contents and methodologies established by the respective teaching guides will be maintained.

An exploratory pilot experience was carried out in the form of a workshop in the subject of Didactics of Language and Literature. This workshop consisted of two face-to-face sessions in which students learned about the main manifestations of the disinformation society. During the second session, the *ChatGPT* artificial intelligence-based chatbot was experimented with. Students will be able to check the ease of use of the program and the possibilities to create false news with a high degree of reliability.

As mentioned above, the following proposal is developed in the subjects taught by the teaching staff of the *Department of Education* of the University of Almeria, *Department of Developmental and Evolutionary Psychology of the* Universities of Almeria and Jaen. Specifically, these activities are proposed in the subjects of *Didactics of Language and Literature* (Primary Education teachers Degree), although work is also carried out in subjects such as *Socio-educational Programs in Childhood, Youth, Adults and the Elderly* (Social Education Degree) and *Family, Society and Social Networks* (Psychology Degree). It is, therefore, an interdisciplinary, interdepartmental and inter university proposal. In either case, the students share a profile as young people between 19 and 21 years old, studying education or behavioral sciences. Each practical group consists of about 30 students. The small class size fosters personalized attention and a dynamic learning environment. The hands-on approach encourages critical thinking and problem-solving skills.

Procedure and Activities

The study will begin with an exhaustive and systematic literature review of empirical research that assesses the level of informational and media competence of citizens on critical issues. For this purpose, the model developed by Codina (2018) and the *PRISMA* (*Preferred Reporting Items for Systematic Reviews and Meta-Analyses*) statement (Page et al., 2021) will be followed. Specifically, a systematic search and review has been performed as it combines the strengths of critical review with a comprehensive search process (Grant and Booth, 2009). This analysis will allow the identification of a reliable assessment and data collection instrument or will facilitate the design of a reliable and valuable instrument to be applied at the end of the different subjects.

Once the instrument has been selected or designed, the different training activities that make up this proposal will be developed following the *Project Based Learning* (PBL) model. The final task will consist of creating (dis)informative content (this should have a specific economic interest more or less hidden) using different discursive strategies typical of these formats, with evidence created from *Artificial Intelligence*. To do so, the following sequence of activities will be followed:

1. *Flipped classroom*: on the web we find illustrative videos about contents based on information disorders such as *fake news*, hoaxes, *deepfakes*, headlines based on *cyber hoaxes*, *post-truth* expressions, etc. In order to receive *feedback* (and not so much as evaluation) and improve student motivation and participation, tools such as *Kahoot!* and *Wooclap* will be used.
2. *Thematic debates*: different real and false contents will be exposed to verify their veracity. It will not be as relevant to identify it as to use the appropriate arguments: feelings generated in the readers, undetermined sources of the article, etc.
3. *Creation of narratives based on information disorders*: in order to deepen the concept of *post-verdict*, students will be asked to create a metafictional story that generates strong feelings (manipulation strategy) in the reader.
4. *Construction of audiovisual evidence*. For this purpose, tools such as *Dall-e*, *Vall-e*, *Stable Diffusion* and *Midjourney* or *Make-A-Video* will be used. The idea of this activity is related to the ease of creating material based on *deepfakes*.
5. *Oral presentation of the project*. Students must present the project, as well as the audiovisual evidence that supports and justifies the news or story. Likewise, they must justify the viralization strategies that would be followed.

The evaluation of this project will be through a rubric in which not only the oral skills of the students will be evaluated, but also the plausibility of the proposal, the adequacy of the evidence, the manipulative nature, as well as the potential for virtualization. This evaluation will be continuous and will be weighted with a self-evaluation and a hetero-evaluation. Once the projects have been completed and presented, the instrument designed to evaluate the students' media and information literacy will be passed around.

Results

The results will depend on the instrument obtained in the systematized search in the two main scientific databases: *Web of Science* (*core collection*) and *SCOPUS*. The following descriptors will be used for the search in the title, abstract, and keywords:

- (Information OR Media OR News OR digital) AND (Literacy OR competence) AND (Level OR Measure* OR Evalua*)
- (((((((TS=(information)) OR TS=(media)) OR TS=(news)) OR TS=(digital)) AND TS=(literacy)) OR TS=(competence)) AND TS=(level)) OR TS=(measure*)) OR TS=(evaluates*))

In addition, the following inclusion and exclusion criteria were applied:

- Regarding the format: only scientific articles indexed in the *WOS* and *Scopus databases* were included. Other materials such as book chapters, doctoral theses, and reports of different natures were excluded.
- Regarding the update of the results, only articles published between 2018 and 2022 were included. Earlier papers were discarded.
- Regarding the type of research: only experimental or quasi-experimental research was considered. Systematic reviews, flexural and theoretical studies, as well as classroom experiences, were excluded.

The following are some of the most important studies:

- "Global Media and Information Literacy Assessment Framework: Country Readiness and Competencies" by UNESCO (2013). This study reviews the main strategies for assessing Media and Information Literacy and their applicability in different contexts.
- "Evaluating Online Media Literacy in Higher Education: Validity and Reliability of the Digital Online Media Literacy Assessment (DOMLA)" by T. Hallaq (2016). This study evaluates the validity and reliability of a media and information literacy assessment instrument.
- "Measuring media and information literacy skills: construction of a test" by Lopes et al. (2018). This study develops and validates a media and information literacy assessment scale for you.
- "Media and Information Literacy in Latin America: Approach, analysis and measurement proposal on the context of Media and Information Literacy in Latin America" by T. Durán (2016). This study analyzes the state of media and information literacy assessment in Latin America and proposes a framework for its evaluation in the region.

Example 1

Title: "Discover the Revolution in Comprehensive Well-being! NutriBoost, the Scientifically-Backed Dietary Supplement Generating Astonishing Testimonials"
Subtitle: "Astounded Users Share Extraordinary Results Achieved with NutriBoost"

(City), (Date) - NutriBoost, the dietary supplement that has taken the well-being industry by storm, continues to amaze experts and users alike. In addition to its scientific backing, the testimonials from individuals who have experienced extraordinary results with NutriBoost have made a significant impact on the community.

One of the most astonishing testimonials comes from Ana Garcia, a successful entrepreneur who claims to have reached unprecedented levels of energy and performance after incorporating NutriBoost into her daily routine. According to Ana, "Since I started taking NutriBoost, I have experienced extraordinary mental clarity and focus. My ability to make quick decisions and stay alert during long workdays is simply mind-blowing. Moreover, I have surpassed my physical limits, achieving exceptional athletic performance."

Ana is not the only one who has shared remarkable testimonials. Other users have reported equally astounding results. Pedro Martinez, a university student, states that NutriBoost has improved his memory and retention abilities, enabling him to achieve outstanding grades in his exams. "I never thought a supplement could make such a difference in my academic performance. NutriBoost has given me a remarkable edge in my studies," Pedro excitedly comments.

These extraordinary testimonials complement the scientific backing that supports the benefits of NutriBoost on physical, emotional, and intellectual health. Clinical studies confirm its positive effects on energy levels, the immune system, emotional well-being, and cognitive function. It is important to note that individual results may vary, and NutriBoost is not intended to provide superhuman abilities. However, testimonials from individuals like Ana and Pedro highlight the transformative potential of this dietary supplement for comprehensive well-being.

If you are seeking a solution that can help you achieve extraordinary results in your daily life, NutriBoost could be the ideal supplement for you. Consult with healthcare professionals to determine if NutriBoost is suitable for your individual needs and discover how it can elevate your well-being to astonishing levels.

Example 2

Title: "Explosive Revelations: The Dark Secret Behind SuperAlpha"
Subtitle: "A story that will make you reconsider your purchases at the popular supermarket chain"
Date: June 26, 2023

In a surprising investigation conducted by an independent team of journalists, alarming evidence has been uncovered that reveals a hidden danger behind the seemingly innocent shelves of SuperAlpha. Our team has uncovered a dark secret that raises serious doubts about the safety of the products sold at the popular supermarket chain.

According to confidential sources within the food industry, SuperAlpha is allegedly involved in an elaborate plan to poison its customers. The store's own-brand products that fill the aisles are the tool used to carry out this deadly conspiracy. The details revealed by our informants are as disturbing as they are revealing.

It is claimed that SuperAlpha's suppliers are using toxic and contaminating ingredients in the production of their own-brand products, which are sold at attractive prices to lure unsuspecting consumers. According to our sources, these ingredients include harmful chemicals and additives that are banned in other countries.

As we delve deeper into our investigation, we have discovered a troubling pattern: the company is allegedly purchasing large quantities of these suspicious ingredients from clandestine suppliers abroad. These suppliers, in turn, operate in the shadows, avoiding any form of rigorous regulation or inspection. But that is not all. Our team has gained access to testimonies from former SuperAlpha employees who claim to have witnessed concerning practices in the company's warehouses. According to these testimonies, insects, and rodents have been found in food storage areas, suggesting a lack of hygiene and quality control in the supply chain.

Considering this alarming situation, we urge consumers to exercise extreme caution when shopping at SuperAlpha. It is strongly recommended to carefully check the ingredients and expiration dates of products before purchasing them. Additionally, considering other supermarket options that can guarantee greater safety and product quality is advised. We have attempted to contact SuperAlpha for a response to these accusations, but as of the time of publication, we have not received any official comment from the company.

This revelatory story is intended to make you reconsider your purchasing decisions and raise awareness about the potential dangers you may face when shopping at SuperAlpha. Remember, your health and well-being are at stake.

ANALYSIS

Some indications that allow us to identify that these texts are false news or contain hoaxes are the following:

- *Lack of credible sources*: The article mentions an "independent team of journalists" and "confidential sources within the food industry" but fails to provide any specific names or affiliations. Without verifiable sources, it is difficult to determine the credibility and reliability of the information presented. No links to specific clinical studies or research are provided to support the purported benefits of the supplement.
- *Exaggerated and Unverified Testimonials*: The testimonials presented in the text are quoted from people named Ana García and Pedro Martínez, but no additional information about these people is provided. There is no way to verify if the testimonials are real or if the people exist.
- *Sensational language and dramatic claims*: The article uses sensational language such as "Explosive Revelations," "dark secret," and "deadly conspiracy." Such language is often employed in fake news articles to grab attention and evoke strong emotional responses from readers, rather than providing objective and factual information.
- *Absence of corroborating evidence*: Although the article asserts that evidence has been uncovered, it fails to provide any concrete proof to support its claims. Claims about toxic ingredients, banned chemicals, clandestine suppliers, and unhygienic practices are made without any supporting documentation or verifiable data.
- *Lack of official response*: While the article mentions attempting to contact *SuperAlpha* for a response, it highlights that no official comment was received from the company. In legitimate journalism, efforts are made to seek and include responses from the parties involved to ensure balanced reporting. The absence of an official response raises questions about the article's credibility.
- *Absence of author information*: The article does not disclose the names or backgrounds of the journalists responsible for the investigation. Transparency regarding the authors' credentials and affiliations is important in determining the credibility of a news piece.
- *Overgeneralization and sweeping statements*: The article makes broad claims about *SuperAlpha* as a whole, suggesting that all its own-brand products are involved in a conspiracy to poison customers. Such sweeping statements without specific evidence or clarification can be a characteristic of false news.
- *Persuasive intent*: The article concludes by urging readers to exercise extreme caution when shopping at *SuperAlpha* and consider alternative supermarkets. The underlying persuasive intent to influence consumer behavior rather than provide unbiased reporting raises suspicions about the article's motives.

In conclusion, the provided text exhibits several characteristics commonly associated with false news, including the lack of credible sources, sensational language, absence of evidence, and a persuasive agenda. It is essential to critically evaluate news articles, seek information from reliable sources, and cross-reference claims before accepting their validity. It is important to seek reliable, evidence-backed information when evaluating health products or supplements.

DISCUSSION AND CONCLUSION

Media and information literacy training has emerged as a crucial competency in today's information-rich environment (Zou'bi, 2022). It equips individuals with the skills to understand, evaluate, and critically engage with the vast amount of information available through media and online sources. Given the prevalence of fake news and misleading content, users must possess the necessary abilities for informed decision-making and problem-solving (García-Roca y Amo, 2023).

It is important to recognize that fake news and misleading content can also impact children (Howard, Neudert, Prakash, 2021). Topics such as war conflicts, terrorist acts, and health recommendations can be particularly dangerous when misrepresented or manipulated. Therefore, addressing media literacy education becomes highly relevant, not only in higher education but also at the primary education level, especially in subjects like Didactics of Language and Literature.

By integrating media literacy into the curriculum, children can develop the skills needed to navigate and critically analyze the information they encounter, enabling them to make informed judgments and protect themselves from potential harm. Moreover, the rise of artificial intelligence (AI) has brought about significant changes in various creative and artistic domains, sparking a new era of possibilities and innovations. AI technologies have demonstrated their capability to revolutionize the way we approach and engage with these fields, opening up new frontiers that were once considered unimaginable. From generating compelling visual art and composing mesmerizing music to crafting captivating storytelling and designing cutting-edge architecture, AI has become an invaluable tool for artists, designers, and creators around the world.

In the realm of visual arts, AI-powered algorithms can analyze vast datasets of artwork from different eras and styles, enabling them to produce stunning visual compositions inspired by the masters of the past while incorporating novel elements. This fusion of tradition and innovation has led to the emergence of mesmerizing art pieces that challenge our perception of creativity itself. Furthermore, AI has empowered artists with the ability to explore and experiment with complex 3D designs and virtual reality creations, redefining the boundaries of artistic expression.

As AI advances, deepfakes become more sophisticated and accessible, raising concerns about their potential impact on society. Deepfakes are manipulated media that convincingly alter videos or images, making it difficult to distinguish between genuine and fabricated content. While they have creative and educational applications, they also pose risks, such as misinformation, political manipulation, and loss of public trust. To address these challenges, individuals need media literacy education to critically assess information sources. Governments and tech companies should collaborate to develop regulations and guidelines for responsible AI and deepfake use, balancing freedom of expression and public protection. Advancements in AI-driven detection tools and digital signatures can also help identify and verify authentic media.

The future of media and AI will be a constant balancing act between creators and deceivers. By fostering an informed society and employing responsible AI practices, we can strive to minimize the harm caused by deepfakes while harnessing their benefits for humanity. Vigilance, adaptability, and proactive measures are essential in navigating this rapidly evolving landscape.

In conclusion, media and information literacy training has become an essential skill set for individuals to effectively navigate the vast sea of information available today. Integrating media literacy education into primary education, particularly in subjects like Didactics of Language and Literature, can equip children with the necessary tools to critically evaluate and engage with media content. Additionally, as AI technologies continue to advance, the potential for deepfake creation highlights the need for media literacy and a critical understanding of the evolving technological landscape. By fostering media literacy skills, we can empower individuals to be discerning consumers and responsible contributors in the digital age.

ACKNOWLEDGMENT AND FINANCING

This publication is part of the R&D project *DEFINERS: Digital language learning of junior language teachers* (*TED2021-129984A-I00*), funded by *MCIN/AEI/10.13039/501100011033/* and by the European Union *NextGenerationEU/PRTR*. In addition, this research is supported by a Margarita Salas postdoctoral contract of the program for the re-qualification of teachers for 2021-2023 called by the University of Almeria (ref. *RR_A_2021_13*) and financed by the Spanish Ministry of Universities and the European Union with *Next Generation* funds.

REFERENCES

Amo, J. M., & Baldrich, K. (2021). Prácticas letradas en el ámbito universitario: los hábitos lectores del alumnado de Grado en Educación Infantil de la Universidad de Almería. *Tonos Digital, 40*(9).

Brørs, G., Norman, C. D., & Norekvål, T. N. (2020). Accelerated importance of eHealth literacy in the COVID-19 outbreak and beyond. *European Journal of Cardiovascular Nursing, 19*(6), 458–461. doi:10.1177/1474515120941307 PMID:32667217

Caro Valverde, M. T. (2018). Argumentative communication in the Knowledge Society, key to distributed leadership for an educational change from professional development. *Journal of Distance Education, 56*, 1–30. https://revistas.um.es/red/article/view/321601

Casacuberta, D. (2003). *Collective creation. On the Internet the creator is the public*. Gedisa.

Cassany, D., & Castellá, J. M. (2013). Approach to critical literacy. *Perspectiva, Florianópolis, 28*(2), 353–374. doi:10.5007/2175-795X.2010v28n2p353

Chartier, R. (2018). *The revolutions of written culture*. Gedisa.

Chesney, R., & Citron, D. (2019). Deepfakes and the New Disinformation War. The Coming Age of Post-Truth Geopolitics. *Foreign Affairs, 98*(1), 147–155.

Chong, Y. Y., Cheng, H. Y., Chan, H. Y. L., Chien, W. T., & Wong, S. Y. S. (2020). COVID-19 pandemic, infodemic and the role of eHealth literacy. *International Journal of Nursing Studies, 108*, 103644. Advance online publication. doi:10.1016/j.ijnurstu.2020.103644 PMID:32447127

Codina, L. (2018). *Systematized reviews for academic papers - 1: Concepts, phases and bibliography*. https://www.lluiscodina.com/revisiones-sistematizadas-fundamentos/

Del-Fresno, M. (2019). Desórdenes informativos: Sobreexpuestos e infrainformados en la era de la posverdad. *El Profesional de la Información, 28*(3). Advance online publication. doi:10.3145/epi.2019.may.02

Destiny, O., & Omar, B. (2020). Fake news and COVID-19: Modelling the predictors of fake news sharing among social media users. *Telematics and Informatics*, *56*, 1–16. doi:10.1016/j.tele.2020.101475 PMID:34887612

Durán, T. (2016). *Alfabetización mediática e informacional en Latinoamérica. Approach, analysis and measurement proposal on the context of media and information literacy in Latin America* [Doctoral dissertation]. Autonomous University of Barcelona.

Eurobarometer. (Ed.). (2018). *Fake news and disinformation online.* doi:10.2759/559993

European Commission. (2018). A multi-dimensional approach to disinformation. Report of the independent High Level Group on Fake News and Online Disinformation. *Directorate-General for Communication Networks, Content and Technology*. http://bit.ly/2Ponxc1

European Commission (Ed.). (2018). *Action Plan against Disinformation. EEAS – European External Action Service – European Union*. Retrieved of https://bit.ly/3wuVw86

García Roca, A. (2021a). The SCP Foundation in the development of academic literacy, reading promotion and creative writing. *Publicaciones*, *51*(1), 15–28. doi:10.30827/publicaciones.v51i1.16488

García Roca, A. (2021b). *Creepypasta* as metafictional textualities: formative opportunities for media and information literacy. *Tonos Digital, 40*, 1-20. http://www.tonosdigital.es/ojs/index.php/tonos/article/view/2715/1195

García Roca, A., & Amo, J. M. (2023). Media and information literacy in a context of disinformation. *Tejuelo*, *37*, 99–128. doi:10.17398/1988-8430.37.99

García Roca, A., Manzano León, A., & Rodríguez Ferrer, J. (in press). Adaptation and training model of the board game untold for narrative and communicative competences. *International Journal of Instruction*.

Golovchenco, Y., Mareike, H., & Adler-Nissen, R. (2018). State, media and civil society in the information warfare over Ukraine: Citizen curators of digital disinformation. *International Affairs*, *94*(5), 975–994. doi:10.1093/ia/iiy148

Grant, M. J., & Booth, A. (2009). A typology of reviews: An analysis of 14 review types and associated methodologies. *Health Information and Libraries Journal*, *26*(2), 91–108. doi:10.1111/j.1471-1842.2009.00848.x PMID:19490148

Hallaq, T. (2016). Evaluating Online Media Literacy in Higher Education: Validity and Reliability of the Digital Online Media Literacy Assessment (DOMLA). *The Journal of Media Literacy Education*, *8*(1), 62–84. doi:10.23860/jmle-8-1-5

Howard, P. N., Neudert, L., & Prakash, N. (2021). Digital misinformation/disinformation and children. UNICEF Office of Global Insight and Policy.

Jenkins, H. (2008). Convergence culture. The culture of media convergence. Barcelona: Paidós.

Jenkins, H., Ford, S., & Green, J. (2013). *Spreadable Media: Creating Value and Meaning in a Networked Culture*. New York University Press.

Jenkins, H., Ito, M., & Boyd, D. (2015). *Participatory culture in a networked era: A conversation on youth, learning, commerce, and politics*. Polity Press.

Knobel, M., & Lankshear, C. (2006). Digital literacy and digital literacies: Policy, pedagogy and research considerations for education. *Nordic Journal of Digital Literacy*, *1*(1), 12–24. doi:10.18261/ISSN1891-943X-2006-01-03

Landow, G. P. (2009). *Hypertext 3.0. Critical theory and new media in the age of globalization*. Planeta.

Livingstone, S. (2008). *Internet literacy: Young people's negotiation of new online opportunities*. MIT Press.

Lopes, P., Costa, P., Araujo, L., & Ávila, P. (2018). Measuring media and information literacy skills: Construction of a test. *Communications*, *43*(4), 1–27. doi:10.1515/commun-2017-0051

López Borrull, A., Vives-Gràcia, J., & Badell, J. (2018). Fake news, ¿amenaza u oportunidad para los profesionales de la información y la documentación? *El Profesional de la Información*, *27*(6), 1346. Advance online publication. doi:10.3145/epi.2018.nov.17

Manzano León, A., Rodríguez Ferrer, J. M., Aguilar, J. M., Fernández Campoy, J. M., Trigueros, R., & Martínez Martínez, A. M. (2022). Play and learn: Influence of gamification and game-based learning in the reading processes of secondary school students. *Revista de Psicodidáctica*, *27*, 38–46. doi:10.1016/j.psicod.2021.07.001

Millán, J. A. (2008). *Reading and the knowledge society*. Government of Navarra / Department of Education.

Naeem, S., & Bhatti, R. (2020). The Covid-19 'infodemic': A new front for information professionals. *Health Information and Libraries Journal*, *37*(3), 233–239. doi:10.1111/hir.12311 PMID:32533803

New London Group. (1996). A Pedagogy of Multiliteracies: Designing Social Futures. *Harvard Educational Review*, *66*(1), 60–93. doi:10.17763/haer.66.1.17370n67v22j160u

Newman, N., Fletcher, R., Schulz, A., Andi, S., Robertson, C. T., & Nielsen, R. K. (2021). *Reuters Institute Digital News Report 2021. 10th Edition.* https://reutersinstitute.politics.ox.ac.uk/sites/default/files/2021- 06/Digital_News_Report_2021_FINAL.pdf

Paakkari, L., & Okan, O. (2020). COVID-19: Health literacy is an underestimated problem. *The Lancet. Public Health*, *5*(5), e249–e250. Advance online publication. doi:10.1016/S2468-2667(20)30086-4 PMID:32302535

Page, M. J., McKenzie, J. E., Bossuyt, P. M., Boutron, I., Hoffmann, T. C., Mulrow, C. D., Shamseer, L., Tetzlaff, J. M., Akl, E. A., Brennan, S. E., Chou, R., Glanville, J., Grimshaw, J. M., Hróbjartsson, A., Lalu, M. M., Li, T., Loder, E. W., Mayo-Wilson, E., McDonald, S., ... Moher, D. (2021). The PRISMA 2020 statement: An updated guideline for reporting systematic reviews. *British Medical Journal*, *372*(71), n71. Advance online publication. doi:10.1136/bmj.n71 PMID:33782057

Pariser, E. (2011). *The Filter Bubble: What the Internet is hiding from you*. The Penguin Press.

Puri, N., Coomes, E. A., Haghbayan, H., & Gunaratne, K. (2020). Social media and vaccine hesitancy: New updates for the era of COVID-19 and globalized infectious diseases. *Human Vaccines & Immunotherapeutics, 16*(11), 2586–2593. Advance online publication. doi:10.1080/21645515.2020.1780846 PMID:32693678

Rodríguez Ferrer, J. M., Manzano León, A., Fernández Jiménez, C., Luque de la Rosa, A., Fernández Campoy, J.M., & Aguilar Parra, J. M. (2023) Shall we play together? Game-based learning for engagement and classroom climate in Spanish socially deprived communities. *Frontiers in Psychology*. doi:10.3389/fpsyg.2023.1163441

Royal Decree 1393/2007, of October 29, 2007, which establishes the organization of official university education. doi:10.3389/fpsyg.2023.1163441

Royal Decree 822/2021, of September 28, establishing the organization of university education and the procedure for quality assurance.

Salavaría, R., Buslón, N., López-Pan, F., León, B., López-Goñi, I., & Erviti, M. (2020). Disinformation in times of pandemic: Typology of Covid-19 hoaxes. *El Profesional de la Información, 29*(3), 1–15. doi:10.3145/epi.2020.may.15

Sánchez-Cruzado, C., Santiago, R., & Sánchez-Compaña, M. T. (2021). Teacher digital literacy: The indisputable challenge after covid-19. *Sustainability (Basel), 13*(4), 1858. Advance online publication. doi:10.3390u13041858

Scolari, C. A. (2015). *Media ecology: environments, evolutions and interpretations*. Gedisa.

Sentell, T., Vamos, S., & Okan, O. (2020). Interdisciplinary Perspectives on Health Literacy Research Around the World: More Important Than Ever in a Time of COVID-19. *International Journal of Environmental Research and Public Health, 17*(9), 3010. doi:10.3390/ijerph17093010 PMID:32357457

Tejedor, S., Cervi, L., Pérez-Escoda, A., & Jumbo, F. T. (2020). Digital literacy and higher education during COVID-19 lockdown: Spain, Italy, and Ecuador. *Publications / MDPI, 8*(4), 48. Advance online publication. doi:10.3390/publications8040048

Tolosa, R., Vera-Rodríguez, R., Fierrez, J., Morales, A., & Ortega-García, J. (2020). Deepfakes and beyond: A Survey of face manipulation and fake detection. *Information Fusion, 64*, 131–148. doi:10.1016/j.inffus.2020.06.014

UNESCO. (2005). *Towards knowledge societies*. UNESCO.

UNESCO. (2013). *Global Media and Information Literacy Assessment Framework: country readiness and competencies*. UNESCO.

UNESCO. (2022). *Global standards on guidelines for the development of media and information literacy curricula*. UNESCO. https://www.unesco.org/sites/default/files/medias/files/2022/02/Global%20Standards%20for%20Media%20and%20Information%20Literacy%20Curricula%20Development%20Guidelines_ES.pdf

Vraga, E. K., Tully, M., & Bode, L. (2020). Empowering Users to Respond to Misinformation about Covid-19. *Media and Communication*, *8*(2), 475–479. doi:10.17645/mac.v8i2.3200

Weiss, A. P., Alwan, A., Garcia, E. P., & García, J. (2020). Surveying fake news: Assessing university faculty's fragmented definition of fake news and its impact on teaching critical thinking. *International Journal for Educational Integrity*, *16*(1), 1–30. doi:10.100740979-019-0049-x

Zou'bi, R. A. (2022). The impact of media and information literacy on students' acquisition of the skills needed to detect fake news. *The Journal of Media Literacy Education*, *28*(2), 58–71. doi:10.23860/JMLE-2022-14-2-5

KEY TERMS AND DEFINITIONS

Artificial Intelligence (AI): Involves the creation of computer systems that possess the ability to carry out tasks traditionally associated with human intelligence, including perception, reasoning, learning, and decision-making.

Clickbait: Encompasses sensational or deceptive online content, predominantly in the form of attention-grabbing headlines or thumbnails, strategically crafted to entice users into clicking on a link, often with the intention of boosting website traffic or generating advertising revenue.

Creative Writing: Involves the skillful utilization of imagination, innovative ideas, and evocative language to craft a wide range of literary compositions, including narratives, poetry, plays, and more, enabling individuals to delve into artistic realms, explore diverse perspectives, and authentically express their thoughts, emotions, and experiences through written form.

Deepfake: Denotes a form of altered media content, commonly videos, employing artificial intelligence techniques to convincingly substitute or overlay an individual's face or voice onto another person's body, giving rise to potentially misleading or deceptive portrayals.

Disinformation: Also known as deliberate misinformation, encompasses the intentional dissemination of fabricated or distorted information, often through various media channels, aiming to deceive, manipulate, or create confusion among individuals, potentially undermining trust, distorting public discourse, and influencing opinions or actions.

Fake News: A form of intentionally deceptive information, encompasses the deliberate creation and widespread distribution of false or misleading content across diverse media platforms, driven by the explicit motive of deceiving, manipulating, or influencing the audience, often resulting in the erosion of trust, polarization of opinions, and disruption of accurate public discourse.

Flipped Classroom: An instructional approach where traditional lecture-based content is delivered to students outside of the classroom, typically through video or online resources, allowing in-class time to be dedicated to interactive activities, discussions, and collaborative learning.

Information and Communication Technology: Refers to the broad range of technologies, systems, and tools used for gathering, storing, managing, and transmitting information electronically, enabling communication, collaboration, and access to digital resources.

Media and Information Literacy: The ability to access, analyze, evaluate, and effectively use various media forms and information sources in a critical and responsible manner.

New Literacy Studies: Refers to an interdisciplinary field of research that examines the social, cultural, and contextual aspects of literacy practices, going beyond traditional notions of reading and writing to encompass a broader understanding of how people engage with and make meaning from texts in diverse contexts.

ENDNOTE

1 Eli Pariser (2011) presents the filter bubble as a potential risk as it acts "like a lens, the filter bubble invisibly transforms the world we experience by controlling what we see and don't see".

Chapter 14
Game-Based Learning to Teach Executive Functions:
A Pilot Study in Higher Education

José M. Rodríguez-Ferrer
https://orcid.org/0000-0003-1811-0040
University of Jaén, Spain

Rubén Camacho Sánchez
University Lleida, Spain

Anastasio García-Roca
https://orcid.org/0000-0002-2277-3034
University of Almería, Spain

Ana Manzano León
https://orcid.org/0000-0001-6966-0355
University of Almería, Spain

ABSTRACT

This pilot study aimed to investigate the effectiveness of game-based learning in teaching executive functions (EF) to preservice teachers. Various games were employed as training tools to improve EF skills, and students' satisfaction and perceptions of the training were evaluated. The results revealed a highly positive response to the utilized methodology, highlighting its efficacy in stimulating students with special educational needs. Participants valued the adaptability of the games and considered the experience enjoyable and innovative. These findings emphasize the importance of providing future educators with practical exposure to active learning methodologies.

DOI: 10.4018/978-1-6684-8861-4.ch014

INTRODUCTION

Executive Functions (EFS)

Regarding human abilities, Executive Functions (EFs) encompass a set of cognitive processes that play a crucial role in our capacity to plan, think, act, move, and reflect on specific actions during learning and behavior. EFs are responsible for vital functions such as creativity, emotional control, and motivation, which are integral to our overall cognitive functioning and well-being.

Over the years, various conceptualizations of EFs have emerged, emphasizing their central role in individual development and involvement in various activities. Lezak (1982) argued that EFs are important for specific tasks and crucial for overall personal growth. Successful performance across different domains requires setting objectives, planning their implementation, and effectively monitoring their progress as they unfold. Therefore, EFs are fundamental in facilitating adaptive behavior and achieving successful goal attainment.

Gilbert and Burgess (2008) highlighted the significance of EFs in decision-making under new or unfamiliar circumstances. They emphasized that individuals with well-developed EFs can make effective decisions and adapt to challenging situations. This ability to respond to novel situations requires flexibility, mental agility, and the capacity to think critically, all skills associated with high-level EFs.

The term "EFs" is often used to refer to a group of cognitive processes in the brain that are triggered by focused attention or deliberate effort. EFs are cognitive processes that require active engagement and conscious control, as they depend on individuals' needs, goals, and motivations. Diamond (2013) noted that EFs are effortful processes that necessitate intentional allocation of cognitive resources and a desire to achieve a specific outcome. This deliberate effort distinguishes EFs from automatic or reflexive behaviors.

It is important to note that EFs can also be referred to as executive cognitive functions, tasks, or executive processes. These terms are often used interchangeably, which can lead to some confusion. Some researchers have used the term "central executive" to specifically describe functions related to working memory, highlighting the crucial role of working memory in coordinating and integrating other cognitive processes (Corso et al., 2013).

EFs are essential for problem-solving and complex decision-making. Miyake et al. (2000) identified three core EFs that are considered foundational to higher-order cognitive abilities. These include flexibility, which refers to switching between objectives or mental tasks; inhibition, which involves suppressing irrelevant or impulsive responses; and working memory, which entails updating and monitoring tasks. These basic EFs provide the cognitive foundation for individuals to engage in more complex problem-solving and cognitive tasks.

The development of EFs throughout an individual's life is directly related to various significant human outcomes. Research has shown that EFs are associated with mental health, physical health, quality of life, social functioning, school readiness, and academic achievement. For example, studies have demonstrated that individuals with well-developed EFs exhibit better mental health outcomes (Fairchild et al., 2009; Taylor Tavares et al., 2007) and improved physical health and overall quality of life (Brown & Landgraf, 2010; Will Crescioni et al., 2011). Additionally, EFs have been found to play a crucial role in social functioning, as they contribute to effective communication, problem-solving, and interpersonal skills (Riggs et al., 2006). EFs are closely tied to school readiness and academic achievement (Morrison et al., 2010; Fuhs et al., 2014).

Researchers and practitioners recognize the importance of developing EFs alongside other cognitive abilities. Diamond and Lee (2011) emphasize the need to foster executive function skills, recognizing their importance alongside other cognitive abilities. In an educational context, students rely on EFs to manage cognitive processes and pursue specific goals. These skills have been found to predict academic success, metacognitive abilities, language acquisition, and theory of mind (Best, 2012; St Clair-Thompson & Gathercole, 2006; Bryce et al., 2015; Fuhs et al., 2014; Carlson et al., 2002).

It has been shown that EFs play an important role in preparing students for school. Several studies have indicated that they can significantly impact mathematics and comprehension (Borella et al., 2010; Duncan et al., 2007; Gathercole et al., 2004). Additionally, children's literacy skills during the early stages of school significantly correlate with their EFs skills during preschool (Blair & Razza, 2007).

Recognizing the significance of EFs, researchers and practitioners have focused on developing strategies and interventions to enhance executive function skills in individuals of all ages. These interventions often involve activities and exercises that target specific EFs, such as working memory training, cognitive flexibility exercises, and inhibitory control tasks (Diamond & Lee, 2011). By strengthening EFs, individuals can improve their ability to plan, organize, regulate emotions, and navigate complex cognitive tasks.

In conclusion, Executive Functions play a vital role in human abilities, encompassing various cognitive processes necessary for planning, thinking, acting, and reflecting on our actions. These functions are central to individual development and are associated with various outcomes, including mental health, physical health, social functioning, school readiness, and academic achievement. Fostering EFs is crucial in educational contexts, as they predict academic success and other important cognitive abilities. By understanding and supporting EFs, we can enhance individuals' overall well-being and success in various domains of life. Efforts to develop EFs through targeted interventions and strategies hold the potential to unlock individuals' cognitive potential and improve their overall functioning.

EFS and Students With Disabilities

In recent years, there have been questions about how far schools are actually adapting inclusive education to meet the needs of children with special educational needs (SEN), which is a key policy objective for our teaching community and society. SEN refers to students with learning, developmental, physical, and emotional disabilities and communication, emotional, and behavioral disorders (Bryant et al., 2019). A substantial amount of attention has previously been paid to the real determinants associated with learning difficulties and the fact that EFS is a powerful predictor of school performance and cognitive development in students with special needs.

It is a complex educational challenge to evaluate and conceptualize differences between children, especially those that pertain to children with special needs. A great deal of research has been conducted on the definition of special needs and learning difficulties in educational settings (Dell'Anna et al., 2019; Sheehy et al., 2009). A variety of theoretical approaches to special needs exist, but one that has significantly impacted the history of special education is the classification of children's learning difficulties according to categories (Terzi, 2005). A successful special education program requires the identification and characterization of children's differences, as well as the determination of which differences are relevant to the special education program. It has also been suggested that special needs may be associated with the limitations of the education system as a whole (Norwich & Kelly, 2006). A report was published by UNESCO (1994) supporting the idea of inclusive education due to the Salamanca World Conference on Special Education.

Students with limited EFS skills are more likely to exhibit behaviors associated with deficits in time management, sustained attention to instruction, and understanding of academic concepts (Bull & Scerif, 2001; Holley et al., 2017). Recent studies indicate that children with mild intellectual disabilities are more likely to have difficulty processing magnitudes (Brankaer et al., 2013). Compared to students with other types of disabilities or the general population, children with mild disabilities have lower mathematical performance (Wei et al., 2013). Undoubtedly, acquiring adequate math and English skills is crucial to achieving a good functional outcome for children with intellectual disabilities, and research indicates that EFS and math skills training is highly effective for children with disabilities. These facts suggest that it may be possible to increase future adaptive functioning (Diamond & Lee, 2011; Jansen et al., 2013; Söderqvist et al., 2012).

GAME-BASED LEARNING AND EFS

In recent years, there has been an increasing interest in exploring innovative and effective approaches to education. One such approach is game-based learning (GBL), which involves using games to facilitate students' learning process and cognitive development. This chapter analyzes the relationship between GBL and the improvement of Essential Future Skills (EFS), and its particular benefits for students with special educational needs.

- Enhancing Essential Future Skills through Game-Based Learning

There is a reciprocal feedback process between learning and improving EFS. As we acquire more knowledge and skills, our EFS improve, and vice versa. By developing and teaching EFS through innovative and effective means, such as game-based learning, we can enhance the learning process and foster the development of essential knowledge for students to become active members of society (Lavega et al., 2013).

Games have been recognized as powerful pedagogical tools that can motivate participants, facilitate tasks of varying complexity, and record player behavior (Plass et al., 2015). A well-designed game-based learning program not only provides students with an enjoyable and playful experience but also guides them towards achieving specific educational objectives (Camacho-Sánchez et al., 2022). Research suggests that games can effectively improve variables such as emotional state, study skills, and cognitive abilities (Roffey, 2009; Charlton et al., 2005; Ke, 2008; Powers et al., 2013).

- Adaptability in Game Design for Different Ages

Game design should consider adaptability as a key component, especially when targeting learners of different ages. Understanding how games can be adaptable to meet the needs of different age groups would allow for the optimization of game-based interventions for enhancing EFS. Obtaining information about adaptability in game design would enable intervention designers to tailor programs to the specific ages and requirements of their students (Zelazo, 2006).

- Empowering Students with Special Needs through Gamification

Gamification, the application of game elements and mechanics to non-game contexts, has shown particular effectiveness in engaging students with special needs and learning difficulties (Jong, 2015; Lan & Hsiao, 2018). Video games have proven to be valuable learning tools for children with hearing disabilities, contributing to improved cognitive abilities and memory (Mawas et al., 2019; Marschark & Hauser, 2008). Research also suggests that collaborative game strategies can positively impact the learning experiences of children with special educational needs, supporting their literacy skills (Cano et al., 2018).

- Challenges in Inclusive Education for Students with Special Needs

A significant percentage of the student population consists of students with special educational needs (SEN). Despite the existence of laws and regulations aimed at protecting these students, several challenges hinder their participation in education. Insufficient resources, inadequate teacher training, and pedagogical deficiencies pose obstacles to inclusive learning (Sakarneh & Al-Swelmyeen, 2020).

- The Role of Game-Based Learning in Overcoming Challenges

Incorporating game-based learning into teaching strategies can address some of the challenges faced in ensuring inclusive education for students with special needs. GBL provides an interactive and engaging learning environment that can accommodate diverse learning styles and promote active participation. By leveraging past approaches, discoveries, and technologies, educators can harness the potential of game-based learning to advance the educational system in the 21st century (de Freitas, 2018; Florian, 2021).

Game-based learning has emerged as a promising approach to education, with the potential to enhance essential future skills and cater to the diverse needs of students, including those with special educational needs. By integrating game elements into the learning process, educators can create an immersive and interactive environment that fosters engagement, motivation, and skill development. The use of games as a pedagogical tool not only provides a fun and enjoyable learning experience but also aligns educational objectives with the gameplay, ensuring that students acquire the necessary knowledge and competencies.

For students with special needs, game-based learning offers unique advantages. The interactive nature of games can provide personalized learning experiences that cater to individual strengths and challenges. Through gamification, educators can adapt game mechanics and elements to accommodate different learning styles and create inclusive environments where every student can thrive.

Research has shown that game-based learning can significantly improve cognitive abilities, memory, problem-solving skills, and social interaction among students with special needs. By presenting information and concepts in a gamified format, complex topics can be simplified and made more accessible, enhancing comprehension and retention.

Furthermore, game-based learning encourages active participation and collaboration, fostering a sense of community and belonging among students. This collaborative aspect is particularly valuable for students with special needs, as it promotes social interaction, communication skills, and teamwork, which are crucial for their overall development.

However, despite the benefits of game-based learning, there are still challenges to overcome. One major challenge is the integration of technology and games into traditional educational settings. Many schools and institutions may face limitations in terms of resources, infrastructure, and teacher training to effectively implement game-based learning approaches.

To address these challenges, it is essential to invest in professional development programs for educators, equipping them with the necessary skills and knowledge to integrate game-based learning effectively. Collaboration between educators, game developers, and researchers can also lead to the creation of more inclusive and accessible games that cater to the specific needs of students with disabilities.

In conclusion, game-based learning has the potential to revolutionize education by enhancing essential future skills and promoting inclusive learning environments. Through interactive gameplay, students can acquire knowledge, develop critical thinking, problem-solving abilities, and collaborate with their peers. For students with special needs, game-based learning offers a unique opportunity to bridge educational gaps and empower them to reach their full potential. By embracing the benefits of game-based learning and addressing the associated challenges, we can pave the way for a more engaging, inclusive, and effective educational system in the 21st century.

METHODOLOGY

The Participants

Students from the Master of Special Education program at the University of Almería participated in this study. A total of 29 students, with an average age of 27 years, participated in the study. There are 29 students in this class, 19 of whom are women and 10 of whom are men. As for the participant profile, the majority are either primary or infant teachers or opponents of primary or infant teaching. A few people from psychology and social education are also present.

Procedure

- Introduction

This section focuses on the analysis and expansion of a theoretical-practical workshop conducted within the context of the Mental Disability subject. The workshop aimed to explore the relationship between executive functions (EFs) and game-based learning. The workshop participants provided informed consent and were divided into groups to engage in practical activities involving games. The content and structure of the workshop were designed to enhance the participants' understanding of EFs and their application in daily and academic activities.

- Theoretical Component of the Workshop

During the theoretical portion of the workshop, participants were provided with comprehensive information about executive functions and their practical relevance. They learned about the various executive functions involved in tasks performed during daily life or academic activities. The objective was to establish a solid foundation of knowledge that would enable participants to identify and analyze the executive functions engaged during gameplay.

Furthermore, the workshop emphasized the specific connections between game mechanics and executive functions. By understanding the relationship between the two, participants were better prepared to explore and evaluate the potential benefits of game-based learning. They learned how different game mechanics could impact executive functions and the cognitive processes associated with them.

- Practical Component of the Workshop

In the practical part of the workshop, participants were given the opportunity to engage in gameplay and apply their theoretical knowledge. They played various games and were then asked to identify the executive functions involved during gameplay. This exercise encouraged participants to observe and analyze their cognitive processes while playing, deepening their understanding of executive functions in a practical context.

Moreover, participants were tasked with determining which types of students could benefit from the game and exploring potential adaptations to make the game accessible to students with specific needs. To aid in this exercise, a stereotyped example of a student in need of adaptations was provided, allowing participants to consider diverse learning requirements and explore inclusive possibilities.

- Participant Feedback

At the conclusion of the workshop, participants were given a questionnaire to gather their feedback and assess their perception of the tasks assigned. This feedback served as valuable input for evaluating the effectiveness of the workshop and gauging the participants' understanding of the relationship between executive functions and gameplay.

- Facilitation and Logistics

To ensure the smooth running of the workshop, the participants were divided into groups of 4-5 individuals, as this group size is conducive to playing most board games effectively. The presence of two facilitators was necessary to efficiently initiate the workshop and provide guidance to the participants. By having at least two copies of each game, the facilitators could explain the rules and mechanics simultaneously to two groups, maximizing efficiency.

Additionally, it was decided that the game would not be changed until participants had played a minimum of two rounds. This decision was justified by the need to grasp the mechanics of the game thoroughly. By allowing participants to engage in multiple rounds, they could gain a deeper understanding of the game's intricacies and its impact on executive functions.

- Conclusion

The theoretical-practical workshop successfully integrated the study of executive functions and game-based learning. By combining theoretical knowledge with practical gameplay, participants gained a deeper understanding of how executive functions are engaged during gameplay. The workshop fostered critical thinking, analysis, and observation skills in participants, allowing them to identify executive functions and consider adaptations for diverse learners.

The workshop's structure, participant consent, and facilitation techniques ensured a conducive learning environment. The feedback obtained from participants through the questionnaire provided valuable insights for future improvements in workshop design and implementation. Overall, the workshop exemplified the potential of game-based learning to enhance the understanding and application of executive functions in educational contexts.

- The games used for this workshop were the following:

Ghosts of Blizt (Devir): Ghosts of Blitz, developed by Devir, is a game that offers exciting possibilities for adapting gameplay to target different executive functions and skills. This game utilizes wooden figures and cards to represent various objects, each distinguished by its unique color (e.g., white ghost, green bottle, red chair). While there are multiple versions of the game, they all share the same fundamental mechanics. The game incorporates two types of cards, each presenting its own cognitive challenges. In the first type, two objects are displayed, with one object sharing a similar color with the figure it represents. For example, a gray ghost and a red armchair might be depicted. It is important to note that the physical figure for the ghost is white, while the chair figure is red. Players are required to quickly identify the congruent figure and retrieve it from the game components. This aspect of the game encourages rapid visual processing and the ability to discern color congruence, fostering skills related to attention to detail and mental flexibility.

The second type of card presents a more complex challenge. In this scenario, two figures are presented, neither of which is congruent in color with the object it represents. The players' objective is to determine which figure and color are not represented in these cards and retrieve the corresponding figure as swiftly as possible. This aspect of the game promotes higher-order thinking skills, such as logical reasoning, deductive thinking, and cognitive flexibility. The versatility of Ghosts of Blitz lies in its adaptability to target specific executive functions and skills. Educators and facilitators can tailor the gameplay to focus on areas such as working memory, cognitive processing speed, attentional control, and decision-making. By modifying the game rules or introducing additional challenges, specific executive functions can be honed and strengthened. For example, to enhance working memory, players can be tasked with memorizing the position and color of figures on the cards and recalling them after a certain period. This variation would require players to actively engage their working memory capacity, facilitating the development of this executive function.

To improve cognitive processing speed, a time constraint can be introduced, encouraging players to identify the congruent or non-represented figure within a limited time frame. This adaptation challenges players to think quickly and make rapid decisions, sharpening their cognitive processing speed. Attentional control can be targeted by introducing distractors or additional objects that are irrelevant to the gameplay. Players must then selectively attend to the relevant objects and ignore the distractions, enhancing their ability to sustain attention and filter out irrelevant information. Decision-making skills can be emphasized by introducing more complex scenarios where players must weigh multiple factors and make strategic choices. This adaptation encourages players to think critically, analyze different options, and consider the potential consequences of their decisions.

In conclusion, Ghosts of Blitz presents an excellent platform for enhancing executive functions and skills through engaging gameplay. By adapting the game mechanics and introducing variations that target specific cognitive processes, educators and facilitators can effectively harness its potential as a versatile tool for cognitive development. Whether working on attention, working memory, cognitive processing speed, or decision-making, Ghosts of Blitz offers an immersive and enjoyable experience that promotes the development of essential cognitive skills.

Enchanted Staircase (Devir): The game described is a simple yet engaging game that accommodates up to four players. Each player has their own designated playing area, consisting of a meeple and a totem of the same color. The game components include meeples, a ladder-like board, and a six-sided die. One side of the die features numbers one to four, while the remaining two sides display a ghost symbol and ghost-shaped hoods that can be used to cover the meeples.

To begin the game, players roll the die from the lowest rung of the ladder. If a numerical value is rolled, the player's meeple advances the corresponding number of rungs on the board. The objective is to reach the top of the ladder first. However, when a ghost symbol is rolled, things take an interesting twist.

When a ghost appears on the die, the player who rolled it takes one of the ghost-shaped hoods and covers their own meeple or another player's meeple with it. As the game progresses, it is possible for all players' meeples to be covered by these hoods, making it impossible to distinguish between them. At this point, if a player rolls a ghost symbol on the die, they have the option to swap the positions of their meeple and another player's meeple, adding an element of strategic decision-making.

The game continues until one meeple reaches the final step of the ladder. At this moment, the first meeple to reach the finish line reveals its color by removing the hood covering it, effectively transforming from a ghost into a recognizable meeple. This adds an exciting element of surprise and suspense, as players eagerly await the revelation of the winning color.

This game provides a delightful combination of luck, strategy, and anticipation. The roll of the die determines the movement of the meeples, while the use of the ghost-shaped hoods introduces a captivating twist that challenges players' memory and observation skills. The element of swapping positions adds an extra layer of strategy and interaction among players. The climactic moment of unveiling the winning color creates a thrilling conclusion to the game. Overall, this simple yet engaging game offers an enjoyable experience for up to four players. With its unique mechanics and elements of surprise, it provides an opportunity for friendly competition, strategic thinking, and a shared sense of excitement as players strive to reach the top of the ladder and reveal the victorious color.

Speed Cubes (Mercury): The game being described is simple and easy to understand. It involves a set of cards and five small cubes, each representing a different color. The objective of the game is to quickly arrange the cubes according to the order indicated by the revealed cards. The cards use simple drawings to depict either a horizontal order, where the cubes are arranged in rows, or a vertical order, where the cubes are stacked to form towers. This game is designed to be accessible and straightforward for players of all ages. It requires minimal setup and can be quickly explained, making it ideal for casual gameplay sessions or introducing new players to the world of board games.

The components of the game include a deck of cards and five small cubes, each assigned a distinct color. These cubes serve as the key elements that players need to arrange correctly during the game. The cards, featuring simple drawings, provide the instructions for the order in which the cubes should be arranged. When a card is revealed, it presents a visual representation of the desired order of the colored cubes. Depending on the specific card, the order can be horizontal or vertical. In the horizontal order, players must arrange the cubes in rows, aligning them according to the indicated sequence. In the vertical order, players stack the cubes on top of each other to form towers, adhering to the specified arrangement.

The gameplay is fast-paced, as players strive to arrange the cubes as quickly as possible in response to the revealed cards. The simplicity of the game mechanics allows for rapid decision-making and encourages players to think on their feet. The game's focus on visual cues and quick ordering fosters skills such as pattern recognition, visual processing, and dexterity. Players must scan the cards, interpret the drawings, and promptly manipulate the cubes to match the desired order. This promotes cognitive agility and enhances hand-eye coordination. Due to its simplicity, the game can accommodate multiple players and encourages friendly competition. Players can challenge each other to complete the ordering tasks in the shortest amount of time or compete to see who can achieve the highest level of accuracy. This adds an element of excitement and engagement to the gameplay experience.

Rhino Hero (Haba): The game described revolves around the construction of a tower using cards as the primary game mechanic. It features two types of cards: roofs, which are dealt to the players, and walls, which are more durable and foldable, allowing for the creation of walls at 90 degrees. Players take turns placing walls and roofs to construct their towers, following specific rules outlined in a table.

In this unique game, the construction of a tower takes center stage, with cards serving as the primary mechanism. The cards are divided into two categories: roofs and walls. Roofs are distributed among the players, while walls, designed to be sturdier, possess the unique characteristic of being foldable, allowing them to create walls at 90-degree angles. The combination of roofs and walls forms the foundation for building the towers. As the game progresses, players take turns placing their walls and roofs strategically to construct their towers. The table mentioned in the description serves as a guide, indicating the specific number of walls and roofs that players can place during their turn. The table ensures fairness and balance, providing clear instructions for tower construction.

However, the gameplay is not solely based on the physical act of building towers. There are additional elements that introduce complexity and excitement to the game. Some roof cards possess special effects that can alter the course of the game. These effects might include skipping a player's turn, changing the order of turns, adding extra roofs to the tower, or requiring the placement of a wooden meeple called a Rhino within a designated space on the card. The inclusion of special effect roof cards adds an element of unpredictability and strategic decision-making to the game. Players must carefully consider the potential consequences and benefits of utilizing these cards, as they can impact both their own towers and those of their opponents. The risk of losing a tower, and thus the game, is ever-present, adding tension and a sense of urgency to each turn. Ultimately, the game reaches its conclusion when a player's tower collapses. This signifies the end of the game, prompting a final evaluation of the towers constructed. The player with the most stable and impressive tower emerges as the winner.

In summary, the game revolving around the construction of towers using cards as the main mechanic offers an engaging and strategic experience. Players strategically place walls and roofs to build their towers, following guidelines outlined in a table. The addition of special effect roof cards introduces an element of unpredictability, adding excitement and complexity to the gameplay. With the looming possibility of tower collapse, players must balance risk and strategy to emerge victorious in this challenging and thrilling game.

Dobble (Asmodee): The game is a card-based matching game that focuses on identifying matching drawings between two cards. The mechanics of the game involve comparing the drawings on each card and determining which drawings match. Depending on the specific game variant, players may release the card, cover the previous card, or add it to their deck.

In this game, the primary components are the cards, which are the sole focus of gameplay. The objective is to find matching drawings between two cards, regardless of the pairings. No matter which two cards are compared, there will always be a matching draw between them. The mechanics of the game are straightforward. Players select two cards and compare the drawings on them. The task is to identify the drawing that appears on both cards, indicating a match. The ability to recognize and remember the drawings is key to succeeding in this game. The gameplay options may vary depending on the specific rules or variations of the game. In some versions, when a match is found, players release the cards by placing them face-up on the table. This allows other players to keep track of the revealed drawings and aids in memory recall. In other variations, players may cover the previously revealed card with a new card, which adds an element of strategy to the game. This forces players to remember the covered cards and recall them when needed.

Additionally, some game variants involve players collecting the matched cards and adding them to their own deck or scoring pile. This introduces a competitive element, as players strive to accumulate the most matched cards throughout the gameplay. The game enhances cognitive skills such as visual perception, concentration, and memory recall. By actively comparing the drawings on the cards and seeking matches, players develop their observational abilities and strengthen their cognitive processes. The simplicity and versatility of the game make it suitable for players of different ages and skill levels. It can be enjoyed as a casual activity or as a more competitive game, depending on the desired level of challenge and engagement.

Dixit (Libelum): Dixit is a well-known narrative game that sparks players' imagination. The game features various elements, including fantasy illustrations on the game cards. With the help of these cards, players can weave stories that can be more or less extensive, depending on the chosen difficulty level. What sets Dixit apart is that there is no defining aspect to the narrative. If all players guess the narrator's card correctly at the end of a round, no additional points are added. Similarly, if no one guesses correctly, no points are awarded. The focus of the game lies in the art of storytelling and stimulating creativity, rather than competing for points.

For some students, Dixit can pose a challenge. This is why using only the cards to create stories can be appealing in many instances. The evocative illustrations on the cards provide ample inspiration for crafting intricate and engaging narratives. The narrative component of Dixit goes beyond mere storytelling. It also encourages active listening, interpretation, and the development of critical thinking skills. As players craft their narratives, they must consider the perspectives and knowledge of their fellow players, using strategic thinking to provide clues that strike a balance between being too obvious and too obscure. Dixit is a beloved game that transports players into a world of wonder and creativity. Through its imaginative illustrations and the power of storytelling, the game offers a captivating experience that transcends traditional gaming norms.

Whether players create elaborate tales or engage in simplified storytelling using the cards alone, Dixit fosters inclusive gameplay, providing opportunities for all players to express their creativity, linguistic abilities, and strategic thinking. In Dixit, the journey of storytelling holds greater significance than the final score. As players immerse themselves in the enchanting world of Dixit, they embark on a collective endeavor to create narratives that inspire, engage, and ignite the imaginations of all who participate. It is a game that sparks creativity and cultivates communication skills while enjoying shared moments of play.

Story Cubes (Asmodee): The game described features nine dice, each side of which is adorned with an icon representing either a specific drawing or an action, depending on the version. These icons serve as the foundation for gameplay, providing players with visual prompts to spark their creativity and storytelling abilities. On each side of the nine dice, a captivating and diverse array of icons is featured. These icons encompass a range of imagery, including objects, characters, animals, symbols, and even actions, depending on the particular version of the game being played. The carefully crafted icons are designed to ignite players' imaginations and inspire them to weave intricate and captivating stories. The gameplay itself revolves around the act of rolling a designated number of dice and utilizing the drawings depicted on the top face of each dice to create a cohesive and engaging story. As the dice tumble across the playing surface, players eagerly observe the resulting icons that are revealed, forming the building blocks of their imaginative narratives.

Each player, in turn, incorporates the icons from the dice into their storytelling, seamlessly integrating the visuals into their imaginative tale. The icons act as catalysts, stimulating the players' creativity and prompting them to devise compelling characters, thrilling plot lines, and captivating settings. The diversity of the icons ensures that each roll of the dice presents unique opportunities for storytelling, enabling endless possibilities and variations in the narratives that unfold. As the story progresses, players may interweave multiple dice icons into their tales, drawing connections and introducing plot twists that captivate both the storyteller and the eager listeners. The game fosters collaboration and engagement, as players actively listen to each other's narratives, offer support, and build upon the unfolding storylines. The act of creating a story from the drawings on the top face of the dice encourages players to think critically, adapt their narratives based on the icons available, and develop their communication skills. It challenges players to connect disparate elements and construct cohesive narratives that captivate listeners and keep the story flowing.

In conclusion, the game involving the nine dice and their illustrative icons presents an engaging and immersive experience that encourages creativity, imagination, and collaborative storytelling. By rolling the dice and incorporating the drawn icons into their narratives, players embark on a journey filled with endless storytelling possibilities. This game not only provides entertainment but also fosters communication skills, critical thinking, and the joy of crafting unique stories together.

Instruments

In order to determine the level of satisfaction of those who participated in the workshop methodology, acceptance of the methodology learned, and perception of the usefulness of the methodology in the classroom by the students participating in the research, an ad hoc questionnaire was prepared with a Likert-type scale (1-5).

Data Analysis

For the data analysis, an average has been made with each factor and shown in the form of a figure to make it more visual.

RESULTS

In general, the results obtained are very high. For each factor measured, an average score of over 4 was obtained, with a maximum score of 5. Figure 1 shows the averages for each factor of the questionnaire carried out for this research. In terms of satisfaction with the methodology used to teach the use of GBL to stimulate Executive Functions, the score obtained is 4.2 out of 5. As for the perception of using the GBL to stimulate executive functions, a score of 4 out of 5 was obtained.

Figure 1. Graphical representation of the maturities for each of the factors

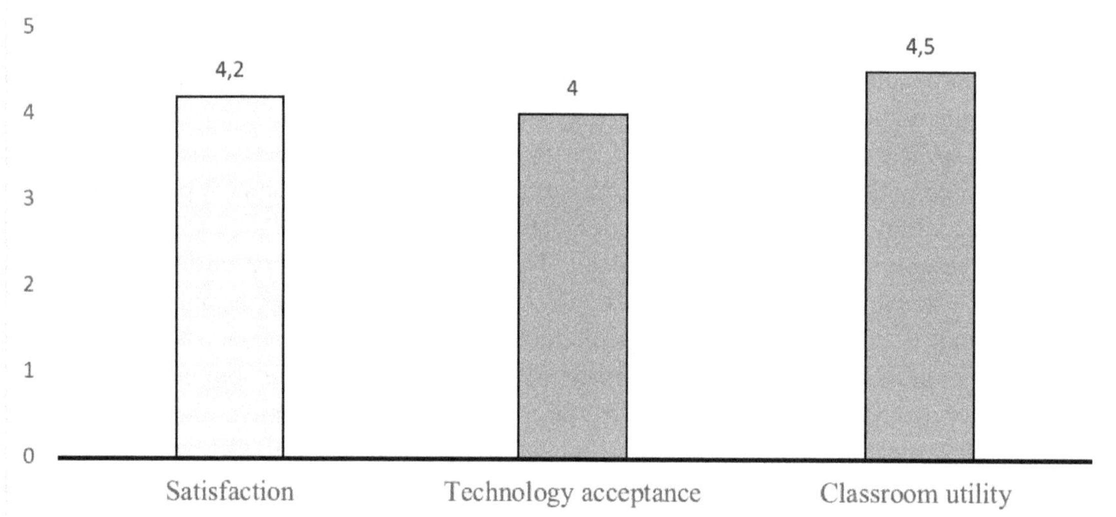

CONCLUSION AND DISCUSSION

In order to assess the effectiveness of a specific methodology to stimulate executive functions, such as game-based learning, students in the Master of Special Education program were asked to conduct a study. The results of the workshop show that the participants highly valued both the methodology used in the workshop (basically, game-based learning) and the use of this methodology for the stimulation of students with special educational needs.

Play strategies are not new in educational settings and are being used more frequently with good results (Abdul & Felicia, 2015; Friehs et al., 2020; Indriasari et al., 2020; Kalogiannakis et al., 2021). People need to be able to participate in their learning process through the use of playful strategies to be more motivated to learn. Psychologists have traditionally considered motivation to be one of the most significant predictors of behavior in terms of its influence on behavior. Therefore, learning methodologies like GBL could be particularly beneficial for students who lack motivation or need to repeat several exercises to learn the material.

There is no doubt that future teachers will gain skills and resources during their initial teacher training, that they will use in their daily work (Ibáez-Córdoba, 2021; Villaverde-Caramés et al., 2021). It is also important to note that these styles provide information on how these people will also apply the same methodologies that have been used with them in the past, which is why it is essential to emphasize that the use of the methodology itself as a vehicle for learning seems to us to be very important since it allows these future teachers to experience first-hand what it is like for their future students.

As a conclusion of this research experience, it is worth mentioning the generally good assessment made during the workshop. Many of the participants commented that it was their first experience with the game as a learning tool and that they were surprised by its adaptability. They also commented that it was fun and innovative to experience the game as a learning methodology, which was a way to connect with the methodology and gain an understanding from their own experience as students.

It is also worth mentioning that although the methodology is highly valued, it is necessary to mention that more teaching staff is needed than usual, at least in the initial sessions in which the students must be introduced to the game and taught how to play each of the games. Once the barrier of the first sessions has been overcome, the students themselves can act as facilitators with classmates who do not know the rules of a game.

REFERENCES

Abdul Jabbar, A. I., & Felicia, P. (2015). Gameplay engagement and learning in game-based learning: A systematic review. *Review of Educational Research*, *85*(4), 740–779. doi:10.3102/0034654315577210

Best, J. R. (2012). Exergaming immediately enhances children's executive function. *Developmental Psychology*, *48*(5), 1501–1510. Advance online publication. doi:10.1037/a0026648 PMID:22148945

Blair, C., & Razza, R. P. (2007). Relating effortful control, executive function, and false belief understanding to emerging math and literacy ability in kindergarten. *Child Development*, *78*(2), 647–663. Advance online publication. doi:10.1111/j.1467-8624.2007.01019.x PMID:17381795

Borella, E., Carretti, B., & Pelegrina, S. (2010). The Specific Role of Inhibition in Reading Comprehension in Good and Poor Comprehenders. *Journal of Learning Disabilities*, *43*(6), 541–552. Advance online publication. doi:10.1177/0022219410371676 PMID:20606207

Brankaer, C., Ghesquière, P., & de Smedt, B. (2013). The development of numerical magnitude processing and its association with working memory in children with mild intellectual disabilities. *Research in Developmental Disabilities*, *34*(10), 3361–3371. Advance online publication. doi:10.1016/j.ridd.2013.07.001 PMID:23911546

Brown, T. E., & Landgraf, J. M. (2010). Improvements in executive function correlate with enhanced per-formance and functioning and health-related quality of life: Evidence from 2 large, double-blind, randomized, placebo-controlled trials in ADHD. *Postgraduate Medicine*, *122*(5), 42–51. Advance online publication. doi:10.3810/pgm.2010.09.2200 PMID:20861587

Bryant, D. P., Bryant, B. R., & Smith, D. D. (2019). Teaching students with special needs in inclusive class-rooms. SAGE Publications Ltd.

Bryce, D., Whitebread, D., & Szűcs, D. (2015). The relationships among executive functions, metacognitive skills and educational achievement in 5 and 7 year-old children. *Metacognition and Learning*, *10*(2), 181–198. Advance online publication. doi:10.100711409-014-9120-4

Bull, R., & Scerif, G. (2001). Executive functioning as a predictor of children's mathematics ability: Inhibition, switching, and working memory. *Developmental Neuropsychology, 19*(3), 273–293. Advance online publication. doi:10.1207/S15326942DN1903_3 PMID:11758669

Camacho-Sánchez, R., Rillo-Albert, A., & Lavega-Burgués, P. (2022). Gamified Digital Game-Based Learning as a Pedagogical Strategy: Student Academic Performance and Motivation. *Applied Sciences (Basel, Switzerland), 12*(21), 11214. doi:10.3390/app122111214

Cano, S., Collazos, C. A., Aristizabal, L. F., Moreira, F., Peñeñory, V. M., & Agredo, V. (2018). Designing Collaborative Strategies Supporting Literacy Skills in Children with Cochlear Implants Using Serious Games. *Advances in Intelligent Systems and Computing, 2*(6), 1317–1326. doi:10.1007/978-3-319-77712-2_126

Carlson, S. M., Moses, L. J., & Breton, C. (2002). How Specific is the Relation between Executive Function and Theory of Mind? Contributions of Inhibitory Control and Working Memory. *Infant and Child Development, 11*(2), 73–92. Advance online publication. doi:10.1002/icd.298

Charlton, B., Williams, R. L., & McLaughlin, T. F. (2005). Educational games: A technique to accelerate the acquisition of reading skills of children with learning disabilities. *International Journal of Special Education, 20*(2), 66–72.

Corso, H. V., Mara, T., de Jou, G. I., & Salles, J. F. (2013). Metacognição e funções executivas: Relações entre os conceitos e implicações para a aprendizagem. *Psicologia: Teoria e Pesquisa (Brasília), 29*(1), 21–29. Advance online publication. doi:10.1590/S0102-37722013000100004

De Freitas, S. (2018). International Forum of Educational Technology & Society Are Games Effective Learning Tools? A Review of Educational Games. *Journal of Educational Technology & Society, 21*(2), 74–84. doi:10.2307/26388380

Dell'Anna, S., Pellegrini, M., & Ianes, D. (2019). Experiences and learning outcomes of students without special educational needs in inclusive settings: A systematic review. *International Journal of Inclusive Education, 25*(1), 944–959. Advance online publication. doi:10.1080/13603116.2019.1592248

Diamond, A. (2013). Executive functions. *Annual Review of Psychology, 64*(1), 135–168. Advance online publication. doi:10.1146/annurev-psych-113011-143750 PMID:23020641

Diamond, A., & Lee, K. (2011). Interventions shown to aid executive function development in children 4 to 12 years old. *Science, 333*(6045), 959–964. Advance online publication. doi:10.1126cience.1204529 PMID:21852486

Duncan, G. J., Dowsett, C. J., Claessens, A., Magnuson, K., Huston, A. C., Klebanov, P., Pagani, L. S., Feinstein, L., Engel, M., Brooks-Gunn, J., Sexton, H., Duckworth, K., & Japel, C. (2007). School Readiness and Later Achievement. *Developmental Psychology, 43*(6), 1428–1446. Advance online publication. doi:10.1037/0012-1649.43.6.1428 PMID:18020822

el Mawas, N., Bratu, M., Caraman, D., & Muntean, C. (2019). Investigating the Learning Impact of Game-based Learning when Teaching Science to Children with Special Learning Needs. *30th Annual Conference of the Society for Information Technology and Teacher Education SITE.*

Emerson, A., Cloude, E. B., Azevedo, R., & Lester, J. (2020). Multimodal learning analytics for game-based learning. *British Journal of Educational Technology*, *51*(5), 1505–1526. doi:10.1111/bjet.12992

Fairchild, G., van Goozen, S. H. M., Stollery, S. J., Aitken, M. R. F., Savage, J., Moore, S. C., & Goodyer, I. M. (2009). Decision Making and Executive Function in Male Adolescents with Early-Onset or Adolescence-Onset Conduct Disorder and Control Subjects. *Biological Psychiatry*, *66*(2), 162–168. Advance online publication. doi:10.1016/j.biopsych.2009.02.024 PMID:19362293

Florian, L. (2021). The Universal Value of Teacher Education for Inclusive Education. In A. Köpfer, J. J. W. Powell, & R. Zahnd (Eds.), *Handbuch Inklusion international / International Handbook of Inclusive Education: Globale, nationale und lokale Perspektiven auf Inklusive Bildung / Global, National and Local Perspectives* (1st ed., pp. 89–106). Verlag Barbara Budrich. doi:10.2307/j.ctv1f70kvj.8

Friehs, M. A., Dechant, M., Vedress, S., Frings, C., & Mandryk, R. L. (2020). Effec-tive gamification of the stop-signal task: Two controlled laboratory experiments. *JMIR Serious Games*, *8*(3), e17810. Advance online publication. doi:10.2196/17810 PMID:32897233

Fuhs, M. W., Nesbitt, K. T., Farran, D. C., & Dong, N. (2014). Longitudinal associations between executive functioning and academic skills across content areas. *Developmental Psychology*, *50*(6), 1698–1709. Advance online publication. doi:10.1037/a0036633 PMID:24749550

Gathercole, S. E., Pickering, S. J., Knight, C., & Stegmann, Z. (2004). Working memory skills and educational attainment: Evidence from national curriculum assessments at 7 and 14 years of age. *Applied Cognitive Psychology*, *18*(1), 1–16. Advance online publication. doi:10.1002/acp.934

Ge, X., & Ifenthaler, D. (2018). Designing engaging educational games and assessing engagement in game-based learning. *Ophthalmology: Breakthroughs in Research and Practice*, 1-18. doi:10.4018/978-1-5225-5198-0.ch001

Gilbert, S. J., & Burgess, P. W. (2008). Executive function. *Current Biology*, *18*(3), 110–114. doi:10.1016/j.cub.2007.12.014 PMID:18269902

Holley, S. R., Ewing, S. T., Stiver, J. T., & Bloch, L. (2017). The Relationship Between Emotion Regulation, Executive Functioning, and Aggressive Behaviors. *Journal of Interpersonal Violence*, *32*(11), 1692–1707. Advance online publication. doi:10.1177/0886260515592619 PMID:26130684

Ibáñez-Córdoba, S. X. (2021). Las prácticas pedagógicas en la formación inicial del profesorado de química y su relación con la construcción de conocimiento profesional: Reflexiones y apuestas desde una revisión documental. *Noria Investigación Educativa*, *2*(8), 35–55. doi:10.14483/25905791.18162

Indriasari, T. D., Luxton-Reilly, A., & Denny, P. (2020). Gamification of student peer review in education: A systematic literature review. *Education and Information Technologies*, *25*(6), 5205–5234. doi:10.100710639-020-10228-x

Jansen, B. R. J., de Lange, E., & van der Molen, M. J. (2013). Math practice and its influence on math skills and executive functions in adolescents with mild to borderline intellectual disability. *Research in Developmental Disabilities*, *34*(5), 1815–1824. Advance online publication. doi:10.1016/j.ridd.2013.02.022 PMID:23511346

Jong, M. S. Y. (2015). Does online game-based learning work in formal education at school? A case study of VISOLE. *Curriculum Journal*, *26*(2), 249–267. Advance online publication. doi:10.1080/09585176.2015.1018915

Kalogiannakis, M., Papadakis, S., & Zourmpakis, A. I. (2021). Gamification in Science Education. A Systematic Review of the Literature. *Education Sciences*, *11*(1), 1–36. doi:10.3390/educsci11010022

Ke, F. (2008). A case study of computer gaming for math: Engaged learning from gameplay? *Computers & Education*, *51*(4), 1609–1620. Advance online publication. doi:10.1016/j.compedu.2008.03.003

Lan, K. Y., & Hsiao, I. (2018). Effective Learning Design of Game-Based 3D Virtual Language Learning Environments for Special Education Students. *Journal of Educational Technology & Society*, *21*(3), 213. https://www.researchgate.net/publication/326250187

Lavega, P., Aráujo, P., & Jaqueira, A. R. (2013). Teaching motor and emotional competencies in university students. *Cultura, Ciencia y Deporte*, *8*(22), 5–15. Advance online publication. doi:10.12800/ccd.v8i22.219

Lezak, M. D. (1982). The problem of assessing executive functions. *International Journal of Psychology*, *17*(1–4), 281–297. Advance online publication. doi:10.1080/00207598208247445

Marschark, M., & Hauser, P. C. (2008). Cognitive underpinnings of learning by deaf and hard-of-hearing students: Differences, diversity, and directions. In M. Marschark & P. C. Hauser (Eds.), *Deaf cognition: Foundations and outcomes* (pp. 3–23). Oxford University Press. doi:10.1093/acprof:oso/9780195368673.003.0001

Miyake, A., Friedman, N. P., Emerson, M. J., Witzki, A. H., Howerter, A., & Wager, T. D. (2000). The Unity and Diversity of Executive Functions and Their Contributions to Complex "Frontal Lobe" Tasks: A Latent Variable Analysis. *Cognitive Psychology*, *41*(1), 49–100. Advance online publication. doi:10.1006/cogp.1999.0734 PMID:10945922

Morrison, F. J., Ponitz, C. C., & McClelland, M. M. (2010). Self-regulation and academic achievement in the transition to school. In S. D. Calkins & M. A. Bell (Eds.), *Child development at the intersection of emotion and cognition* (pp. 203–224). American Psychological Association. doi:10.1037/12059-011

Müller, U., & Kerns, K. (2015). The Development of Executive Function development of EF and its neural. Handbook of Child Psychology and Developmental Science: Cognitive Processes.

Norwich, B., Kelly, N., & Not Available, E. P. T. (2006). Evaluating children's Participation in SEN procedures: Lessons for educa-tional psychologists. *Educational Psychology in Practice*, *22*(3), 255–271. Advance online publication. doi:10.1080/02667360600845836

Plass, J. L., Homer, B. D., & Kinzer, C. K. (2015). Foundations of Game-Based Learning. *Educational Psychologist*, *50*(4), 258–283. Advance online publication. doi:10.1080/00461520.2015.1122533

Powers, K. L., Brooks, P. J., Aldrich, N. J., Palladino, M. A., & Alfieri, L. (2013). Effects of video-game play on information processing: A meta-analytic investigation. *Psychonomic Bulletin and Review, 20*(6). doi:10.3758/s13423-013-0418-z

Riggs, N. R., Jahromi, L. B., Razza, R. P., Dillworth-Bart, J. E., & Mueller, U. (2006). Executive function and the promotion of social-emotional competence. *Journal of Applied Developmental Psychology*, *27*(4), 300–309. Advance online publication. doi:10.1016/j.appdev.2006.04.002

Roffey, S. (2009). Promoting social and emotional learning with games: "It's fun and we learn things.". *Simulation & Gaming*, *40*(5), 626–644. Advance online publication. doi:10.1177/1046878109333793

Sakarneh, M. A., & Al-Swelmyeen, M. B. (2020). The extent to which the Jordanian inclusive basic school teachers use the constructivism theory in teaching. *Journal of Educational and Social Research*, *10*(1), 182. Advance online publication. doi:10.36941/jesr-2020-0017

Sheehy, K., Rix, J., Collins, J., Hall, K., Nind, M., & Wearmouth, J. (2009). *A systematic review of whole class, subject based, pedagogies with reported outcomes for the academic and social inclusion of pupils with special educational needs in mainstream classrooms. EPPI-Centre, Social Science Research Unit, Institute of Education*. University of London.

Söderqvist, S., Nutley, S. B., Ottersen, J., Grill, K. M., & Klingberg, T. (2012). Computerized training of non-verbal reasoning and working memory in children with intellectual disability. *Frontiers in Human Neuroscience*, *6*. Advance online publication. doi:10.3389/fnhum.2012.00271 PMID:23060775

St Clair-Thompson, H. L., & Gathercole, S. E. (2006). Executive functions and achievements in school: Shifting, updating, inhibition, and working memory. *Quarterly Journal of Experimental Psychology*, *59*(4), 745–759. Advance online publication. doi:10.1080/17470210500162854 PMID:16707360

Taylor Tavares, J., Clark, L., Cannon, D. M., Erickson, K., Drevets, W. C., & Sahakian, B. J. (2007). Distinct Profiles of Neurocognitive Function in Unmedicated Unipolar Depression and Bipolar II Depression. *Biological Psychiatry*, *62*(8), 917–924. Advance online publication. doi:10.1016/j.biopsych.2007.05.034 PMID:17825802

Terzi, L. (2005). Beyond the dilemma of difference: The capability approach to disability and Special Edu-cational Needs. *Journal of Philosophy of Education*, *39*(3), 443–459. Advance online publication. doi:10.1111/j.1467-9752.2005.00447.x

UNESCO. (1994). The Salamanca statement and framework for action on special needs education. In *The Salamanca statement and framework for action on special needs education*. UNESCO.

Villaverde-Caramés, E. J., Fernández-Villarino, M. Á., Toja-Reboredo, B., & González Valeiro, M. Á. (2021). Revisión de la literatura sobre las características que definen a un buen docente de Educación Física: Consideraciones desde la formación del profesorado [A literature review of the characteristics that define a good physical education teacher considera-tions]. *Retos*, (41), 471–479. doi:10.47197/retos.v0i41.84421

Wei, X., Lenz, K. B., & Blackorby, J. (2013). Math Growth Trajectories of Students with Disabilities: Disa-bility category, gender, racial, and Socioeconomic Status Differences From Ages 7 to 17. *Remedial and Special Education*, *34*(3), 154–165. Advance online publication. doi:10.1177/0741932512448253

Will Crescioni, A., Ehrlinger, J., Alquist, J. L., Conlon, K. E., Baumeister, R. F., Schatschneider, C., & Dutton, G. R. (2011). High trait self-control predicts positive health behaviors and success in weight loss. *Journal of Health Psychology*, *16*(5), 750–759. Advance online publication. doi:10.1177/1359105310390247 PMID:21421645

Zelazo, P. D. (2006). The Dimensional Change Card Sort (DCCS): A method of assessing executive function in children. *Nature Protocols*, *1*(1), 297–301. Advance online publication. doi:10.1038/nprot.2006.46 PMID:17406248

KEY TERMS AND DEFINITIONS

Disabilities: Physical, cognitive, sensory, or developmental impairments that may affect a person's ability to perform certain tasks, participate in activities, or interact with their environment. Disabilities can vary widely in their nature, severity, and impact on an individual's daily life.

Executive Functions: Executive functions refer to a set of cognitive processes that enable individuals to plan, organize, initiate, monitor, and control their thoughts, actions, and behaviors. These functions are responsible for managing higher-order mental processes and play a crucial role in goal-directed behavior, problem-solving, decision-making, and self-regulation.

Game-Based Learning: Refers to an educational approach that incorporates elements of gameplay into the learning process. It involves using games, either digital or physical, as a medium for delivering instructional content, engaging learners, and promoting active participation and problem-solving skills.

Higher Education: Higher education refers to the level of education that follows secondary education (high school) and provides advanced academic and professional knowledge and skills. It typically includes undergraduate programs, such as bachelor's degrees, and postgraduate programs, such as master's degrees and doctoral degrees. Higher education institutions include universities, colleges, and professional schools.

Inclusive Education: Is an approach to education that aims to ensure that all students, regardless of their abilities, disabilities, or other characteristics, can fully participate in and benefit from the educational environment. It emphasizes providing equitable access, opportunities, and support to all learners, promoting diversity, and fostering a sense of belonging for every student.

Special Needs: Is an umbrella term used in the field of education to refer to students who require additional support and accommodations to access and benefit from the educational environment. These students may have physical, cognitive, sensory, emotional, or behavioral challenges that require specialized interventions and individualized educational plans.

Compilation of References

Abdul Jabbar, A. I., & Felicia, P. (2015). Gameplay engagement and learning in game-based learning: A systematic review. *Review of Educational Research*, *85*(4), 740–779. doi:10.3102/0034654315577210

Abramovich, S., Schunn, C., & Higashi, R. M. (2013). Are badges useful in education? It depends upon the type of badge and expertise of learner. *Educational Technology Research and Development*, *61*(2), 217–232. doi:10.100711423-013-9289-2

Abu-Al-Aish, A., & Love, S. (2013). Factors influencing students' acceptance of M-learning: An investigation in higher education. *International Review of Research in Open and Distance Learning*, *14*(5), 82–107. doi:10.19173/irrodl.v14i5.1631

Agreda-Montoro, M., Ortíz-Colón, A. M., & Rodríguez, J. (2020). Gamificación y competencia digital Escape Rooms para la formación de los profesionales educativos. La tecnología como eje del cambio metodológico, 693-696.

Aguilera-Hermida, A. P. (2020). Uso y aceptación de estudiantes universitarios del aprendizaje en línea de emergencia debido a la COVID-19. *Revista internacional de investigación educativa abierta*. doi:10.1016/j.ijedro.2020.100011

Akbulut, Y., Odabaşı, H., & Kuzu, A. (2011). Perceptions of preservice teachers regarding the integration of information and communication technologies in Turkish education faculties. *The Turkish Online Journal of Educational Technology*, *10*(3), 175–184.

Akman, Y. (2021). Dijital okuryazarlık, çevrim içi öğrenme ve akademik isteklilik arasındaki ilişkinin incelenmesi. *The Journal of Turkish Educational Sciences*, *19*(2), 1012–1036. doi:10.37217/tebd.982846

Akour, H. (2009). *Determinants of mobile learning acceptance: an empirical investigation in higher education* [Unpublished PhD dissertation]. Oklahoma State University.

Al-Adwan, A., Al-Adwan, A., & Smedley, J. (2018). Exploring students' acceptance of e-learning using Technology Acceptance Model in Jordanian universities. *International Journal of Education and Development Using Information and Communication Technology*, *9*(2), 4–18.

Aldossary, K. (2021). Online distance learning for translation subjects: Tertiary level instructors' and students' perceptions in Saudi Arabia. *Turkish Online Journal of Distance Education*, *22*(6), 96–109. doi:10.17718/tojde.961821

Alegre, M., & Silva, C. (2020). Teaching practices in Higher Education: Didactic analysis of institutional conferences at the IES of Puerto Tirol—Chaco, Argentina. *Magazine of Studies and Experiences in Education*, *19*(39), 287–304. doi:10.21793/rexe.20201939alegre17

Aleixandre-Benavent, R., Castelló-Cogollos, L., & Valderrama-Zurián, J.C. (2020). Information and communication during the first months of Covid-19. Timeline, infodemic and misinformation, fake news, ongoing investigations, and the role of information specialist. *The Information Professional*, e290408. doi:10.3145/epi.2020.jul.08

Al-Emran, M., Elsherif, H. M., & Shaalan, K. (2016). Investigating attitudes towards the use of mobile learning in higher education. *Computers in Human Behavior*, *56*, 93–102. doi:10.1016/j.chb.2015.11.033

Alimohamadi, Y., Hosamirudsari, H., Hesari, E., & Sepandi, M. (2022). Global COVID-19 vaccine acceptance rate: A systematic review and meta-analysis. *Zeitschrift für Gesundheitswissenschaften*, ●●●, 1–13. doi:10.100710389-022-01757-5 PMID:36188446

Al-Jarrah, J. M., Talafhah, R. H., & Al-Jarrah, T. M. (2019). ESL teacher perceptions of using educational mobile applications to develop the language skills of ESL elementary school students. *European Journal of Foreign Language Teaching.*, *4*(1), 65–86.

Aljawarneh, S. A. (2020). Reviewing and exploring innovative ubiquitous learning tools in higher education. *Journal of Computing in Higher Education*, *32*(1), 57–73. doi:10.100712528-019-09207-0

Allo, M. D. G. (2020). Is the online learning good in the midst of Covid-19 pandemic? The case of EFL learners. *Journal Sinestesia*, *10*(1), 1–10. https://sinestesia.pustaka.my.id/journal/article/view/24

Ally, M. (2004). Foundations of educational theory for online learning. In T. Anderson (Ed.), *The theory and practice of online learning* (pp. 3–31). Athabasca University.

Almaiah, M. A. (2018). Acceptance and usage of a mobile information system services in University of Jordan. *Education and Information Technologies*, *23*(5), 1873–1895. doi:10.100710639-018-9694-6

Almaiah, M. A., & Alismaiel, O. A. (2019). Examination of factors influencing the use of mobile learning system: An empirical study. *Education and Information Technologies*, *24*(1), 885–909. doi:10.100710639-018-9810-7

Almaiah, M. A., Al-Khasawneh, A., & Althunibat, A. (2020). Exploring the critical challenges and factors influencing the e-learning system usage during COVID-19 pandemic. *Education and Information Technologies*, *25*(6), 5261–5280. doi:10.100710639-020-10219-y PMID:32837229

Almaiah, M. A., Jalil, M. A., & Man, M. (2016). Extending the TAM to examine the effects of quality features on mobile learning acceptance. *Journal of Computers in Education*, *3*(4), 453–485. doi:10.100740692-016-0074-1

Al-Mamary, Y. H. S. (2022). Why do students adopt and use Learning Management Systems?: Insights from Saudi Arabia. *International Journal of Information Management Data Insights*, *2*(2), 100088. doi:10.1016/j.jjimei.2022.100088

Alowayr, A. (2022). Determinants of mobile learning adoption: Extending the unified theory of acceptance and use of technology (UTAUT). *The International Journal of Information and Learning Technology*, *39*(1), 1–12. doi:10.1108/IJILT-05-2021-0070

Al-Qerem, W. A., & Jarab, A. S. (2021). COVID-19 vaccination acceptance and its associated factors among a middle eastern population. *Frontiers in Public Health*, *9*, 632914. Advance online publication. doi:10.3389/fpubh.2021.632914 PMID:33643995

Al-Rahmi, A. M., Al-Rahmi, W. M., Alturki, U., Aldraiweesh, A., Almutairy, S., & Al-Adwan, A. S. (2022). Acceptance of mobile technologies and M-learning by university students: An empirical investigation in higher education. *Education and Information Technologies*, *27*(6), 7805–7826. doi:10.100710639-022-10934-8

Al-Seghayer, K. (2014). The four most common constraints affecting English teaching in Saudi Arabia. *International Journal of English Linguistics*, *4*(5), 17–26. doi:10.5539/ijel.v4n5p17

Al-Shehri, S. (2014). Mobile learning in the Arab world: Contemporary and future implications. In X. Xu (Ed.), *Interdisciplinary mobile media and communications: social, political, and economic implications* (pp. 48–62). IGI Global. doi:10.4018/978-1-4666-6166-0.ch003

AlShurman, B. A., Khan, A. F., Mac, C., Majeed, M., & Butt, Z. A. (2021). What demographic, social, and contextual factors influence the intention to use COVID-19 vaccines: A scoping review. *International Journal of Environmental Research and Public Health*, *18*(17), 17. Advance online publication. doi:10.3390/ijerph18179342 PMID:34501932

Alsina M. & Farrés, I. (2021). ¿Jugar o aprender? El aprendizaje lúdico en la formación musical del maestro. *Revista electrónica complutense de investigación en educación musical, 18*, 83-96. . doi:10.5209/reciem.67853

Alvarez-Herrero, J. F., & Fernández-Herrero, J. (2020). *Aprendizaje online versus presencial. Percepciones del alumnado del grado de Magisterio en Educación Infantil*. Octaedro.

Amo, J. M., & Baldrich, K. (2021). Prácticas letradas en el ámbito universitario: los hábitos lectores del alumnado de Grado en Educación Infantil de la Universidad de Almería. *Tonos Digital, 40*(9).

Amponsah, S., Torto, B. A., & Badu-Nyarko, S. K. (2018). Ghanaian mature students' motivation to pursue degree programmes through distance education. *International Review of Education*, *64*(5), 585–606. doi:10.100711159-018-9728-8

Amutha, D. (2020). The Role and impact of ICT in improving the quality of education. SSRN *Electronic Journal*. doi:10.2139/ssrn.3585228

Anderson, N. (2009). *Equity and information communication technology (ICT) in education*. Peter Lang.

Angermeyer, M. C., & Matschinger, H. (2005). Labelling -Stereotype- Discrimination. An Investigation of the Stigma Process. *Social Psychiatry and Psychiatric Epidemiology*, *40*, 391–395. doi:10.100700127-005-0903-4 PMID:15902409

Arat, T., & Bakan, Ö. (2014). Uzaktan Eğitim ve Uygulamaları. *Selçuk Üniversitesi Sosyal Bilimler Meslek Yüksekokulu Dergisi*, *14*(1-2), 363–374.

Araya, S., Rodríguez, A., Badilla, N., & Marchena, K. (2021). The flipped classroom as a teaching resource in the Costa Rican context: Case study on its implementation in a secondary school. *Education Magazine*, *46*(1), 103–119. doi:10.15517/revedu.v46i1.44333

Aristovnik, A., Keržiˇ, D., Ravšelj, D., Tomaževiˇ, N., & Umek, L. (2020). Impacts of the COVID-19 Pandemic on Life of Higher Education Students: A Global Perspective. *Sustainability, 12*, 1-34. https://doi.org/https://doi.org/10.3390/su12208438

Armando, N., Almeida, R., Fernandes, J. M., Silva, J. S., & Boavida, F. (2021). End-to-end experimentation of a 5G vertical within the scope of blended learning. *Discover Internet of Things*, *1*(9), 9. Advance online publication. doi:10.100743926-021-00008-5

Armie, M., & Membrive Pérez, V. (2022). *Escaping boredom in the classroom: Breakouts, breakout boxes and escape rooms*. Cambridge Scholars Publishing.

Arnold, B. J. (2014). Gamification in education. *Proceedings of the American Society of Business and Behavioral Sciences*, *21*(1), 32–39.

Arsenault, C., Harper, S., Nandi, A., Rodríguez, J. M. M., Hansen, P. M., & Johri, M. (2017). Monitoring equity in vaccination coverage: A systematic analysis of demographic and health surveys from 45 Gavi-supported countries. *Vaccine*, *35*(6), 951–959. doi:10.1016/j.vaccine.2016.12.041 PMID:28069359

Ausín, V., Abella, V., Delgado, V., & Hortigüela, D. (2016). Project-based learning through ICT: An experience of teaching innovation from university classrooms. *Formación Universitaria*, *9*(3), 31–38. doi:10.4067/S0718-50062016000300005

Aydın, B. (2017). Three birds with a stone: Technology integration in language education with reverse mentoring model. *Journal of Teacher Education and Educators*, *6*(2), 177–190.

Aydın, S. (2013). Teachers' perceptions about the use of computers in EFL teaching and learning: The case of Turkey. *Computer Assisted Language Learning*, *26*(3), 214–233. doi:10.1080/09588221.2012.654495

Aytekin, A., Özköse, H., & Ayaz, A. (2022). Unified theory of acceptance and use of technology (UTAUT) in mobile learning adoption: Systematic literature review and bibliometric analysis. *COLLNET Journal of Scientometrics and Information Management*, *16*(1), 75–116. doi:10.1080/09737766.2021.2007037

Aznar-Díaz, I., Romero-Rodríguez, J., & Rodríguez-García, A. M. (2018). La tecnología móvil de realidad virtual en educación: Una revisión del estado de la literatura científica en España. *EDMETIC*, *7*(1), 256–274. doi:10.21071/edmetic.v7i1.10139

Aznar-Lou, I., Serrano-Blanco, A., & Luciano, J. V. (2017). Effectiveness of the "what's up!" intervention to reduce stigma and psychometric properties of the Youth Program Questionnaire (YPQ): Results from a cluster non-randomized controlled trial conducted in Catalan high schools. *Frontiers in Psychology*, *8*, 1608. doi:10.3389/fpsyg.2017.01608 PMID:28959228

Azzouz, N., & Gutierrez-Colon Plana, M. (2020). Effect of Gamification on students' motivation and learning achievement in Second Language Acquisition within higher education: a literature review 2011-2019. *The EuroCALL Review*, *28*, 40. . doi:10.4995/eurocall.2020.12974

Baba, Y., Nemoto, T., Tsujino, N., Yamaguchi, T., Katagiri, N., & Mizuno, M. (2017). Stigma toward psychosis and its formulation process: Prejudice and discrimination against early stages of schizophrenia. *Comprehensive Psychiatry*, *73*, 181–186. doi:10.1016/j.comppsych.2016.11.005 PMID:28040575

Baiocco, R., Pezzella, A., Pistella, J., Kouta, C., Rousou, E., Rocamora-Perez, P., López-Liria, R., Dudau, V., Doru, A. M., Kuckert-Wöstheinrich, A., Ziegler, S., Nielsen, D., Bay, L. T., & Papadopoulos, I. (2022). LGBT+ training needs for health and social care professionals: A cross-cultural comparison among seven european countries. *Sexuality Research & Social Policy*, *19*(1), 22–36. doi:10.100713178-020-00521-2

Baladrón Pazos, A. J., Correyero Ruiz, B., & Manchado Pérez, B. (2020). La transformación digital de la docencia universitaria en comunicación durante la crisis de la COVID-19 en España: una aproximación desde la perspectiva del alumnado. *Revista latina de comunicación social, 78*, 265-287. https://www.doi.org/10.4185/RLCS-2020-1477

Bandura, A. (1977). Self-efficacy: Toward a unifying theory of behavioral change. *Psychological Review*, *84*(2), 191–215. doi:10.1037/0033-295X.84.2.191 PMID:847061

Bano, M., Zowghi, D., Kearney, M., Schuck, S., & Aubusson, P. (2018). Mobile learning for science and mathematics school education: A systematic review of empirical evidence. *Computers & Education*, *121*, 30–58. doi:10.1016/j.compedu.2018.02.006

Barahona, A. (2020). Gamification for classroom management: An implementation, using ClassDojo. *Sustainability (Basel)*, *12*(22), 9371. Advance online publication. doi:10.3390u12229371

Baran, E. (2014). A review of research on mobile learning in teacher education. *Journal of Educational Technology & Society*, *17*(4), 17–32.

Barata, G., Gama, S., Jorge, J., & Gonçalvez, D. (2015). Gamification for smarter learning: tales from the trenches. *Smart Learning Environments, 2*(10).

Barbour, M. K., LaBonte, R., Kelly, K., Hodges, C., Moore, S., Lockee, B., Trust, T., Bond, A., & Hill, P. (2020). *Understanding pandemic pedagogy: Differences between emergency remote, remote, and online teaching*. A special report of the State of the Nation: K-12 E-Learning in Canada Project. Canadian eLearning Network.

Bárcena-Toyos, P. (2022). Teacher identity in CLIL: A case study of in-service teachers. *Latin American Journal of Content and Language Integrated Learning*, *15*(2), e1516. doi:10.5294/laclil.2022.15.1.6

Barnes, J. A., Jr. (2018). *Teachers' perceptions of implementing m-learning using pedagogical approaches* [Unpublished doctoral dissertation]. The University of Memphis, Memphis, TN.

Barral, A., Ardi-Pastores, V., & Simmons, R. (2018). Student Learning in an Accelerated Introductory Biology Course is Significantly Enhanced by a Flipped Learning Environment. *CBE Life Sciences Education*, *17*(3), 1–9. doi:10.1187/cbe.17-07-0129 PMID:30040530

Barry, C. L., McGinty, E. E., Pescosolido, B. A., & Goldman, H. H. (2014). Stigma, discrimination, treatment effectiveness, and policy: Public views about drug addiction and mental illness. *Psychiatric Services (Washington, D.C.)*, *65*(10), 1269–1272. doi:10.1176/appi.ps.201400140 PMID:25270497

Başal, A. (2015). English language teachers and technology education. *Journal of Theory and Practice in Education.*, *1*(4), 1496–1511.

Bates, A. W. (2015). *Teaching in a Digital Age: Guidelines for Designing Teaching and Learning*. Tony Bates Associates Ltd.

Bautista, J. M., Espigares, M. J., & Hernández, R. M. (2017). Aprendizaje Basado en Proyectos (ABP) ante el reto de una nueva enseñanza de las Ciencias. Rev. Bras. *Ensino Ciènc. Tecnol.*, *10*, 43–60.

Bawa'aneh, M. S. (2021). Distance learning during COVID-19 pandemic in UAE public schools: Student satisfaction, attitudes and challenges. *Contemporary Educational Technology*, *13*(3), ep304. Advance online publication. doi:10.30935/cedtech/10872

Bax, S. (2003). CALL—Past, present and future. *System*, *31*(1), 13–28. doi:10.1016/S0346-251X(02)00071-4

Bax, S. (2011). Normalisation revisited: The effective use of technology in language education. *International Journal of Computer-Assisted Language Learning and Teaching*, *1*(2), 1–15. doi:10.4018/ijcallt.2011040101

Bay, D. N. (2021). Digital literacy levels of preschool teacher candidates. *Mustafa Kemal University Journal of the Faculty of Education*, *5*(7), 172–187.

Beach, K. (2003). Consequential transitions: A developmental view of knowledge propagation through social organizations. In T. Tuomi-Gröhn & Y. Engeström (Eds.), *Between school and work: New perspectives on transfer and boundary-crossing* (pp. 39–62). Earli.

Beck, A., Lakkaraju, K., & Rai, V. (2017). Small is big: Interactive trumps passive information in breaking information barriers and impacting behavioral antecedents. *PLoS One*, *12*(1), e0169326. doi:10.1371/journal.pone.0169326 PMID:28099478

Beck, C., & Kosnik, C. (2006). *Innovations in teacher education: a social constructivist approach*. SUNY Press. doi:10.1353/book5061

Bedford, H., Attwell, K., Danchin, M., Marshall, H., Corben, P., & Leask, J. (2018). Vaccine hesitancy, refusal and access barriers: The need for clarity in terminology. *Vaccine*, *36*(44), 6556–6558. doi:10.1016/j.vaccine.2017.08.004 PMID:28830694

Bell, S. (2010). Project-based learning for the 21st century: Skills for the future. *The Clearing House: A Journal of Educational Strategies, Issues and Ideas*, *83*(2), 39–43. doi:10.1080/00098650903505415

Beltrán, A., & Romero-Riaño, E. (2020). El papel de la gamificación en la conciencia ambiental: Una revisión bibliométrica. Evolución de la estructura intelectual. *Prisma Social*, *30*, 161–185. https://revistaprismasocial.es/article/ view/3764

BeluceA. C.OliveiraK. L. D. (2015). Students' motivation for learning in virtual learning environments. *Paidéia (Ribeirão Preto)*, *25*, 105-113. doi:10.1590/1982-43272560201513

Bentham, C., Daunt, A., Taylor, S., & Simmons, M. (2013). Mental health workshops delivered by medical students in Cambridge secondary schools: An evaluation of learning. *Psychiatria Danubina*, *25*, 224–230. PMID:23995181

Beres, D. (2011). Mobile-assisted language learning from the student perspective: Encouraging effective language learning strategies outside of the classroom. In B. R. Facer & M. Abdous (Eds.), *Academic podcasting and mobile assisted language learning: applications and outcomes* (pp. 93–110). IGI Global. doi:10.4018/978-1-60960-141-6.ch006

Bergmann, J., & Sams, A. (2014). *Our story: How to create an "upside down class"?* SM Editions.

Best, J. R. (2012). Exergaming immediately enhances children's executive function. *Developmental Psychology*, *48*(5), 1501–1510. Advance online publication. doi:10.1037/a0026648 PMID:22148945

Black, P., & Wiliam, D. (2009). Developing the theory of formative assessment. *Educational Assessment, Evaluation and Accountability*, *21*(1), 5–31. doi:10.100711092-008-9068-5

Blair, C., & Razza, R. P. (2007). Relating effortful control, executive function, and false belief understanding to emerging math and literacy ability in kindergarten. *Child Development*, *78*(2), 647–663. Advance online publication. doi:10.1111/ j.1467-8624.2007.01019.x PMID:17381795

Blau, I., & Hameiri, M. (2010). Implementing technological change at schools: The impact of online communication with families on teacher interactions through learning management system. *Interdisciplinary Journal of E-Learning and Learning Objects*, *6*, 245–257. doi:10.28945/1313

Borella, E., Carretti, B., & Pelegrina, S. (2010). The Specific Role of Inhibition in Reading Comprehension in Good and Poor Comprehenders. *Journal of Learning Disabilities*, *43*(6), 541–552. Advance online publication. doi:10.1177/0022219410371676 PMID:20606207

Borg, S. (2005). *Teacher cognition in language teaching*. doi:10.1057/9780230523470_10

Borras, O. (2015). *Fundamentos de la gamificación*. Universidad Politécnica de Madrid.

Botte, B., Bakkes, S., & Veltkamp, R. (2020). Motivation in gamification: Constructing a correlation between gamification achievements and self-determination theory. *Games and Learning Alliance: 9th International Conference, GALA 2020, Laval, France, December 9-10, 2020 Proceedings*, *9*, 157–166.

Bozkurt, A. (2020). A global outlook to the interruption of education due to COVID-19 pandemic: Navigating in a time of uncertainty and crisis. *Asian Journal of Distance Education*, *15*(1), 1–126. doi:10.5281/zenodo.3878572

Bradley, V. M. (2021). Learning Management System (LMS) use with online instruction. *International Journal of Technology in Education*, *4*(1), 68–92. doi:10.46328/ijte.36

Brandtzæg, P. B., Heim, J., & Karahasanović, A. (2011). Understanding the new digital divide—A typology of Internet users in Europe. *International Journal of Human-Computer Studies*, *69*(3), 123–138. doi:10.1016/j.ijhcs.2010.11.004

Brankaer, C., Ghesquière, P., & de Smedt, B. (2013). The development of numerical magnitude processing and its association with working memory in children with mild intellectual disabilities. *Research in Developmental Disabilities*, *34*(10), 3361–3371. Advance online publication. doi:10.1016/j.ridd.2013.07.001 PMID:23911546

Brearley, L., Eggers, R., Steinglass, R., & Vandelaer, J. (2013). Applying an equity lens in the Decade of Vaccines. *Vaccine, 31*, 103–107. doi:10.1016/j.vaccine.2012.11.088 PMID:23598470

Brørs, G., Norman, C. D., & Norekvål, T. N. (2020). Accelerated importance of eHealth literacy in the COVID-19 outbreak and beyond. *European Journal of Cardiovascular Nursing, 19*(6), 458–461. doi:10.1177/1474515120941307 PMID:32667217

Brown, H. D. (2007). *Principles of language learning and teaching*. Pearson Education.

Brown, T. E., & Landgraf, J. M. (2010). Improvements in executive function correlate with enhanced per-formance and functioning and health-related quality of life: Evidence from 2 large, double-blind, randomized, placebo-controlled trials in ADHD. *Postgraduate Medicine, 122*(5), 42–51. Advance online publication. doi:10.3810/pgm.2010.09.2200 PMID:20861587

Brown, T. H., & Mbati, L. S. (2015). Mobile learning: Moving past the myths and embracing the opportunities. *International Review of Research in Open and Distance Learning, 16*(2), 115–135. doi:10.19173/irrodl.v16i2.2071

Brusilovsky, P., & Millán, E. (2007). User models for adaptive hypermedia and adaptive educational systems. In *The Adaptive Web* (pp. 3–53). Springer. doi:10.1007/978-3-540-72079-9_1

Bryant, D. P., Bryant, B. R., & Smith, D. D. (2019). Teaching students with special needs in inclusive class-rooms. SAGE Publications Ltd.

Bryant, A., & Charmaz, K. (Eds.). (2007). *The SAGE handbook of grounded theory*. Sage Publications. doi:10.4135/9781848607941

Bryce, D., Whitebread, D., & Szűcs, D. (2015). The relationships among executive functions, metacognitive skills and educational achievement in 5 and 7 year-old children. *Metacognition and Learning, 10*(2), 181–198. Advance online publication. doi:10.100711409-014-9120-4

Bryman, A. (2012). *Social research methods* (4th ed.). Oxford University Press.

Bucy, E. P., & Newhagen, J. E. (2004). *Media access: Social and psychological dimensions of new technology use*. Psychology Press.

Bulanda, J. J., Bruhn, C., Byro-Johnson, T., & Zentmyer, M. (2014). Addressing mental health stigma among young adolescents: Evaluation of a youth-led approach. *Health & Social Work, 39*(2), 73–80. doi:10.1093/hsw/hlu008 PMID:24946423

Bull, R., & Scerif, G. (2001). Executive functioning as a predictor of children's mathematics ability: Inhibition, switching, and working memory. *Developmental Neuropsychology, 19*(3), 273–293. Advance online publication. doi:10.1207/S15326942DN1903_3 PMID:11758669

Bustamante, J. C., Segura-Berges, M., Lizalde-Gil, M., & Peñarrubia-Lozano, C. (2022). análisis cualitativos de la implementación del e-learning y la enseñanza híbrida durante la pandemia del COVID-19 en las universidades españolas. *Sostenibilidad, 14*(19), 1–19. doi:10.3390u141912003

Byram, M., Nichols, A., & Stevens, D. (2001). Introduction. In M. Byram, A. Nichols, & D. Stevens (Eds.), *Doing intercultural competence in practice* (pp. 1–8). Multilingual Matters Ltd.

Cabero, J., & Llorente, C. (2020). Covid-19: radical transformation of digitization in university institutions. *Virtual Campus, 9*(2), 25-34. (www.revistacampusvirtuales.es)

Calvo Roselló, V., & López Rodríguez, M. I. (2021). Gamificación en el aula universitaria: Una experiencia de "escape room inversa.". *REDU Revista de Docencia Universitaria, 19*(1), 45–74. doi:10.4995/redu.2021.15666

Camacho-Sánchez, R., Rillo-Albert, A., & Lavega-Burgués, P. (2022). Gamified Digital Game-Based Learning as a Pedagogical Strategy: Student Academic Performance and Motivation. *Applied Sciences (Basel, Switzerland), 12*(21), 11214. doi:10.3390/app122111214

Campus Explorer. (2019). *Different Types of Distance Learning*. Retrieved December 12, 2022, from https://www.campusexplorer.com/college-advice-tips/7021E31E/Different-Types-of-Distance-Learning/

Cangalaya-Sevillano, L. M., Casazola-Cruz, O. D., & Aguilar, J. A. F. (2022). Gamificación en el proceso de enseñanza y aprendizaje de estudiantes universitarios. *Horizontes revista de investigación en ciencias de la educación, 6*(23), 637-647. . doi:10.33996/revistahorizontes.v6i23.364

Cangas, A. J., & Galván, B. (2020). Estudio preliminar de la utilidad de un programa de realidad virtual contra el estigma en salud mental. *International Journal of Developmental and Educational Psychology, 1*(1), 45–52. doi:10.17060/ijodaep.2020.n1.v1.1760

Cangas, A. J., Navarro, N., Aguilar-Parra, J. M., Trigueros, R., Gallego, J., Zárate, R., & Gregg, M. (2019). Analysis of the Usefulness of a Serious Game to Raise Awareness about Mental Health Problems in a Sample of High School and University Students: Relationship with Familiarity and Time Spent Playing Video Games. *Journal of Clinical Medicine, 8*(10), 1504. doi:10.3390/jcm8101504 PMID:31547017

Cangas, A. J., Navarro, N., Parra, J., Ojeda, J. J., Cangas, D., Piedra, J. A., & Gallego, J. (2017). Stigma-Stop: A serious game against the stigma toward mental health in educational settings. *Frontiers in Psychology, 8*, 1385. doi:10.3389/fpsyg.2017.01385 PMID:28878702

Cano, S., Collazos, C. A., Aristizabal, L. F., Moreira, F., Peñeñory, V. M., & Agredo, V. (2018). Designing Collaborative Strategies Supporting Literacy Skills in Children with Cochlear Implants Using Serious Games. *Advances in Intelligent Systems and Computing, 2*(6), 1317–1326. doi:10.1007/978-3-319-77712-2_126

Çapar, M., & Keçik, İ. (2017). Generation types view and perspectives of EFL instructors on their expectations related to professional development programs. *ELT Research Journal, 5*(2).

Caparrós, M. (2017). *Gamificación en educación: Una guía práctica*. Retrieved from https://www.trespuntoelearning.com/gamificacion-en-educacion-guia-practica/

Carlson, S. M., Moses, L. J., & Breton, C. (2002). How Specific is the Relation between Executive Function and Theory of Mind? Contributions of Inhibitory Control and Working Memory. *Infant and Child Development, 11*(2), 73–92. Advance online publication. doi:10.1002/icd.298

Caro Valverde, M. T. (2018). Argumentative communication in the Knowledge Society, key to distributed leadership for an educational change from professional development. *Journal of Distance Education, 56*, 1–30. https://revistas.um.es/red/article/view/321601

Casacuberta, D. (2003). *Collective creation. On the Internet the creator is the public*. Gedisa.

Casañas, R., Arfuch, V. M., Castellví, P., Gil, J. J., Torres, M., Pujol, A., & Caussa, A. (2018). EspaiJove. net"-a school-based intervention programme to promote mental health and eradicate stigma in the adolescent population: Study protocol for a cluster randomized controlled trial. *BMC Public Health, 18*(1), 939. doi:10.118612889-018-5855-1 PMID:30064404

Cassany, D., & Castellá, J. M. (2013). Approach to critical literacy. *Perspectiva, Florianópolis, 28*(2), 353–374. doi:10.5007/2175-795X.2010v28n2p353

Castellano Gil, J. M., Almagro, J., & Fajardo, Á. B. (2021). Percepción estudiantil sobre la educación online en tiempos de COVID-19: Universidad de Almería (España). *Revista Scientific*, *6*(19), 185–207. doi:10.29394/Scientific.issn.2542-2987.2021.6.19.9.185-207

Castells, M. (2002). *The Internet galaxy: Reflections on the Internet, business, and society*. Oxford University Press. doi:10.1093/acprof:oso/9780199255771.001.0001

Cataldi, J. R., Sevick, C., Pyrzanowski, J., Wagner, N., Brewer, S. E., Narwaney, K. J., Shoup, J. A., Resnicow, K., Glanz, J., Dempsey, A., & Kwan, B. M. (2019). Addressing personal parental values in decisions about childhood vaccination: Measure development. *Vaccine*, *37*(38), 5688–5697. doi:10.1016/j.vaccine.2019.08.009 PMID:31421930

Çavuş, N., & İbrahim, D. (2017). Learning English using children's stories in mobile devices: Children's stories in mobile devices. *British Journal of Educational Technology*, *48*(2), 625–641. doi:10.1111/bjet.12427

Centres for Disease Control and Prevention (CDC). (2022). *Answers to your most common questions about childhood vaccines*. https://www.cdc.gov/vaccines/parents/FAQs.html

Centro IDEA-UMinho. (2020). *Quais são os objetivos do Centro IDEA-UMinho?* Retrieved December 12, 2022, from https://idea.uminho.pt/pt/quem-somos/Paginas/missao.aspx

Chan, S. S., So, W. K., Wong, D. C., Lee, A. C. K., & Tiwari, A. (2007). Improving older adults' knowledge and practice of preventive measures through a telephone health education during the SARS epidemic in Hong Kong: A pilot study. *International Journal of Nursing Studies*, *244*(7), 1120–1127. doi:10.1016/j.ijnurstu.2006.04.019 PMID:16857203

Chao, C. C. (2015). Rethinking transfer: Learning from CALL teacher education as consequential transition. *Language Learning & Technology*, *19*(1), 102–118.

Charlton, B., Williams, R. L., & McLaughlin, T. F. (2005). Educational games: A technique to accelerate the acquisition of reading skills of children with learning disabilities. *International Journal of Special Education*, *20*(2), 66–72.

Charmaz, K. (2006). *Constructing grounded Theory: a practical guide through qualitative analysis*. Sage Publications.

Chartier, R. (2018). *The revolutions of written culture*. Gedisa.

Chavoshi, A., & Hamidi, H. (2019). Social, individual, technological and pedagogical factors influencing mobile learning acceptance in higher education: A case from Iran. *Telematics and Informatics*, *38*, 133–165. doi:10.1016/j.tele.2018.09.007

Chehaibar, L. (2020). Curriculum flexibility. Tensions in times of pandemic. In H. Casanova (Coord.), Education and pandemic. An academic vision (pp. 83-91). UNAM.

Chen, C. H., Wang, K. C., & Lin, Y. H. (2015). The Comparison of solitary and collaborative modes of game-based learning on Students' science learning and motivation. *Journal of Educational Technology & Society*, *18*(2), 237–248.

Chen, C.-M., Li, M.-C., & Chen, T.-C. (2020). A web-based collaborative reading annotation system with gamification mechanisms to improve reading performance. *Computers & Education*, *144*, 103697. doi:10.1016/j.compedu.2019.103697

Cheng, M., Tao, D., Xie, S., Cao, X., & Yuen, A. H. (2022). Exploring students' learning management system acceptance patterns: Antecedents and consequences of profile membership. *Computers in Human Behavior*, *135*, 107374. doi:10.1016/j.chb.2022.107374

Cheng, M., & Yuen, A. H. K. (2018). Student continuance of learning management system use: A longitudinal exploration. *Computers & Education*, *120*, 241–253. doi:10.1016/j.compedu.2018.02.004

Chen, J. C. (2017). Nontraditional adult learners: The neglected diversity in postsecondary education. *SAGE Open*, *7*(1), 2158244017697161. doi:10.1177/2158244017697161

Chen, L., Lin, T., & Tang, S. (2021). A qualitative exploration of nursing undergraduates' perceptions towards scaffolding in the flipped classroom of the Fundamental Nursing Practice Course: A qualitative study. *BMC Family Practice*, *22*(1), 245–255. doi:10.118612875-021-01597-4 PMID:34903171

Chen, Z., Chen, W., Jia, J., & An, H. (2020). The effects of using mobile devices on language learning: A meta-analysis. *Educational Technology Research and Development*, *68*(4), 1769–1789. doi:10.100711423-020-09801-5

Chesney, R., & Citron, D. (2019). Deepfakes and the New Disinformation War. The Coming Age of Post-Truth Geopolitics. *Foreign Affairs*, *98*(1), 147–155.

Chiva-Bartoll, O., & Fernández-Rio, J. (2021). Advocating for service-learning as a pedagogical model in physical education: Towards an activist and transformative approach. *Physical Education and Sport Pedagogy*. Advance online publication. doi:10.1080/17408989.2021.1911981

Cho, K., Lee, S., Joo, M., & Becker, B. (2018). The effects of using mobile devices on student achievement in language learning: A meta-analysis. *Education Sciences*, *8*(3), 1–16. doi:10.3390/educsci8030105

Chong, Y. Y., Cheng, H. Y., Chan, H. Y. L., Chien, W. T., & Wong, S. Y. S. (2020). COVID-19 pandemic, infodemic and the role of eHealth literacy. *International Journal of Nursing Studies*, *108*, 103644. Advance online publication. doi:10.1016/j.ijnurstu.2020.103644 PMID:32447127

Chou, Y. (2015). *Gamification & behavioral design. Learn how to use gamification to make a positive impact on your work and life.* Retrieved on September 17, 2018 from https://yukaichou.com/gamification-examples/octalysis-complete-gamification-framework/

Clarke, B., & Svanaes, S. (2014). *An updated literature review on the use of tablets in education*. Family Kids and Youth.

Clark, R. C., & Mayer, R. E. (2008). *E-learning and the science of instruction: Proven guidelines for consumers and designers of multimedia learning*. Wiley.

Clement, S., Schauman, O., Graham, T., Maggioni, F., Evans-Lacko, S., Bezborodovs, N., & Thornicroft, G. (2015). What is the impact of mental health-related stigma on help-seeking? A systematic review of quantitative and qualitative studies. *Psychological Medicine*, *45*(1), 11–27. doi:10.1017/S0033291714000129 PMID:24569086

Coates, H., James, R., & Baldwin, G. (2005). A critical examination of the effects of learning management systems on university teaching and learning. *Tertiary Education and Management*, *11*(1), 19–36. doi:10.1080/13583883.2005.9967137

Code, J., Ralph, R., & Forde, K. (2020). Pandemic designs for the future : perspectives of technology education teachers during. *Information and Learning Sciences, 121*(5/6), 419-431. https://doi.org/https://doi.org/10.1108/ILS-04-2020-0112

Codina, L. (2018). *Systematized reviews for academic papers - 1: Concepts, phases and bibliography*. https://www.lluiscodina.com/revisiones-sistematizadas-fundamentos/

Cohen, D., & Crabtree, B. (2006). *Semi-structured interviews*. Retrieved from http://www.qualres.org/HomeSemi-3629.html

Comas-Quinn, A., Mardomingo, R., & Valentine, C. (2009). Mobile blogs in language learning: Making the most of informal and situated learning opportunities. *ReCALL*, *21*(1), 96–112. doi:10.1017/S0958344009000032

Cormier, D., McAuley, A., Siemens, G., & Stewart, B. (2010). *What is a MOOC?* [Video file]. YouTube. https://www.youtube.com/watch?v=eW3gMGqcZQc

Cornellà, P., & Meritxell, B. (2020). Gamificación y aprendizaje basado en juegos. *Enseñanza de las ciencias de la tierra, 28*(1), 5-19.

Corrigan, P., Markowitz, F. E., Watson, A., Rowan, D., & Kubiak, M. A. (2003). An attribution model of public discrimination towards persons with mental illness. *Journal of Health and Social Behavior*, *44*(2), 162–179. doi:10.2307/1519806 PMID:12866388

Corso, H. V., Mara, T., de Jou, G. I., & Salles, J. F. (2013). Metacognição e funções executivas: Relações entre os conceitos e implicações para a aprendizagem. *Psicologia: Teoria e Pesquisa (Brasília)*, *29*(1), 21–29. Advance online publication. doi:10.1590/S0102-37722013000100004

Coşkun Karademir, T., & Filiz, O. (2021). Öğretim Elemanlarının Uzaktan Eğitim Sürecinde Ölçme Değerlendirmeye Yönelik Yaşadığı Sorunlar ve Çözüm Önerileri. In Uzaktan Eğitimde Güncel Sorunlar ve Çözüm Önerileri, ASBÜ yayınları.

Courville, K. A., Lowe, A., Dannelley, B., Sellers, J., & Deal, B. (2021). Learning-by teaching and service learning to promote bleeding control education: An academic-community partnership. *Nurse Educator*. Advance online publication. doi:10.1097/NNE.0000000000001038 PMID:33958559

Crawford, J., Butler-Henderson, K., Rudolph, J., Malkawi, B., Glowatz, M., Burton, R., Magni, P. A., & Lam, S. (2020). COVID-19: 20 countries' higher education intra-period digital pedagogy responses. *Journal of Applied Learning & Teaching*, *3*(1), 9–28. https://doi.org/https://doi.org/10.37074/jalt.2020.3.1.7

Cronbach, L. J., & Snow, R. E. (1977). Aptitudes and instructional methods: A handbook for research on interactions. Academic Press.

CRUE Universidades Españolas. (2020). *Comunicado de la CRUE a los rectores de las universidades españolas de 30 de marzo de 2020*. CRUE Universidades Españolas.

Cueva Gaibor, D. A. (2020). Transformación digital en la universidad actual. *Conrado*, *16*(77), 483–489.

Da Rocha Seixas, L., Gomes, A. S., & de Melo Filho, I. J. (2016). Effectiveness of gamification in the engagement of students. *Computers in Human Behavior*, *58*, 48–63. doi:10.1016/j.chb.2015.11.021

Dainton, C., & Wong, J. (2022). Repairing our broken relationship with the vaccine hesitant: Empathy, compassion, and humility are needed. *Canadian Family Physician Medecin de Famille Canadien*, *211-213*(3), 211–213. Advance online publication. doi:10.46747/cfp.6803211 PMID:35292461

Dalsgaard, C., & Paulsen, M. F. (2009). Transparency in cooperative online education. *International Review of Research in Open and Distance Learning*, *10*(3). Advance online publication. doi:10.19173/irrodl.v10i3.671

Danish, J., & Hmelo-Silver, C. E. (2020). On activities and affordances for mobile learning. *Contemporary Educational Psychology*, *60*(101829), 1–5. doi:10.1016/j.cedpsych.2019.101829

Darling-Hamond, L. & Bransford, J. (2005). *Preparing Teachers Education for a Changing World*. Jossey-Bass Publishing Co.

Darmi, R., & Albion, P. (2014, February). A review of integrating mobile phones for language learning. *International Conference on Mobile Learning*, Madrid, Spain.

Das, K. (2019). *The role and impact of ICT in improving the quality of education: an overview.* https://www.kdsepathsala.com/2021/09/the-role-and-impact-of-ict-in-improving.html

Dashtestani, R. (2016). Moving bravely towards mobile learning: Iranian students' use of mobile devices for learning English as a foreign language. *Computer Assisted Language Learning*, *29*(4), 815–832. doi:10.1080/09588221.2015.1069360

Datosmacro.com. (n.d.). *COVID-19—Administred Vaccines 2023.* https://datosmacro.expansion.com/otros/coronavirus-vacuna

Davis, L. (2020). Don't hide the hurt: Using storytelling for trauma-informed teaching. In A. Ellsworth (Ed.), *Cases on emotionally responsive teaching and mentoring* (pp. 13–26). IGI Global., doi:10.4018/978-1-7998-2971-3.ch002

Daza, Caridad, & Fernández-Sánchez. (2019). Gamificando el aula universitaria. Análisis de una experiencia de Escape Room en educación superior. *Revista de estudios y experiencias en educación, 18*(36), 105-115. doi:10.21703/rexe.20191836sierra15

Daza, C. (2021). Teaching skills in the 21st century facing the virtuality of education on the occasion of COVID-19. *Redipe Bulletin, 10*(5), 177–188.

De Freitas, S. (2018). International Forum of Educational Technology & Society Are Games Effective Learning Tools? A Review of Educational Games. *Journal of Educational Technology & Society, 21*(2), 74–84. doi:10.2307/26388380

De Miguel, M. (Ed.). (2005). *Metodologías de enseñanza para el desarrollo de competencias. Orientaciones para el profesorado universitario ante el Espacio Europeo de Educación Superior.* Alianza.

De Paepe, L., Zhu, C., & DePryck, K. (2018). Drop-out, retention, satisfaction and attainment of online learners of Dutch in adult education. *International Journal on E-Learning, 17*(3), 303–323.

De Paepe, L., Zhu, C., & Depryck, K. (2018). Online Dutch L2 learning in adult education: Educators' and providers' viewpoints on needs, advantages and disadvantages. *Open Learning, 33*(1), 18–33. doi:10.1080/02680513.2017.1414586

Dehganzadeh, H., & Dehganzadeh, H. (2020). Investigating effects of digital gamification-based language learning: A systematic review. *J. Eng. Lang. Teach. Learn., 12*, 53–93.

Dehghanzadeh, H., Fardanesh, H., Hatami, J., Talaee, E., & Noroozi, O. (2019). Using gamification to support learning English as a second language: A systematic review. *Computer Assisted Language Learning, 34*(7), 1–24. doi:10.1080/09588221.2019.1648298

Del Barrio, V. (2010). Salud mental infanto-juvenil: Requisitos para su evaluación, diagnóstico precoz y prevención. Infocoponline. *Revista de Psicología*, 13–19.

Del-Fresno, M. (2019). Desórdenes informativos: Sobreexpuestos e infrainformados en la era de la posverdad. *El Profesional de la Información, 28*(3). Advance online publication. doi:10.3145/epi.2019.may.02

Dell'Anna, S., Pellegrini, M., & Ianes, D. (2019). Experiences and learning outcomes of students without special educational needs in inclusive settings: A systematic review. *International Journal of Inclusive Education, 25*(1), 944–959. Advance online publication. doi:10.1080/13603116.2019.1592248

Denzin, N. K., & Lincoln, Y. S. (2005). Introduction: the discipline and practice of qualitative research. In N. K. Denzin & Y. S. Lincoln (Eds.), The Sage Handbook of Qualitative Research (3rd ed.). SAGE Publications, Inc.

Denzin, N. K. (1978). *The research act: A theoretical introduction to sociological methods.* McGraw-Hill.

Destiny, O., & Omar, B. (2020). Fake news and COVID-19: Modelling the predictors of fake news sharing among social media users. *Telematics and Informatics, 56*, 1–16. doi:10.1016/j.tele.2020.101475 PMID:34887612

Deterding, S., Dixon, D., Khaled, R., & Nacke, L. (2011). From game design elements to gamefulness: defining "gamification". *Proceedings of the 15th International Academic MindTrek Conference: Envisioning Future Media Environments*, 9-15. 10.1145/2181037.2181040

Devi, S., Lakshmi, V., & Aparna, M. (2020). Moodle: an effective learning management system for 21st century learners. *Alochana Chakra Journal, 9*(6).

Dewey, J. (1938). *Experience and education*. Simon and Schuster.

Dhawan, S. (2020). Online learning: A panacea in the time of COVID-19 crisis. *Journal of Educational Technology Systems, 49*(1), 5–22. doi:10.1177/0047239520934018

Diamond, A. (2013). Executive functions. *Annual Review of Psychology, 64*(1), 135–168. Advance online publication. doi:10.1146/annurev-psych-113011-143750 PMID:23020641

Diamond, A., & Lee, K. (2011). Interventions shown to aid executive function development in children 4 to 12 years old. *Science, 333*(6045), 959–964. Advance online publication. doi:10.1126cience.1204529 PMID:21852486

Dias, S. B., & Diniz, J. A. (2014). Towards an enhanced learning management system for blended learning in higher education incorporating distinct learners' profiles. *Journal of Educational Technology & Society, 17*(1), 307–319.

Diaz, A. (2021). Pandemic times: Knowledge management and technology in service learning. *Revista Virtual Universidad Católica Del Norte, 63*, 1–4. doi:10.35575/rvucn.n63a1

Díaz-Noguera, M. D, Hervás-Gómez, C., De la Calle-Cabrera, A. M. y López-Meneses, E. (2022). La autonomía, la motivación y la pedagogía digital son factores clave en las percepciones de los estudiantes españoles de educación superior sobre el aprendizaje en línea durante la pandemia de COVID-19. *Revista internacional de investigación ambiental y salud pública, 19*(2), 654. doi:10.3390/ijerph19020654

Dicheva, D., Dichev, C., Agre, G., & Angelova, G. (2015). Gamification in education: A systematic mapping study. *Journal of Educational Technology & Society, 18*(3), 75–88. doi:10.1109/EDUCON.2014.6826129

Díez-Gutiérrez, E. J., & Gajardo K. (2021). Evaluación online en educación superior en tiempos de coronavirus ¿Qué piensan los estudiantes? *Bordón: Revista de pedagogía, 73*(1), 39-58.

Dillon, M. B., Radley, K. C., Tingstrom, D. H., Dart, E. H., & Barry, C. T. (2019). The effects of tootling via ClassDojo on student behavior in elementary classrooms. *School Psychology Review, 48*(1), 18–30. doi:10.17105/SPR-2017-0090.V48-1

Dindar, M., Ren, L., & Järvenoja, H. (2020). An experimental study on the effects of gamified cooperation and competition on English vocabulary learning. *Britiish J. Educ. Technol., 2020*(52), 142–159.

Doherty, M., Schmidt-Ott, R., Santos, J. I., Stanberry, L. R., Hofstetter, A. M., Rosenthal, S. L., & Cunningham, A. L. (2016). Vaccination of special populations: Protecting the vulnerable. *Vaccine, 34*(52), 6681–6690. doi:10.1016/j.vaccine.2016.11.015 PMID:27876197

Dolphin, L., & Hennessy, E. (2017). Labelling effects and adolescent responses to peers with depression: An experimental investigation. *BMC Psychiatry, 17*(1), 228. doi:10.118612888-017-1389-9 PMID:28646873

Dos Santos, M. J., & Dos Santos, C. A. (2019). Virtual reality in the classroom: Geography teaching practice. *Geosaberes, 10*(22), 72–80. doi:10.26895/geosaberes.v10i22.814

Duart, J. M. (2011). La Red en los procesos de enseñanza de la Universidad. *Comunicar, 37*(19), 10–13. doi:10.3916/C37-2011-02-00

Duhigg, C. (2012). The power of habit: Why we do what we do in life and business (Vol. 34). Random House.

Duncan, G. J., Dowsett, C. J., Claessens, A., Magnuson, K., Huston, A. C., Klebanov, P., Pagani, L. S., Feinstein, L., Engel, M., Brooks-Gunn, J., Sexton, H., Duckworth, K., & Japel, C. (2007). School Readiness and Later Achievement. *Developmental Psychology, 43*(6), 1428–1446. Advance online publication. doi:10.1037/0012-1649.43.6.1428 PMID:18020822

Dunn, L. (2005). The importance of reflexivity in the changing field of qualitative research. *Forum Qualitative Sozialforschung/Forum: Qualitative Social Research*, *6*(3).

Durán, C., Páez, D., & Nolasco, C. (2021). Profile, challenges and challenges of the university student in the 21st century. *REDIPE Bulletin Magazine, 10*(5), 189-198. https://revista.redipe.org/index.php/1/article/view/1296

Durán, T. (2016). *Alfabetización mediática e informacional en Latinoamérica. Approach, analysis and measurement proposal on the context of media and information literacy in Latin America* [Doctoral dissertation]. Autonomous University of Barcelona.

Dvalidze, N. (2017). Teaching English language through storytelling. *Journal of Teacher Education*, *6*(2), 53–58.

Dzandu, M., Pathak, B., & Cesare, S. (2022). Acceptability of the COVID-19 contact-tracing app—Does culture matter? *Government Information Quarterly*, *39*(4), 101750. Advance online publication. doi:10.1016/j.giq.2022.101750 PMID:35909915

Eastin, M. S., Cicchirillo, V., & Mabry, A. (2015). Extending the digital divide conversation: Examining the knowledge gap through media expectancies. *Journal of Broadcasting & Electronic Media*, *59*(3), 416–437. doi:10.1080/08838151.2015.1054994

Economic Commission for Latin America and the Caribbean. (2020). *Education in times of the COVID- 19 pandemic*. https://www.cepal.org/es/comunicados/la-cepal-la-unesco-publican-documento-que-analiza-desafios-la-educacion-que-ha-brido-la

Ede, S. (2022). Gamification and Motivation. *Issues and Trends in Learning Technologies, 10*(1).

Edwards, E., Lumsden, J., Rivas, C., Steed, L., Edwards, L., Thiyagarajan, A., Sohanpal, R., Caton, H., Griffiths, C.J., Munafo, M.R., Taylor, S., & Walton, R. (2016). Gamification for health promotion: Systematic review of behaviour change techniques in smartphone apps. *BMJ Open*, 6. doi:10.1136/bmjopen-2016-012447

Eisner, E. W. (1991). *The enlightened eye: qualitative inquiry and the enhancement of educational practice*. Macmillan.

Eissa, A., Lofters, A., Akor, N., Prescod, C. & Nnorom, O. (2021). Increasing SARS-CoV-2 vaccination rates among Black people in Canada. *CMAJ: Canadian Medical Association Journal*, E1220-E1221. doi:10.1503/cmaj.210949

el Mawas, N., Bratu, M., Caraman, D., & Muntean, C. (2019). Investigating the Learning Impact of Game-based Learning when Teaching Science to Children with Special Learning Needs. *30th Annual Conference of the Society for Information Technology and Teacher Education SITE.*

Elaish, M. M., Shuib, L., Ghani, N. A., & Yadegaridehkordi, E. (2019). Mobile english language learning (MELL): A literature review. *Educational Review*, *71*(2), 257–276. doi:10.1080/00131911.2017.1382445

Ellis, G., & Brewster, J. (2014). *Tell it Again! The storytelling handbook for primary teachers* (3rd ed.). British Council.

Emelyanova, N., & Voronina, E. (2014). Introducing a learning management system at a Russian university: Students' and teachers' perceptions. *International Review of Research in Open and Distance Learning*, *15*(1), 272–289. doi:10.19173/irrodl.v15i1.1701

Emerson, A., Cloude, E. B., Azevedo, R., & Lester, J. (2020). Multimodal learning analytics for game-based learning. *British Journal of Educational Technology*, *51*(5), 1505–1526. doi:10.1111/bjet.12992

Ernest, P., Catasús, M. G., Hampel, R., Heiser, S., Hopkins, J., Murphy, L., & Stickler, U. (2013). Online teacher development: Collaborating in a virtual learning environment. *Computer Assisted Language Learning*, *26*(4), 311–333. doi:10.1080/09588221.2012.667814

Espinales, A. (2018). Gamificación en el desarrollo de la competencia matemática Plantear y Resolver Problemas. *La revista científica del ITSUP, 12*(1), 1390-9770.

Espinoza Mina, M. A., & Gallegos Barzola, D. (2020). Soft skills in education and business: Systematic Mapping. *UISRAEL Scientific Magazine, 7*(2), 39-56. doi:10.35290/rcui.v7n2.2020.245

Eurobarometer. (Ed.). (2018). *Fake news and disinformation online.* doi:10.2759/559993

European Centre for Disease Control and Prevention (ECDC). (2016). *Let's talk about protection: Enhancing childhood vaccination uptake.* https://www.ecdc.europa.eu/en/publications-data/lets-talk-about-protection-enhancing-childhood-vaccination-uptake

European Centre for Disease, Prevention, and Control (ECDC). (2015). *Vaccine hesitancy among healthcare workers and their patients in Europe. A qualitative study.* https://www.ecdc.europa.eu/en/publications-data/vaccine-hesitancy-among-healthcare-workers-and-their-patients-europe

European Commission (Ed.). (2018). *Action Plan against Disinformation. EEAS – European External Action Service – European Union.* Retrieved of https://bit.ly/3wuVw86

European Commission. (2018). A multi-dimensional approach to disinformation. Report of the independent High Level Group on Fake News and Online Disinformation. *Directorate-General for Communication Networks, Content and Technology.* http://bit.ly/2Ponxc1

Eveleigh, B. (2020). *Embracing virtual and blended learning post COVID-19.* Pamoja. https://pamojaeducation.com/blog/embracing-virtual-and-blended-learning-post-covid-19

Fadel, C., Bialink, M., & Trilling, B. (2016). *Education in four dimensions.* Editorial Granika.

Fairchild, G., van Goozen, S. H. M., Stollery, S. J., Aitken, M. R. F., Savage, J., Moore, S. C., & Goodyer, I. M. (2009). Decision Making and Executive Function in Male Adolescents with Early-Onset or Adolescence-Onset Conduct Disorder and Control Subjects. *Biological Psychiatry, 66*(2), 162–168. Advance online publication. doi:10.1016/j.biopsych.2009.02.024 PMID:19362293

Fairhead, J., Leach, M., & Small, M. (2006). Public engagement with science? Local understandings of a vaccine trial in the Gambia. *Journal of Biosocial Science, 38*(1), 103–116. doi:10.1017/S0021932005000945 PMID:16266443

Falagas, M. E., & Zarkadoulia, E. (2008). Factors associated with suboptimal compliance to vaccinations in children in developed countries: A systematic review. *Current Medical Research and Opinion, 24*(6), 1719–1741. doi:10.1185/03007990802085692 PMID:18474148

Fasce, P. (2018). Portare il gioco a scuola, non la gamification. *Educazione Aperta,* 4.

Faul, F., Erdfelder, E., Lang, A. G., & Buchner, A. (2007). G* power 3: A flexible statistical power analysis program for the social, behavioral, and biomedical sciences. *Behavior Research Methods, 39*(2), 175–191. doi:10.3758/BF03193146 PMID:17695343

Fernández Torres, M. J., Chamizo Sánchez, R. & Sánchez Villarrubia, R. (2021). Universidad y pandemia: la comunicación en la educación a distancia. *Revista internacional de comunicación, 52,* 156-174. . doi:10.12795/Ambitos.2021.i52.10

Fernandez-Marquez, E., Vazquez-Cano, E., Lopez-Meneses, E., & Sirignano, F. M. (2020). The digital competence of student of different European universities. *Revista ESPACIOS, 41*(13). https://www.revistaespacios.com/a20v41n13/20411315.html

Fernández-Portero, I., & Castillo, C. (2022). Gamification in the English language class: Analysis of pre-service teachers' perceptions. *Computer Assisted Language Learning Electronic Journal, 23* (1), 425-444.

Ferrara, J. (2013). Games for persuasion argumentation, procedurality, and the lie of gamification. *Games and Culture, 8*(4), 289–304. doi:10.1177/1555412013496891

Fidalgo, P. (2012). *Learning networks and moodle use in online courses: a social network analysis study* [Unpublished doctoral dissertation]. Universidade Nova de Lisboa, Lisbon.

Fidalgo-Blanco, Á., Sein-Echaluce, ML, & García-Peñalvo, F. (2020). *Real advantages in the application of the Aula Invertida-Flipped Classroom method.* GRAIL Group.

Fidalgo, P., Thormann, J., Kulyk, O., & Lencastre, J. A. J. A. (2020). Students' perceptions on distance education: A multinational study. *International Journal of Educational Technology in Higher Education, 17*(1), 18. doi:10.118641239-020-00194-2

Fidan, M. (2020). Covid-19 belirsizliğinde eğitim: İlkokulda zorunlu uzaktan eğitime ilişkin öğretmen görüşleri. *Uşak Üniversitesi Eğitim Araştırmaları Dergisi, 6*(2), 24–43. doi:10.29065/usakead.736643

Fierro-Suero, S., Almagro, B.J., & Sáenz-López, P. (2019). Necesidades psicológicas, motivación e inteligencia emocional en Educación Física. *Revista electrónica interuniversitaria de formación profesorado, 22*, 167-186.

Figueroa Flores, J. F. (2015). Using gamification to enhance second language learning. *Digital Education Review, 27*, 32–54.

Fishbein, M., & Cappella, J. (2006). The role of theory in developing effective health communications. *Journal of Communication, 56*(suppl_1), S1–S17. doi:10.1111/j.1460-2466.2006.00280.x

Florian, L. (2021). The Universal Value of Teacher Education for Inclusive Education. In A. Köpfer, J. J. W. Powell, & R. Zahnd (Eds.), *Handbuch Inklusion international / International Handbook of Inclusive Education: Globale, nationale und lokale Perspektiven auf Inklusive Bildung / Global, National and Local Perspectives* (1st ed., pp. 89–106). Verlag Barbara Budrich. doi:10.2307/j.ctv1f70kvj.8

Fombona, J., & Pascual, M. Á. (2017). La producción científica sobre Realidad Aumentada, un análisis de la situación educativa desde la perspectiva SCOPUS. *EDMETIC, 6*(1), 39–61. doi:10.21071/edmetic.v6i1.5807

Francisco Amat, A., & Moliner Miravet, L. (2010). El aprendizaje servicio en la Universidad: una estrategia en la formación de ciudadanía crítica. *Revista electrónica interuniversitaria de formación del profesorado, 13*(4), 69-77.

Friehs, M. A., Dechant, M., Vedress, S., Frings, C., & Mandryk, R. L. (2020). Effec-tive gamification of the stop-signal task: Two controlled laboratory experiments. *JMIR Serious Games, 8*(3), e17810. Advance online publication. doi:10.2196/17810 PMID:32897233

Fuentes, G., Moreno, L., Rincón, D., & Silva, M. (2020). Assessment of soft skills in higher education. *University Formation, 14*(4), 49–60. doi:10.4067/S0718-50062021000400049

Fuhs, M. W., Nesbitt, K. T., Farran, D. C., & Dong, N. (2014). Longitudinal associations between executive functioning and academic skills across content areas. *Developmental Psychology, 50*(6), 1698–1709. Advance online publication. doi:10.1037/a0036633 PMID:24749550

Galeana de la O. L. (n.d.). *Aprendizaje basado en proyectos.* Retrieved from http://ceupromed.ucol.mx/revista/PdfArt/1/27.pdf

Galletta, A. (2013). *Mastering the semi-structured interview and beyond: From research design to analysis and publication.* NYU Press.

Gamage, S. H. P. W., Ayres, J. R., & Behrend, M. B. (2022). A systematic review on trends in using Moodle for teaching and learning. *International Journal of STEM Education*, *9*(1), 9. doi:10.118640594-021-00323-x PMID:35096513

Gandía, J. B. (2012). La rentabilidad del cómic en la enseñanza de la cultura en E/LE. *Foro de profesores en E/LE*, 8, 1-8.

Garagui, E. (2021). *Las profesiones del futuro, la nueva generación de empleos*. https://www.linkedin.com/pulse/las-profesiones-del-futuro-la-nueva-generaci%C3%B3n-de-empleos-everylgbt?trk=public_post-content_share-article

García Peñalvo, F. J., Corell, A., Abella García, V. y Grande de Prado, M. (2020). La evaluación online en la educación superior en tiempos de la COVID-19. *Education in the Knowledge Society: EKS,* 21. . doi:10.14201/eks.23013

García Prieto, F. J., López-Aguilar, D. & Delgado-García, M. (2022). Competencia digital del alumnado universitario y rendimiento académico en tiempos de COVID-19. *Revista de medios y educación Pixel-Bil, 64*, 1-41 doi:10.12795/pixelbit.91862

García Roca, A. (2021b). *Creepypasta* as metafictional textualities: formative opportunities for media and information literacy. *Tonos Digital, 40*, 1-20. http://www.tonosdigital.es/ojs/index.php/tonos/article/view/2715/1195

García Roca, A. (2021a). The SCP Foundation in the development of academic literacy, reading promotion and creative writing. *Publicaciones*, *51*(1), 15–28. doi:10.30827/publicaciones.v51i1.16488

García Roca, A., & Amo, J. M. (2023). Media and information literacy in a context of disinformation. *Tejuelo*, *37*, 99–128. doi:10.17398/1988-8430.37.99

García Roca, A., Manzano León, A., & Rodríguez Ferrer, J. (in press). Adaptation and training model of the board game untold for narrative and communicative competences. *International Journal of Instruction.*

Garcia-Alonso, A., Mateos-Aparicio, I., Pérez-Rodríguez, M. L., Sánchez-Paniagua, M., Rodriguez-Rodriguez, E., Moreno-Guzmán, M., Sanz Landaluce, J., Rueda, C., Gallo, A., & De la Peña-Armada, R. (2023). Gamificación y desarrollo de contenidos virtuales para la dinamización del aula y mejora del aprendizaje universitario. *Actas INTED*, 788–791. Advance online publication. doi:10.21125/inted.2023.0252

García-Peñalvo, F. J. y Corell, A. (2020). La COVID-19: ¿enzima de la transformación digital de la docencia o reflejo de una crisis metodológica y competencial en la educación superior? *Campus virtuales, 9*(2), 83-98.

García-Varcálcel, A., & Gómez-Pablos, V. (2017). Project-based learning (ABP): evaluation from the perspective of primary school students. *Educational Research Journal, 1*(35), 113. https://digitum.um.es/digitum/bitstream/10201/52389/1/246811-963851-1-PB.pdf

Garrison, D. R., & Anderson, T. (2003). *E-learning in the 21st century: A framework for research and practice*. Psychology Press.

Garrison, D. R., & Kanuka, H. (2004). Blended learning: Uncovering its transformative potential in higher education. *The Internet and Higher Education*, *7*(2), 95–105. doi:10.1016/j.iheduc.2004.02.001

Gathercole, S. E., Pickering, S. J., Knight, C., & Stegmann, Z. (2004). Working memory skills and educational attainment: Evidence from national curriculum assessments at 7 and 14 years of age. *Applied Cognitive Psychology*, *18*(1), 1–16. Advance online publication. doi:10.1002/acp.934

GAVI Alliance. (2013). *Summary of definitions of mission and strategic goal level indicators in GAVI Alliance Strategy 2011-2015*. https://www.gavi.org/programmes-impact/our-impact/measuring-our-performance/2011-2015-indicators

Gaviria Bustamante, K., Carriazo Diaz, C., & Perez Reyes, M. (2020). Educational planning as a fundamental tool for quality education. *Utopia and Latin American Praxis, 25*(3), 87-95. https://www.redalyc.org/articulo.oa?id=27963600007

Gaviria, D., Arango, J., Valencia, A., & Bran, L. (2019). Perception of the flipped classroom strategy in university settings. *Mexican Journal of Educational Research, 24*(81), 1–15. https://www.redalyc.org/journal/140/14062583011/html/

Gaytan, J., & McEwen, B. C. (2007). Effective online instructional and assessment strategies. *American Journal of Distance Education, 21*(3), 117–132. doi:10.1080/08923640701341653

Ge, X., & Ifenthaler, D. (2018). Designing engaging educational games and assessing engagement in game-based learning. *Ophthalmology: Breakthroughs in Research and Practice*, 1-18. doi:10.4018/978-1-5225-5198-0.ch001

Gee, J. (2003). *What video games have to teach us about learning and literacy*. Palgrave Macmillan. doi:10.1145/950566.950595

Geertz, C. (1973). *The interpretation of cultures: Selected essays*. Basic Books.

Geoghegan, S., O'Callaghan, K. P., & Offit, P. A. (2020). Vaccine safety: Myths and misinformation. *Frontiers in Microbiology, 11*, 372. doi:10.3389/fmicb.2020.00372 PMID:32256465

Georgakakou-Koutsonikou, N., & Williams, J. M. (2017). Children and young people's conceptualizations of depression: A systematic review and narrative meta-synthesis. *Child: Care, Health and Development, 43*(2), 161–181. doi:10.1111/cch.12439 PMID:28090667

Gilbert, S. J., & Burgess, P. W. (2008). Executive function. *Current Biology, 18*(3), 110–114. doi:10.1016/j.cub.2007.12.014 PMID:18269902

Gilster, P. (1997). *Digital literacy*. John Wiley.

Gil-Villa, F., Urchaga, J. D., & Sánchez-Fernández, A. (2020). Proceso de digitalización y adaptación a la enseñanza no presencial motivada por la pandemia de COVID-19: análisis de la percepción y repercusiones en la comunidad universitaria. *Revista latina de comunicación social, 78*, 99-119. https://www.doi.org/10.4185/RLCS-2020-1470

Giray, G. (2021). An assessment of student satisfaction with e-learning: An empirical study with computer and software engineering undergraduate students in Turkey under pandemic conditions. *Education and Information Technologies, 26*(6), 6651–6673. doi:10.100710639-021-10454-x PMID:33686329

Godwin-Jones, R. (2008). Emerging technologies: Web-writing 2.0: Enabling, documenting, and assessing writing online. *Language Learning & Technology, 12*(2), 7–13.

Golovchenco, Y., Mareike, H., & Adler-Nissen, R. (2018). State, media and civil society in the information warfare over Ukraine: Citizen curators of digital disinformation. *International Affairs, 94*(5), 975–994. doi:10.1093/ia/iiy148

Gómez Contreras, J. L., Bonilla Torres, C. A., & Esteban Ojeda, Y. C. (2022). Uso de TIC y TAC en la educación superior: Un análisis bibliométrico. *Revista complutense de educación, 33*(3), 601-613.

Gómez-Gómez, M., Hijón-Neira, R., Santacruz-Valencia, L., & Pérez-Marín, D. (2022). Impacto del proceso de enseñanza y aprendizaje remoto de emergencia en la competencia digital y el estado de ánimo en la formación del profesorado. *Education in the Knowledge Society, 23*. Advance online publication. doi:10.14201/eks.27037

González Juárez, L., & Noreña Peña, A. L. (2011). Comunicación intercultural como medio para favorecer el cuidado culturalmente aceptable. *Enfermería Universitaria, 8*(1), 55–60.

Gonzalez, T., De la Rubia, M. A., Hincz, K. P., Comas-Lopez, M., Subirats, L., Fort, S., & Sacha, G. M. (2020). Influence of COVID-19 confinement on students' performance in higher education. *PLoS ONE, 15*(10), 1-23. https://doi.org/https://doi.org/10.1371/journal.pone.0239490

González, D., Jeong, J., Gallego, A., & Cañada, F. (2018). Influence of the flipped methodology on the emotions felt by students of the Primary Education Degree in science classes depending on the baccalaureate studied. *Chemical Educator (Boise, Idaho)*, *29*(1), 77–88. doi:10.22201/fq.18708404e.2018.1.63698

González, M., & Abad, E. (2020). The flipped classroom: A challenge for university teaching. *VEC Virtuality. Education in Science*, *11*(20), 75–91. https://revistas.unc.edu.ar/index.php/vesc/article/view/27449

González, M., & Huerta, P. (2019). Flipped classroom experience to promote prosumer students of the higher level. *RIED. Iberoamerican Journal of Distance Education*, *22*(2), 245–263. doi:10.5944/ried.22.2.23065

González, T., de la Rubia, M., Hincz, K., López, M. C., Subirats, L., Fort, S., & Sacha, G. M. (2020). Influencia del confinamiento por COVID-19 en el desempeño de los estudiantes en la educación superior. *PLoS One*, *15*(10). Advance online publication. doi:10.1371/journal.pone.0239490 PMID:33035228

Goss, M. D., Temte, J. L., Barlow, S., Temte, E., Bell, C., Birstler, J., & Chen, G. (2020). An assessment of parental knowledge, attitudes, and beliefs regarding influenza vaccination. *Vaccine*, *38*(6), 1565–1571. doi:10.1016/j.vaccine.2019.11.040 PMID:31776028

Governo da República Portuguesa. (2019). *Portaria n.o 359/2019. Diário da República n.o 193/2019, Série I de 2019-10-08.* Retrieved December 12, 2022, from https://dre.pt/home/-/dre/125085420/details/maximized

Governo da República Portuguesa. (2020). *Comunicado do Conselho de Ministros de 19 de março de 2020.* Comunicados do Conselho de Ministros.

Granda Asencio, L. Y., Espinoza Freire, E. E., & Mayon Espinoza, S. E. (2019). Las TICs como herramientas didácticas del proceso de enseñanza-aprendizaje. *Revista Conrado*, *15*(66), 104–110.

Grande de Prado, M., García-Peñalvo, F. J., Corell, A. & Abella García, V. (2021). Evaluación en Educación Superior durante la pandemia de la COVID-19. *Campus virtuales*, *1*(10), 49-58.

Grant, M. J., & Booth, A. (2009). A typology of reviews: An analysis of 14 review types and associated methodologies. *Health Information and Libraries Journal*, *26*(2), 91–108. doi:10.1111/j.1471-1842.2009.00848.x PMID:19490148

Gregg, L., & Tarrier, N. (2007). Virtual reality in mental health: A review of the literature. *Social Psychiatry and Psychiatric Epidemiology*, *42*(5), 343–354. doi:10.100700127-007-0173-4 PMID:17431528

Greyling, T., & Rossouw, S. (2022). Positive attitudes towards COVID-19 vaccines: A cross-country analysis. *PLoS One*, *17*(3), e0264994. doi:10.1371/journal.pone.0264994 PMID:35271637

Guarino, I., Aceto, G., Ciuonzo, D., Montieri, A., Persico, V., & Pescapé, A. (2022). Contextual counters and multimodal Deep Learning for activity-level traffic classification of mobile communication apps during COVID-19 pandemic. *Computer Networks*, *21*(9), 109452. Advance online publication. doi:10.1016/j.comnet.2022.109452 PMID:36447639

Guichon, N. (2012). *Versl'intégration des TIC dansl'enseignement des langues.* Didier.

Gülbahar, Y. (2008). ICT usage in higher education: A case study on preservice teachers and instructions. *The Turkish Online Journal of Educational Technology*, *7*(1), 32–36.

Gunawardena, C., McIsaac, M., & Jonassen, D. (2006). Distance education. In D. H. Jonassen (Ed.), *Handbook of Research for Educational Communications and Technology* (pp. 355–396). Association for Educational Communications and Technology.

Gust, D. A., Darling, N., Kennedy, A., & Schwartz, B. (2008). Parents with doubts about vaccines: Which vaccines and reasons why. *Pediatrics*, *122*(4), 718–725. doi:10.1542/peds.2007-0538 PMID:18829793

Haines, K. (2016). Expanding the knowledge base of teachers' use of communication tools for language learning. *System*, *62*, 102–112. doi:10.1016/j.system.2016.07.008

Hallaq, T. (2016). Evaluating Online Media Literacy in Higher Education: Validity and Reliability of the Digital Online Media Literacy Assessment (DOMLA). *The Journal of Media Literacy Education*, *8*(1), 62–84. doi:10.23860/jmle-8-1-5

Hamadi, M., El-Den, J., Azam, S. & Sriratanaviriyakul, N. (2022). Integrating social media as cooperative learning tool in higher education classrooms: An empirical study. *Journal of King Saud University—Computer and Information Sciences, 34*(6), 3722-3731. doi:10.1016/j.jksuci.2020.12.007

Hamari, J., & Koivisto, J. (2013). Social motivations to use gamification: An empirical study of gamifying exercise. *Proceedings of the 21st European Conference on Information Systems.*

Hamari, J., Koivisto, J., & Sarsa, H. (2014). Does gamification work? A literature review of empirical studies on gamification. *Proceedings of the 47th Hawaii International Conference on System Sciences.* 10.1109/HICSS.2014.377

Hamborsky, J., Kroger, A., & Wolfe, S. (2015). *Epidemiology and prevention of vaccine-preventable diseases.* Centres for Disease Control and Prevention (CDC). https://www.cdc.gov/vaccines/pubs/pinkbook/index.html

Hamidi, H., & Jahanshaheefard, M. (2019). Essential factors for the application of education information system using mobile learning: A case study of students of the university of technology. *Telematics and Informatics*, *38*, 207–224. doi:10.1016/j.tele.2018.10.002

Hamutoğlu, N. B., Canan Güngören, Ö., Kaya Uyanık, G., & Gür Erdoğan, D. (2017). Adapting digital literacy scale into Turkish. *Ege Journal of Education*, *18*(1), 408–429.

Hanson-Smith, E. (2016). Teacher education and technology. The Routledge handbook of language learning and technology, 210-222.

Hargittai, E. (2002). Second-level digital divide: Differences in people's online skills. *First Monday*, *7*(4). Advance online publication. doi:10.5210/fm.v7i4.942

Hargittai, E. (2010). Digital Na(t)ives? Variation in internet skills and uses among members of the "net Generation.". *Sociological Inquiry*, *80*(1), 92–113. doi:10.1111/j.1475-682X.2009.00317.x

Hart, L. C. (2002). Preservice teachers' beliefs and practice after participating in an integrated content/methods course. *School Science and Mathematics*, *102*(1), 4–15. doi:10.1111/j.1949-8594.2002.tb18191.x

Hebebci, M. T., Bertiz, Y., & Alan, S. (2020). Investigation of Views of Students and Teachers on Distance Education Practices during the Coronavirus (COVID-19) Pandemic. *International Journal of Technology in Education and Science*, *4*(4), 267–282. doi:10.46328/ijtes.v4i4.113

Hernández Angulo, J., Garbey Savigne, E., Enríquez O'Farrill, I. J., Hernández Angulo, J., Garbey Savigne, E., & Enríquez O'Farrill, I. J. (2021). Importancia de la perspectiva afectiva e intercultural en la comunicación profesional en tiempos de crisis. *Revista Habanera de Ciencias Médicas*, *20*(1). http://scielo.sld.cu/scielo.php?script=sci_abstract&pid=S1729-519X2021000100018&lng=es&nrm=iso&tlng=es

Hernández, C. A. H., Prada, R. P., & Espinel, G. A. E. (2022). Competencias TIC del docente en época de no presencialidad: Una mirada desde los estudiantes de comunicación social. *Revista Saber. Ciencia y Libertad*, *17*(1), 432–453. doi:10.18041/2382-3240aber.2022v17n1.7825

Hernández-Silva, C., & Tecpan Flores, S. (2017). Flipped classroom mediated by the use of virtual platforms: A case study in the training of physics teachers. *Studies Pedagogical (Valdivia)*, *43*(3), 193–204.

Herrington, J., Reeves, T. C., & Oliver, R. (2005). Online learning as information delivery: Digital myopia. *Journal of Interactive Learning Research*, *16*(4), 353–367.

Hew, K. (2009). Use of audio podcast in K-12 and higher education: A review of research topics and methodologies. *Educational Technology Research and Development*, *57*(3), 333–357. doi:10.100711423-008-9108-3

Hew, K., Huang, B., Wah Samuel Chu, K., & Chiu, D. (2016). Engaging Asian students through game mechanics: Findings from two experiment studies. *Computers & Education*, *92-93*, 221–236. doi:10.1016/j.compedu.2015.10.010

Hockly, N. (2013). Mobile learning. *ELT Journal*, *67*(1), 80–84. doi:10.1093/elt/ccs064 PMID:23923690

Hodges, C., Moore, S., Lockee, B., Trust, T., & Bond, A. (2020). La diferencia entre la enseñanza remota de emergencia y el aprendizaje en línea. *EDUCAUSE Review*, *27*, 1–12. https://www.educacionyfp.gob.es/prensa/actualidad/2020/04/20200415-sectorial.html

Hodgson, T. R., Cunningham, A., McGee, D., Kinne, L., & Murphy, T. J. (2017). Assessing behavioral engagement in flipped and non-flipped mathematics classrooms: Teacher abilities and other potential factors. *International Journal of Education in Mathematics Science and Technology*, *5*(4), 248–261. doi:10.18404/ijemst.296538

Holley, S. R., Ewing, S. T., Stiver, J. T., & Bloch, L. (2017). The Relationship Between Emotion Regulation, Executive Functioning, and Aggressive Behaviors. *Journal of Interpersonal Violence*, *32*(11), 1692–1707. Advance online publication. doi:10.1177/0886260515592619 PMID:26130684

Horspool, A., & Lange, C. (2012). Applying the scholarship of teaching and learning: Student perceptions, behaviours and success online and face-to-face. *Assessment & Evaluation in Higher Education*, *37*(1), 73–88. doi:10.1080/02602938.2010.496532

Houlden, S., & Veletsianos, G. (2020). *Coronavirus pushes universities to switch to online classes — but are they ready?* Retrieved December 11, 2022, from https://theconversation.com/coronavirus-pushes-universities-to-switch-to-online-classes-but-are-they-ready-132728

Hoven, D., & Palalas, A. (2011). (Re)conceptualizing design approaches for mobile language learning. *CALICO Journal*, *28*(3), 699–720. doi:10.11139/cj.28.3.699-720

Howard, P. N., Neudert, L., & Prakash, N. (2021). Digital misinformation/disinformation and children. UNICEF Office of Global Insight and Policy.

Hsu, L. (2016). Examining EFL teachers' technological pedagogical content knowledge and the adoption of mobile-assisted language learning: A partial least square approach. *Computer Assisted Language Learning*, *29*(8), 1287–1297. doi:10.1080/09588221.2016.1278024

Huang, J. H., Lin, Y. R., & Chuang, S. T. (2007). Elucidating user behavior of mobile learning: A perspective of the extended technology acceptance model. *The Electronic Library*, *25*(5), 585–598. doi:10.1108/02640470710829569

Hue, L. T., & Ab Jalil, H. (2013). Attitudes towards ICT integration into curriculum and usage among university lecturers in Vietnam. *International Journal of Instruction*, *6*(2), 53–66.

Hung, C. Y., Sun, J. C. Y., & Liu, J. Y. (2019). Effects of flipped classrooms integrated with MOOCs and game-based learning on the learning motivation and outcomes of students from different backgrounds. *Interactive Learning Environments*, *27*(8), 1028–1046. doi:10.1080/10494820.2018.1481103

Huotari, K., & Hamari, J. (2017). A definition for gamification: Anchoring gamification in the service marketing literature. *Electronic Markets*, *27*(1), 21–31. doi:10.100712525-015-0212-z

Huttar, C. M., & Brintzenhofeszoc, K. (2019). Virtual Reality and Computer Simulation in Social Work Education: A Systematic Review. *Journal of Social Work Education*, •••, 1–11. doi:10.1080/10437797.2019.1648221

Hwang, G., & Fu, Q. (2019). Trends in the research design and application of mobile language learning: A review of 2007-2016 publications in selected SSCI journals. *Interactive Learning Environments*, *27*(4), 567–581. doi:10.1080/10494820.2018.1486861

Hwang, W., Chen, H., Shadiev, R., Huang, Y., & Chen, C. (2014). Improving English as a foreign language writing in elementary schools using mobile devices in familiar situational contexts. *Computer Assisted Language Learning*, *27*(5), 359–378. doi:10.1080/09588221.2012.733711

Ibáñez-Córdoba, S. X. (2021). Las prácticas pedagógicas en la formación inicial del profesorado de química y su relación con la construcción de conocimiento profesional: Reflexiones y apuestas desde una revisión documental. *Noria Investigación Educativa*, *2*(8), 35–55. doi:10.14483/25905791.18162

Ibarra Sáiz, M. S. & Rodríguez-Gómez, G. (2010). Los procedimientos de evaluación como elementos de desarrollo de la función orientadora en la universidad. *Revista española de orientación y psicopedagogía, 21*(2), 443-461. . doi:10.5944/reop.vol.21.num.2.2010.11558

Ilari, B., Chen-Hafteck, L., & Crawford, L. (2013). Singing and cultural understanding: A music education perspective. *International Journal of Music Education*, *31*(2), 202–216. doi:10.1177/0255761413487281

Indriasari, T. D., Luxton-Reilly, A., & Denny, P. (2020). Gamification of student peer review in education: A systematic literature review. *Education and Information Technologies*, *25*(6), 5205–5234. doi:10.100710639-020-10228-x

INE. (2019). *Encuesta sobre equipamiento y uso de la tecnología de la información y comunicación en los hogares*. INE. https://www.ine.es/dyngs/INEbase/es/operacion.htm?c=estadistica_C&cid=1254736176741&menu=ultiDatos&idp=1254735976608

Jansen, B. R. J., de Lange, E., & van der Molen, M. J. (2013). Math practice and its influence on math skills and executive functions in adolescents with mild to borderline intellectual disability. *Research in Developmental Disabilities*, *34*(5), 1815–1824. Advance online publication. doi:10.1016/j.ridd.2013.02.022 PMID:23511346

Jarrett, C., Wilson, R., O'Leary, M., Eckersberger, E., & Larson, H. J. (2015). Strategies for addressing vaccine hesitancy—A systematic review. *Vaccine*, *33*(34), 4180–4190. doi:10.1016/j.vaccine.2015.04.040 PMID:25896377

Jenkins, H. (2008). Convergence culture. The culture of media convergence. Barcelona: Paidós.

Jenkins, H., Ford, S., & Green, J. (2013). *Spreadable Media: Creating Value and Meaning in a Networked Culture*. New York University Press.

Jenkins, H., Ito, M., & Boyd, D. (2015). *Participatory culture in a networked era: A conversation on youth, learning, commerce, and politics*. Polity Press.

Jin, Q., Raza, S. H., Yousaf, M., Munawar, R., Shah, A. A., Hassan, S., Shaikh, R. S., & Ogadimma, E. C. (2022). Ingraining polio vaccine acceptance through public service advertisements in the digital era: The moderating role of misinformation, disinformation, fake news, and religious fatalism. *Vaccines*, *10*(10), 10. Advance online publication. doi:10.3390/vaccines10101733 PMID:36298598

Johnson, K. E., & Golombek, P. R. (2018). Making L2 teacher education matter through Vygotskian-inspired pedagogy and research. In The Routledge handbook of sociocultural theory and second language development (pp. 443-456). Taylor and Francis. doi:10.4324/9781315624747-28

Johnson, K. E. (2009). *Second language teacher education: a sociocultural perspective*. Routledge. doi:10.4324/9780203878033

Johnson, K. E., & Golombek, P. R. (2020). Informing and transforming language teacher education pedagogy. *Language Teaching Research*, *24*(1), 116–127. doi:10.1177/1362168818777539

Johnson, L., Smith, R., Willis, H., Levine, A., & Haywood, K. (2011). *The 2011 horizon report*. The New Media Consortium.

Jonassen, D. H. (2003). *Learning to solve problems with technology: A constructivist perspective*. Prentice Hall.

Jong, M. S. Y. (2015). Does online game-based learning work in formal education at school? A case study of VISOLE. *Curriculum Journal*, *26*(2), 249–267. Advance online publication. doi:10.1080/09585176.2015.1018915

Juan-Lázaro, O., & Área, M. (2021). Gamificación superficial en e-learning: evidencias sobremotivación y autorregulación [Thin layer gamification in e-learning: evidence on motivation and self-regulation]. *Pixel-Bit. Revista de Medios y Educación*, *62*, 146–181.

Jung, S., & Huh, J. H. (2019). An efficient LMS platform and its test bed. *Electronics (Basel)*, *8*(2), 154. doi:10.3390/electronics8020154

Jurgelaitis, M., Čeponienė, L., Čeponis, J., & Drungilas, V. (2019). Implementing gamification in a university-level UML modeling course: A case study. *Computer Applications in Engineering Education*, *27*(2), 332–343. doi:10.1002/cae.22077

Kaden, U. (2020). COVID-19 school closure-related changes to the professional life of a K–12 teacher. *Education and Information Technologies*, *10*(6), 1–13.

Kalimeri, K. G., Beiró, M., Urbinati, A., Bonanomi, A., Rosina, A., & Cattuto, C. (2019). Human values and attitudes towards vaccination in social media. *Companion Proceedings of the 2019 World Wide Web Conference*, 248-254. 10.1145/3308560.3316489

Kalogiannakis, M., Papadakis, S., & Zourmpakis, A. I. (2021). Gamification in Science Education. A Systematic Review of the Literature. *Education Sciences*, *11*(1), 1–36. doi:10.3390/educsci11010022

Kang, D., & Park, M. (2022). Interaction and online courses for satisfactory university learning during the COVID-19 pandemic. *International Journal of Management Education*, *20*(3), 100678. Advance online publication. doi:10.1016/j.ijme.2022.100678

Kapp, K. (2012). *The Gamification of Learning and Instruction: game-based methods and strategies for training and education*. Pfeiffer.

Kapp, K. (2012). *The gamification of learning and ınstruction: Game-based methods and strategies for training and education*. Pfeiffer.

Kapur, R. (2019). *Use of ICT in Improving Quality of Education*. Retrieved from https://www.researchgate.net/publication/333446961_Use_of_ICT_in_Improving_Quality_of_Education

Kara, M., Erdoğdu, F., Kokoç, M., & Cagiltay, K. (2019). Challenges Faced by Adult Learners in Online Distance Education: A Literature Review. *Open Praxis*, *11*(1), 5. doi:10.5944/openpraxis.11.1.929

Karatepe, F., Küçükgençay, N., & Peker, B. (2020). Öğretmen Adayları Senkron Uzaktan Eğitime Nasıl Bakıyor? Bir Anket Çalışması. *Journal of Social and Humanities Sciences Research*, *7*(53), 1262–1274.

Kattoua, T., Al-Lozi, M., & Alrowwad, A. (2016). A review of literature on e-learning systems in higher education. *International Journal of Business Management and Economic Research*, *7*(5).

Kavlu, A. (2017). Implementation of project-based learning (PBL) in EFL (English as a foreign language) classrooms in Fezalar Educational Institutions (Iraq). *International Journal of Social Sciences and Educational Studies*, *4*(2), 67–79.

Kaya-Özgül, B., Aktaş, N., & Çetinkaya-Özdemir, E. (2023). Examination of digital literacy levels of primary school teachers and primary school teacher candidates according to various variables. *Cumhuriyet International Journal of Education*, *12*(1), 204–221.

Kedrova, G., & Potemkin, S. (2015, October). New trends in implementation of ICT in higher education. In *2015 9th International Conference on Application of Information and Communication Technologies (AICT) (pp. 531-535)*. IEEE. 10.1109/ICAICT.2015.7338617

Keegan, D. (2002). *The Future of Learning: From eLearning to mLearning*. https://files.eric.ed.gov/fulltext/ED472435.pdf

Ke, F. (2008). A case study of computer gaming for math: Engaged learning from gameplay? *Computers & Education*, *51*(4), 1609–1620. Advance online publication. doi:10.1016/j.compedu.2008.03.003

Kentnor, H. E. (2015). Distance education and the evolution of online learning in the United States. In D. J. Flinders & C. M. Moroye (Eds.), *Curriculum and Teaching Dialogue* (pp. 21–34). American Association for Teaching & Curriculum.

Keskin, N. O., & Metcalf, D. (2011). The current perspectives, theories and practices of mobile learning. *The Turkish Online Journal of Educational Technology*, *10*(2), 202–208.

Keskin, S., & Yurdugül, H. (2020). Factors affecting students' preferences for online and blended learning: Motivational vs. cognitive. *European Journal of Open Distance and E-Learning*, *22*(2), 72–86. doi:10.2478/eurodl-2019-0011

Kessler, G., & Hubbard, P. (2017). Language Teacher Education and Technology. In C. A. Chapelle & S. Sauro (Eds.), *The handbook of technology and second language teaching and learning* (pp. 278–292). Wiley. doi:10.1002/9781118914069.ch19

Kim, H., & Kwon, Y. (2012). Exploring smartphone applications for effective mobile-assisted language learning. *Multimedia-Assisted Language Learning*, *15*(1), 31–57. doi:10.15702/mall.2012.15.1.31

Kingston, S. (2018). Project-based learning & student achievement: What does the research tell us? *PBL Evidence Matters*, *1*(1), 1–11.

Klopfer, E. (2008). *Augmented learning: research and design of mobile educational games*. MITPress. doi:10.7551/mitpress/9780262113151.001.0001

Knobel, M., & Lankshear, C. (2006). Digital literacy and digital literacies: Policy, pedagogy and research considerations for education. *Nordic Journal of Digital Literacy*, *1*(1), 12–24. doi:10.18261/ISSN1891-943X-2006-01-03

Koehler, M. J., & Mishra, P. (2009). What is technological pedagogical content knowledge? *Contemporary Issues in Technology & Teacher Education*, *9*(1), 60–70.

Koehler, M. J., Mishra, P., & Yahya, K. (2007). Tracing the development of teacher knowledge in a design seminar: Integrating content, pedagogy and technology. *Computers & Education*, *49*(3), 740–762. doi:10.1016/j.compedu.2005.11.012

Köksal, D. (2004). To Kill the Blackboard? Technology in Language Teaching and Learning. *Turkish Online Journal of Educational Technology-TOJET*, *3*(3), 62–72.

Kolb, A. Y., & Kolb, D. A. (2009). The learning way: Meta-cognitive aspects of experiential learning. *Simulation & Gaming*, *40*(3), 297–327. doi:10.1177/1046878108325713

Kondo, M., Ishikawa, Y., Smith, C., Sakamoto, K., Shimomura, H., & Wada, N. (2012). Mobile Assisted Language Learning in university EFL courses in Japan: Developing attitudes and skills for self-regulated learning. *ReCALL*, *24*(2), 169–187. doi:10.1017/S0958344012000055

Kong, S. C. (2018). Parents' perceptions of e-learning in school education: Implications for the partnership between schools and parents. *Technology, Pedagogy and Education*, *27*(1), 15–31. doi:10.1080/1475939X.2017.1317659

Kosnik, C., Menna, L., Dharamshi, P., & Beck, C. (2018). Constructivism as a framework for literacy teacher education courses: The cases of six literacy teacher educators. *European Journal of Teacher Education*, *41*(1), 105–119. doi:10.1080/02619768.2017.1372742

Koster. (2004). *A theory of fun*. Paraglyph Press.

Krajcik, J. S., & Blumenfeld, P. C. (2005). Project-based learning. In R. K. Sawyer (Ed.), *The Cambridge handbook of learning sciences* (pp. 317–334). Cambridge University Press. doi:10.1017/CBO9780511816833.020

Kramsch, C. (1998). *Language and culture*. Oxford University Press.

Krasulia, A., & Saks, K. (2020, July). Students' perceptions towards mobile learning in an English as a foreign language class. *20th International Conference on Advanced Learning Technologies*. 10.1109/ICALT49669.2020.00078

Kreps, S. E., Goldfarb, J. L., Brownstein, J. S., & Kriner, D. L. (2021). The relationship between us adults' misconceptions about COVID-19 vaccines and vaccination preferences. *Vaccines*, *9*(8), 901. doi:10.3390/vaccines9080901 PMID:34452025

Krippendorff, K. (2013). *Content analysis: an introduction to its methodology* (3rd ed.). Sage Publications.

Kroger, A., Bahta, L., & Hunter, P. (2022). *General best practice guidelines for immunization. Best practices guidance of the Advisory Committee on Immunization Practices*. ACIP.

Kuang, J. F. (2007). Developing students' cultural awareness through foreign language teaching. *SinoUS English Teaching*, *4*(12), 74–81.

Kukulska-Hulme, A. (2009). Will mobile learning change language learning? *ReCALL*, *21*(2), 157–165. doi:10.1017/S0958344009000202

Kukulska-Hulme, A. (2012). Smart devices or people? A mobile learning quandary. *International Journal of Learning and Media*, *4*(3-4), 73–77. doi:10.1162/IJLM_a_00105

Kukulska-Hulme, A., Lee, H., & Norris, L. (2017). Mobile learning revolution: Implications for language pedagogy. In C. A. Chapelle & S. Sauro (Eds.), *The handbook of technology and second language teaching and learning* (pp. 217–233). Wiley. doi:10.1002/9781118914069.ch15

Kukulska-Hulme, A., & Traxler, J. (2005). *Mobile learning: A handbook for educators and trainers*. Routledge.

Kumar, B. A., & Chand, S. (2019). Mobile learning adoption: A systematic review. *Education and Information Technologies*, *24*(1), 471–487. doi:10.100710639-018-9783-6

Kurt, S. (2011). Use of constructivist approach in architectural education. *Procedia: Social and Behavioral Sciences*, *15*, 3980–3988. doi:10.1016/j.sbspro.2011.04.402

Kuru Gönen, S. İ. (2019). A qualitative study on a situated experience of technology integration: Reflections from pre-service teachers and students. *Computer Assisted Language Learning*, *32*(3), 163–189. doi:10.1080/09588221.2018.1552974

Kvale, S., & Brinkmann, S. (2015). *Interviews: learning the craft of qualitative research interviewing* (3rd ed.). Sage Publications.

Lakhal, S., Khechine, H., & Pascot, D. (2007). Evaluation of the effectiveness of podcasting in teaching and learning. In G. Richards (Ed.), *Proceedings of World Conference on E-Learning in Corporate, Government, Healthcare, and Higher Education* (pp. 6181-6188). AACE.

Lall, S., & Singh, N. (2020). COVID-19: Unmasking the new face of education. *International Journal of Research in Pharmaceutical Sciences*, *11*(1), 48–53. doi:10.26452/ijrps.v11iSPL1.2122

Lamanauskas, V., & Makarskaitė-Petkevičienė, R. (2021). Distance lectures in university studies: Advantages, disadvantages, improvement. *Contemporary Educational Technology*, *13*(3), ep309. Advance online publication. doi:10.30935/cedtech/10887

Landers, R. N. (2014). Developing a theory of gamified learning: Linking serious games and gamification of learning. *Simulation & Gaming*, *45*(6), 752–768. doi:10.1177/1046878114563660

Landers, R. N., & Armstrong, M. B. (2017). Enhancing instructional outcomes with gamification: An empirical test of the technology-enhanced training effectiveness model. *Computers in Human Behavior*, *71*, 499–507. doi:10.1016/j.chb.2015.07.031

Landers, R. N., Bauer, K. N., Callan, R. C., & Armstrong, M. B. (2015). Psychological theory and the gamification of learning. In T. Reiners & L. Wood (Eds.), *Gamification in education and business* (pp. 165–186). Springer. doi:10.1007/978-3-319-10208-5_9

Landow, G. P. (2009). *Hypertext 3.0. Critical theory and new media in the age of globalization*. Planeta.

Landrum, R. E. (2000). Motivational differences among traditional and nontraditional students enrolled in metropolitan universities. *Metropolitan Universities*, *11*(1), 87–92.

Lan, K. Y., & Hsiao, I. (2018). Effective Learning Design of Game-Based 3D Virtual Language Learning Environments for Special Education Students. *Journal of Educational Technology & Society*, *21*(3), 213. https://www.researchgate.net/publication/326250187

Lankshear, C., & Knobel, M. (2008). Introduction: Digital literacies - Concepts, policies and practices. In C. Lankshear & M. Knobel (Eds.), *Digital literacies: concepts, policy and practice* (pp. 1–17). Peter Lang.

Lansu, T. A. (2018). Burn! How implicit and explicit attitudes predict early adolescents "hot sauce" aggression toward classroom peers. *Journal of Experimental Child Psychology*, *167*, 423–432. doi:10.1016/j.jecp.2017.11.006 PMID:29241999

Lan, Y. F., & Sie, Y. S. (2010). Using RSS to support mobile learning based on media richness theory. *Computers & Education*, *55*(2), 723–732. doi:10.1016/j.compedu.2010.03.005

Laro-González, E. (2020). Innovar enseñando: La educación del futuro. Las TICs como factor motivador en la enseñanza. *Revista Jurídica de Investigación e Innovación Educativa*, *21*, 11–23. doi:10.24310/REJIE.2020.v0i21.7530

Lasco, G., & Yu, V. G. (2021). Communicating COVID-19 vaccines: Lessons from the dengue vaccine controversy in the Philippines. *BMJ Global Health*, *005422*(3), e005422. Advance online publication. doi:10.1136/bmjgh-2021-005422 PMID:33653732

Laurillard, D., & Pachler, N. (2007). Pedagogical forms of mobile learning: Framing. In N. Pachler (Ed.), Mobile learning: Towards a research agenda (pp. 33-54). WLE Centre, IoE.

Lavega, P., Aráujo, P., & Jaqueira, A. R. (2013). Teaching motor and emotional competencies in university students. *Cultura, Ciencia y Deporte*, *8*(22), 5–15. Advance online publication. doi:10.12800/ccd.v8i22.219

Leask, J., Kinnersley, P., Jackson, C., Cheater, F., Bedford, H., & Rowles, G. (2012). Communicating with parents about vaccination: A framework for health professionals. *BMC Pediatrics*, *12*(1), 154. doi:10.1186/1471-2431-12-154 PMID:22998654

Ledford, V., & Nan, X. (2023). Boosting vaccination: Accelerating vaccine uptake through communication science. *Metode Science Studies Journal*, *0*(13), 103–109. doi:10.7203/metode.13.23759

Lee, J., & Hammer, J. (2011). Gamification in education: What, how, why bother? *Academic Exchange Quarterly*, *15*, 1–5.

Legard, R., Keegan, J., & Ward, K. (2003). In-depth interviews. In Qualitative research practice: A guide for social science students and researchers (pp. 138-169). Academic Press.

Leung, E., & Pluswik, E. (2018). Effectiveness of Gamification Activities in a Project-based Learning Classroom. *Asee Annual Conference and Exposition*. 10.18260/1-2--30361

Leung, S. O. A., Akinwunmi, B., Elias, K. M., & Feldman, S. (2019). Educating healthcare providers to increase Human Papillomavirus (HPV) vaccination rates: A Qualitative Systematic Review. *Vaccine: X*, *3*, 100037. doi:10.1016/j.jvacx.2019.100037 PMID:31463471

Levy, M., & Kennedy, C. (2005). Learning Italian via Mobile SMS. In A. Kukulska-Hulme & J. Traxler (Eds.), *Mobile learning: a handbook for educators and trainers* (pp. 76–83). Routledge.

Lezak, M. D. (1982). The problem of assessing executive functions. *International Journal of Psychology*, *17*(1–4), 281–297. Advance online publication. doi:10.1080/00207598208247445

Li, M., Luo, Y., Watson, R., Zheng, Y., Ren, J., Tang, J., & Chen, Y. (2021). Healthcare workers' (HCWs) attitudes and related factors towards COVID-19 vaccination: A rapid systematic review. *Postgraduate Medical Journal*. . doi:10.1136/postgradmedj-2021-140195

Liaw, S. S., & Huang, H. M. (2012). A case of study of investigating users' acceptance toward mobile learning. In F. L. Gaol (Ed.), *Recent progress in data engineering and internet technology* (pp. 299–306). Springer. doi:10.1007/978-3-642-28798-5_41

Liberati, A., Altman, D. G., Tetzlaff, J., Mulrow, C., Gøtzsche, P. C., & Loannidis, J. P. (2009). La declaración Prisma para informar revisiones sistemáticas y metaanálisis de estudios que evalúan las intervenciones de atención médica: Explicación y elaboración. *PLoS Medicine*, *6*(7), 1–28. PMID:19621070

Lin, C., Mullen, J., Smith, D., Kotarba, M., Kaplan, S. J., & Tu, P. (2021). Healthcare providers' vaccine perceptions, hesitancy, and recommendation to patients: A systematic review. *Vaccines*, *9*(7), 713. doi:10.3390/vaccines9070713 PMID:34358132

Ling, R. (2010). Texting as a life phase medium. *Journal of Computer-Mediated Communication*, *15*(2), 277–292. doi:10.1111/j.1083-6101.2010.01520.x

Lin, H., Hwang, G., Chang, S., & Hsu, Y. (2021). Facilitating critical thinking in decision making-based professional training: An online interactive peer-review approach in flipped learning context. *Computers & Education*, *173*, 104266. Advance online publication. doi:10.1016/j.compedu.2021.104266

Lin, X., & Gao, L. (2020). Students' sense of community and perspectives of taking synchronous and asynchronous online courses. *Asian J. Distance Educ.*, *15*, 169–179. doi:10.5281/zenodo.3881614

Li, R. (2022). Effects of mobile-assisted language learning on EFL/ESL reading comprehension. *Journal of Educational Technology & Society*, *25*(3), 15–29.

Lister, C., West, J. H., Cannon, B., Sax, T., & Brodegard, D. (2014). Just a fad? Gamification in health and fitness apps. *JMIR Serious Games*, *2*(2), e9. doi:10.2196/games.3413 PMID:25654660

Liu, T. Y. (2017). Developing an English mobile learning attitude scale for adult learners. *Journal of Educational Technology Systems*, *45*(3), 424–435. doi:10.1177/0047239516658448

Livingstone, S. (2008). *Internet literacy: Young people's negotiation of new online opportunities*. MIT Press.

Longhurst, R. (2003). Semi-structured interviews and focus groups. *Key Methods in Geography, 3*, 143-156.

Looyestyn, J., Kernot, J., Boshoff, K., Ryan, J., Edney, S., & Maher, C. (2017). Does gamification increase engagement with online programs? A systematic review. *PLoS One*, *12*(3), e0173403. doi:10.1371/journal.pone.0173403 PMID:28362821

Lopes, P., Costa, P., Araujo, L., & Ávila, P. (2018). Measuring media and information literacy skills: Construction of a test. *Communications*, *43*(4), 1–27. doi:10.1515/commun-2017-0051

López Borrull, A., Vives-Gràcia, J., & Badell, J. (2018). Fake news, ¿amenaza u oportunidad para los profesionales de la información y la documentación? *El Profesional de la Información*, *27*(6), 1346. Advance online publication. doi:10.3145/epi.2018.nov.17

López, P. (2018). E*l Aprendizaje Basado en Proyectos en la enseñanza de la historia. Propuesta de dos unidades didácticas para la Educación Secundaria Obligatoria* [Master's thesis]. Retrieved from http://file:///C:/Users/danij/Downloads/tfm_2017-18_MFPR_plm970_1987.pdf

Lopez, Lira, L., & Uribe Lopez, A. (2022). Emerging pedagogies developed in higher education from the confinement by covid-19. *Opening*, *14*(1), 114–131. doi:10.32870/Ap.v14n1.2149

López, M., Laviana, M., Fernández, L., López, A., Rodríguez, A. M., & Aparicio, A. (2008). La lucha contra el estigma y la discriminación en salud mental: Una estrategia compleja basada en la información disponible. *Revista de la Asociación Española de Neuropsiquiatría*, *101*(1), 43–83. doi:10.4321/S0211-57352008000100004

Lord, G., & Lomicka, L. (2011). Calling on educators: Paving the way for the future of technology and CALL. In N. Arnold & L. Ducate (Eds.), *Present and future promises of CALL: From theory and research to new directions in language teaching* (pp. 441–470). CALICO.

Lorenzo-Álvarez, R., Rudolphi-Solero, T., Ruiz-Gómez, M. J., & Sendra-Forte, F. (2019). Medical Student Education for Abdominal Radiographs in a 3D Virtual Classroom Versus Traditional Classroom: A Randomized Controlled Trial. *AJR. American Journal of Roentgenology*, *213*(3), 644–650. doi:10.2214/AJR.19.21131 PMID:31287725

Lozano, A. A., & Izquierdo, J. (2019). Technology in Second Language Education: Overcoming the Digital Divide. *Emerging Trends in Education*, *2*(3), 52–70. doi:10.19136/etie.a2n3.3250

Lozano-Díaz, A., Fernández-Prados, J. S., Figueredo Canosa, V., & Martínez-Martínez, A. M. (2020). Impactos del confinamiento por el COVID-19 entre universitarios: Satisfacción vital, resiliencia y capital social online. *RISE*, *9*(1), 79–104. doi:10.17583/rise.2020.5925

Luque Vilca, O. M., Bolivar Espinoza, N., Achahui Ugarte, V. E., & Gallegos Ramos, J. R. (2022). Estrés académico en estudiantes universitarios frente a la educación virtual asociada al covid-19. *Puriq*, *4*, e200. doi:10.37073/puriq.4.1.200

Maatuk, A. M., Elberkawi, E. K., Aljawarneh, S., Rashaideh, H., & Alharbi, H. (2022). The COVID-19 pandemic and E-learning: Challenges and opportunities from the perspective of students and instructors. *Journal of Computing in Higher Education*, *34*(1), 21–38. doi:10.100712528-021-09274-2 PMID:33967563

Mac Callum, K. M., Jeffrey, L., & Kinshuk. (2014b). Comparing the role of ICT literacy and anxiety in the adoption of mobile learning. *Computers in Human Behavior*, *39*, 8–19. doi:10.1016/j.chb.2014.05.024

Mac Callum, K., Jeffrey, L., & Na, K. (2014a). Factors impacting teachers' adoption of mobile learning. *Journal of Information Technology Education*, *13*, 141–162. doi:10.28945/1970

MacIntyre, C. R. (2013). Elderly vaccination—The glass is half full. *Health (Irvine, Calif.)*, *5*(12), 80–85. doi:10.4236/health.2013.512A011

MacIntyre, C. R., Menzies, R., Kpozehouen, E., Chapman, M., Travaglia, J., Woodward, M., ... Adair, T. (2016). Equity in disease prevention: Vaccines for the older adults–a national workshop, Australia 2014. *Vaccine*, *34*(46), 5463–5469. doi:10.1016/j.vaccine.2016.09.039 PMID:27686835

Madewell, Z. J., Chacón-Fuentes, R., Jara, J., Mejía-Santos, H., Molina, I.-B., Alvis-Estrada, J. P., Ortiz, M.-R., Coello-Licona, R., & Montejo, B. (2021). Knowledge, attitudes, and practices of seasonal influenza vaccination in healthcare workers, Honduras. *PLoS One*, *16*(2), e0246379. Advance online publication. doi:10.1371/journal.pone.0246379 PMID:33539428

Madigan, E. M., Goodfellow, M., & Stone, J. A. (2007). Gender, perceptions, and reality: Technological literacy among first-year students. *SIGCSE Bulletin*, *39*(1), 410–414. doi:10.1145/1227504.1227453

Mahdi, H. S. (2018). Effectiveness of mobile devices on vocabulary learning: A meta-analysis. *Journal of Educational Computing Research*, *56*(1), 134–154. doi:10.1177/0735633117698826

Mahlangu, V. P. (2018). The Good, the Bad, and the Ugly of Distance Learning in Higher Education. In M. Sinecen (Ed.), *Trends in E-learning* (pp. 17–29). IntechOpen. doi:10.5772/intechopen.75702

Mahoney, K. (2017). *The assessment of emergent bilinguals: supporting English language learners*. Multilingual Matters.

Major, B., McCoy, S., Kaiser, C., & Quinton, W. (2003). Prejudice and self-esteem: A transactional model. *European Review of Social Psychology*, *14*(1), 77–104. doi:10.1080/10463280340000027

Malaby, T. M. (2007). Beyond play: A new approach to games. *Games and Culture*, *2*(2), 95–113. doi:10.1177/1555412007299434

Manovich, L. (2001). *The language of new media*. The MIT Press.

Manzano León, A., Arrifano Tadeu, P. J. (2022). Escape Rooms educativos: una experiencia en una universidad portuguesa. *International journal of developmental and educational psychology INFAD revista de psicología, 1*(1), 281-288. . doi:10.17060/ijodaep.2022.n1.v1.2374

Manzano León, A., Rodríguez Ferrer, J. M., Aguilar, J. M., Fernández Campoy, J. M., Trigueros, R., & Martínez Martínez, A. M. (2022). Play and learn: Influence of gamification and game-based learning in the reading processes of secondary school students. *Revista de Psicodidáctica*, *27*, 38–46. doi:10.1016/j.psicod.2021.07.001

Manzano León, A., Sánchez, M., Trigueros Ramos, R., Álvarez Hernández, J., & Aguilar-Parra, J. M. (2020). Gamificación y Breakout Edu en formación profesional. El programa "Grey Place" en integración social. *EDMETIC*, *9*(1), 3–20. doi:10.21071/edmetic.v9i1.12067

Marcelo, C., Yot, C., & Mayor, C. (2015). Enseñar con tecnologías digitales en la Universidad. *Comunicar*, *45*(23), 117–124.

Marinoni, G., Van't Land, H., & Jensen, T. (2020). *The impact of Covid-19 on higher education around the world.* Retrieved November 17, 2022, from https://www.iau-aiu.net/IAU-Global-Survey-on-the-Impact-of-COVID-19-on-Higher-Education-around-the

Marín, V., & Muñoz, V. P. (2018). Trabajar el cuerpo humano con realidad aumentada en educación infantil. *Revista Tecnología. Ciência & Educação (Bauru), 9*, 148–158.

Markauskaite, L. (2007). Exploring the structure of trainee teachers' ICT literacy: The main components of, and relationships between, general cognitive and technical capabilities. *Educational Technology Research and Development, 55*(6), 547–572. doi:10.100711423-007-9043-8

Marley, J. L. (2007). Gender Differences and Distance Education: Major Research Findings and Implications for LIS Education. *Journal of Education for Library and Information Science, 48*(1), 13–20. https://doi.org/http://www.jstor.org/stable/40324317

Marschark, M., & Hauser, P. C. (2008). Cognitive underpinnings of learning by deaf and hard-of-hearing students: Differences, diversity, and directions. In M. Marschark & P. C. Hauser (Eds.), *Deaf cognition: Foundations and outcomes* (pp. 3–23). Oxford University Press. doi:10.1093/acprof:oso/9780195368673.003.0001

Martin, A. (2008). Digital literacy and the "digital society". In C. Lankshear & M. Knobel (Eds.), *Digital literacies: Concepts, policies, and practices* (pp. 151–176). Peter Lang.

Martínez, E., Carbonell, V., Álvarez, J., Flórez, M., & Montoya, E. (2023). Let's escape! the use of a digital Escape Room to increase student's motivation and learning on kinematics concepts. *INTED2023 Proceedings*. 10.21125/inted.2023.0236

Martínez-Baena, A., & Faus-Boscá, J. (2018). Acoso escolar y Educación Física: Una revisión sistemática. *Retos, 34*, 338–345.

Mayer, R. E. (2020). Where is the learning in mobile technologies for learning? *Contemporary Educational Psychology, 60*(101824), 1–3. doi:10.1016/j.cedpsych.2019.101824

Mei, B., Brown, G. T., & Teo, T. (2018). Toward an understanding of preservice English as a Foreign Language teachers' acceptance of computer-assisted language learning 2.0 in the People's Republic of China. *Journal of Educational Computing Research, 56*(1), 74–104. doi:10.1177/0735633117700144

Mellor, C. (2014). School-based interventions targeting stigma of mental illness: Systematic review. *Psychiatric Bulletin, 38*(4), 164–171. doi:10.1192/pb.bp.112.041723 PMID:25237538

Mercader, C., & Gairín, J. (2017). ¿Cómo utiliza el profesorado universitario las tecnologías digitales en sus aulas? *Revista de Docencia Universitaria, 15*(2), 257–274. doi:10.4995/redu.2017.7635

Mercer, N., & Fisher, E. (1997). The importance of talk. In R. Wegerif & P. Scrimshaw (Eds.), *Computers and talk in the primary classroom* (pp. 13–21). Multilingual Matters.

Meskill, C. (2009). CMC in language teacher education: Learning with and through instructional conversations. *Innovation in Language Learning and Teaching, 3*(1), 51–63. doi:10.1080/17501220802655474

Meyer, B. (2015). Learning through telepresence with iPads: Placing schools in local/global communities. *Interactive Technology and Smart Education, 12*(4), 270–284. doi:10.1108/ITSE-09-2015-0027

Mihaylova, M., Gorin, S., Reber, T. P., & Rothen, N. (2022). A meta-analysis on mobile assisted language learning applications: Benefits and risks. *Psychologica Belgica, 62*(1), 252–271. doi:10.5334/pb.1146 PMID:36186897

Millán, J. A. (2008). *Reading and the knowledge society.* Government of Navarra / Department of Education.

Miller, C. (2013). The gamification of education. *Developments in Business Simulation and Experiential Learning: Proceedings of the Annual ABSEL conference,* 40.

Miller, E. C., Reigh, E., Berland, L., & Krajcik, J. (2021). Supporting equity in virtual science instruction through Project-Based Learning: Opportunities and challenges in the era of COVID-19. *Journal of Science Teacher Education, 32*(6), 642–663. Advance online publication. doi:10.1080/1046560X.2021.1873549

Ministerio de Educación y Formación Profesional. (2020). *El Ministerio de Educación y Formación Profesional y las CCAA acuerdan mantener la duración del curso escolar hasta junio.* https://www.educacionyfp.gob.es/prensa/actualidad/2020/04/20200415-sectorial.html

Ministry of Education and Science of Ukraine. (2017). *Програма курсу "Інформатика" для 5–9 класів загальноосвітніх навчальних закладів* [Program of the Course "Informatics" for Grades 5–9 of Secondary Schools]. Retrieved November 12, 2022, from https://mon.gov.ua/storage/app/media/zagalna seredny a/programy-5-9-klas/onovlennya-12-2017/programainformatika-5-9-traven-2015.pdf

Ministry of Education and Science of Ukraine. (2020). *Letter No. 1/9-154.* Retrieved November 12, 2022, from https://mon.gov.ua/storage/app/media/news/Новини/2020/03/11/1_9-154.pdf

Ministry of Education of the United Arab Emirates. (2020a). *Confirmed options for re-opening schools in the academic year 2020-21.* Retrieved November 13, 2022, from https://u.ae/en/information-and-services/education/confirmed-options-for-reopening-schools-in-the-academic-year-2020-21

Ministry of Education of the United Arab Emirates. (2020b). *Partial re-opening of higher education institutes in the academic year 2020-21.* Retrieved November 13, 2022, from https://u.ae/en/information-and-services/education/higher-education/partial-re-opening-of-higher-education-institutes-in-the-academic-year-2020-21

Ministry of Education of the United Arab Emirates. (2020c). *Distance learning in times of COVID-19.* Retrieved November 13, 2022, from https://u.ae/en/information-and-services/education/distance-learning-in-times-of-covid-19

Ministry of Universities of Spain. (2023, July). *UNIVBase.* Retrieved July 4, 2023, from https://estadisticas.universidades.gob.es/jaxiPx/Datos.htm?path=/Universitaria/Alumnado/EEU_2023/Master/Matriculados//l0/&file=1_1_Mat_Master_Sex_Edad(2)_Rama.px

Minović, M., García-Peñalvo, F. J., & Kearney, N. A. (2016). Gamification ecosystems in engineering education. *International Journal of Engineering Education, 32*(1B), 308–309.

Miyake, A., Friedman, N. P., Emerson, M. J., Witzki, A. H., Howerter, A., & Wager, T. D. (2000). The Unity and Diversity of Executive Functions and Their Contributions to Complex "Frontal Lobe" Tasks: A Latent Variable Analysis. *Cognitive Psychology, 41*(1), 49–100. Advance online publication. doi:10.1006/cogp.1999.0734 PMID:10945922

Mohamed, K., Rzymski, P., Islam, M. S., Makuku, R., Mushtaq, A., Khan, A., Ivanovska, M., Makka, S. A., Hashem, F., Marquez, L., Cseprekal, O., Filgueiras, I. S., Fonseca, D. L. M., Mickael, E., Ling, I., Arero, A. G., Cuschieri, S., Minakova, K., Rodríguez-Román, E., ... Rezaei, N. (2022). COVID-19 vaccinations: The unknowns, challenges, and hopes. *Journal of Medical Virology, 94*(4), 1336–1349. doi:10.1002/jmv.27487 PMID:34845731

Molina, P., & Ruiz, M. C. (2023). *La gamificación como herramienta motivadora en la empresa.* https://hdl.handle.net/10953.1/19582

Molinero Bárcenas, M., & Chávez Morales, U. (2019). Technological tools in the teaching-learning process in higher education students. *RIDE. Ibero-American Journal for Educational Research and Development, 10*(19). Advance online publication. doi:10.23913/ride.v10i19.494

Montrieux, H., Grove, F. D., & Schellens, T. (2014). Mobile learning in secondary education: Teachers' and students' perceptions and acceptance of tablet computers. *International Journal of Mobile and Blended Learning*, *6*(1), 26–40. doi:10.4018/ijmbl.2014040103

Moore, M., & Kearsley, G. (2012). *Distance education: a systems view of online learning* (3rd ed.). Wadsworth.

Moreira, F., Ferreira, M. J., & Cardoso, A. (2018, November). ECLECTIC approach applied to data transformation curriculum course. In *Proceedings. 11th Annual International Conference of Education, Research and Innovation (ICERI 2018),* (pp. 6021-6029). 10.21125/iceri.2018.2411

Moreno Fuentes, E., Perales Molada, R., & Hidalgo Navarrete, J. (2019). *Aula de encuentro: revista de investigación y comunicación de experiencias educativas*. Academic Press.

Moreno Oliver, V., Hernández-Leo, D., & Peig Olivé, E. (2022). Impacto pre-post pandemia en escenarios educativos online e híbridos: El caso de la Escuela de Ingeniería de la Universitat Pompeu Fabra. *Revista Actas de las Jenui*, *7*, 223–229.

Moreno, E. y Lopezosa, M.D. (2020). Gamificación a través de un proyecto de aprendizaje-servicio: diseñando un Breakout educativo desde la universidad para el alumnado de primaria. *Revista científica electrónica de Educación y Comunicación en la Sociedad del Conocimiento, 20*(1).

Moreno, E., & Lopezosa, M.D. (2020). Gamificación a través de un proyecto de aprendizaje-servicio: diseñando un breakout educativo desde la universidad para el alumnado de primaria. *Revista científica electrónica de Educación y Comunicación en la Sociedad del Conocimiento, 20*(1). . doi:10.30827/eticanet.v20i1.15524

Moreno, N., Leiva, J., Galván, M., López, E., & García, F. (2017). Realidad aumentada y realidad virtual para la enseñanza-aprendizaje del inglés desde un enfoque comunicativo e intercultural. In J. Ruiz Palmero, J. Sánchez-Rodríguez, & E. Sánchez-Rivas (Eds.), *Innovación docente y uso de las TIC en educación* (pp. 1–11). UMA Editorial.

Moreno-Rodríguez, R. (2020) Reflexiones en torno al Impacto del Covid-19 sobre la Educación Universitaria: Aspectos a Considerar acerca de los Estudiantes con Discapacidad. *Revista internacional de educación para la justicia social, 9*(3), 1-6. https://revistas.uam.es/ riejs/article/view/1222

Morrison, F. J., Ponitz, C. C., & McClelland, M. M. (2010). Self-regulation and academic achievement in the transition to school. In S. D. Calkins & M. A. Bell (Eds.), *Child development at the intersection of emotion and cognition* (pp. 203–224). American Psychological Association. doi:10.1037/12059-011

Morrow, S. L. (2005). Quality and trustworthiness in qualitative research in counseling psychology. *Journal of Counseling Psychology*, *52*(2), 250–260. doi:10.1037/0022-0167.52.2.250

Mossberger, K., Tolbert, C. J., & Stansbury, M. (2003). *Virtual inequality: beyond the digital divide*. Georgetown University Press.

Moustakas, C. (1994). Phenomenological research methods. *Sage (Atlanta, Ga.)*.

Moya, S., & Camacho, M. (2021). Identifying the key success factors for the adoption of mobile learning. *Education and Information Technologies*, *26*(4), 3917–3945. doi:10.100710639-021-10447-w

Muhammed, A., & Kainat, A. (2020). Online learning amid the COVID-19 pandemic: Students' perspectives. *Online Submission*, *2*(1), 45–51.

Muijen, M. (2012). The state of psychiatry in Europe: Facing the challenges, developing consensus. *International Review of Psychiatry (Abingdon, England)*, *24*(4), 274–277. doi:10.3109/09540261.2012.697264 PMID:22950764

Müller, U., & Kerns, K. (2015). The Development of Executive Function development of EF and its neural. Handbook of Child Psychology and Developmental Science: Cognitive Processes.

Nachega, J. B., Sam-Agudu, N. A., Masekela, R., van der Zalm, M. M., Nsanzimana, S., Condo, J., Ntoumi, F., Rabie, H., Kruger, M., Wiysonge, C. S., Ditekemena, J. D., Chirimwami, R. B., Ntakwinja, M., Mukwege, D. M., Noormahomed, E., Paleker, M., Mahomed, H., Tamfum, J.-J. M., Zumla, A., & Suleman, F. (2021). Addressing challenges to rolling out COVID-19 vaccines in African countries. *The Lancet. Global Health*, *9*(6), 746–748. doi:10.1016/S2214-109X(21)00097-8 PMID:33713633

Naeem, S., & Bhatti, R. (2020). The Covid-19 'infodemic': A new front for information professionals. *Health Information and Libraries Journal*, *37*(3), 233–239. doi:10.1111/hir.12311 PMID:32533803

Nah, K. C., White, P., & Sussex, R. (2008). The potential of using a mobile phone to access the Internet for learning EFL listening skills within a Korean context. *ReCALL*, *20*(3), 331–347. doi:10.1017/S0958344008000633

Nair, S., & Mathew, J. (2019). A theoretical framework for gamified learning. *International Journal of Innovative Technology and Exploring Engineering*, *8*(12).

Naismith, L., Lonsdale, P., Vavoula, G., & Sharples, M. (2004). *Literature review in mobile technologies and learning: Report 11*. Futurelab Series.

Navarro-Mateos, C., Pérez-López, I., & Marzo, P. (2021). La gamificación en el ámbito educativo español: revisión sistemática [Gamification in the Spanish educational field: a systematic review]. *Retos*, *42*, 507–516. doi:10.47197/retos.v42i0.87384

Negre i Walczak, C. (2017). *'Breakout Edu', microgamificación y aprendizaje significativo.* https://www.educaweb.com/noticia/2017/07/26/breakoutedu-microgamificacion-aprendizaje-significativo-15068/

Negre, C., & Carrión, S. (2020). *Desafío en el aula. Manual práctico para llevar los juegos de escape educativo a clase.* PAIDÓS Educación.

New London Group. (1996). A Pedagogy of Multiliteracies: Designing Social Futures. *Harvard Educational Review*, *66*(1), 60–93. doi:10.17763/haer.66.1.17370n67v22j160u

Newman, N., Fletcher, R., Schulz, A., Andi, S., Robertson, C. T., & Nielsen, R. K. (2021). *Reuters Institute Digital News Report 2021. 10th Edition.* https://reutersinstitute.politics.ox.ac.uk/sites/default/files/2021- 06/Digital_News_Report_2021_FINAL.pdf

Ng, W. (2015). Sustaining mobile learning at a personal level: Mobile digital literacy. In W, Ng, & T. M. Cumming (Eds.), Sustaining mobile learning: theory, research and practice (pp. 84-107). Routledge.

Nguyen, M. H. (2017). Negotiating contradictions in developing teacher identity during the EAL practicum in Australia. *Asia-Pacific Journal of Teacher Education*, *45*(4), 399–415. doi:10.1080/1359866X.2017.1295132

Nguyen, M. H. (2019). *English language teacher education: a sociocultural perspective on preservice teachers' learning in the professional experience*. Springer. doi:10.1007/978-981-13-9761-5

Nguyen, M. H., & Brown, J. (2016). Influences on preservice writing instruction during the secondary English as an additional language practicum in Australia. *The Australian Journal of Teacher Education*, *41*(8), 84–101. doi:10.14221/ajte.2016v41n8.5

Ng, W. (2012). Can we teach digital natives digital literacy? *Computers & Education*, *59*(3), 1065–1078. doi:10.1016/j.compedu.2012.04.016

Norwich, B., Kelly, N., & Not Available, E. P. T. (2006). Evaluating children's Participation in SEN procedures: Lessons for educa-tional psychologists. *Educational Psychology in Practice*, *22*(3), 255–271. Advance online publication. doi:10.1080/02667360600845836

O'Donovan, S., Gain, J., & Marais, P. (2013). A case study in the gamification of a university-level games development course. In *Proceedings of SAICSIT 2013* (pp. 242-251). ACM Press.

Ocak, G., & Karakuş, G. (2019). An investigation of digital literacy self-efficacy skills of pre-service teachers in terms of different variables. *Afyon Kocatepe University Journal of Social Sciences*, *21*(1), 129–147.

Ocampo, F., De Luna, R., & Pérez, M.G. (2018). Factores de relación entre la inteligencia emocional y el rendimiento académico en estudiantes de la ESCOM del IPN. *Revista electrónica cuerpos académica grupos investigación,* 5.

Ochieng, C., Anand, S., Mutwiri, G., Szafron, M., & Alphonsus, K. (2021). Factors associated with COVID-19 vaccine hesitancy among visible minority groups from a global context: A scoping review. *Vaccines*, *9*(12), 1445. doi:10.3390/vaccines9121445 PMID:34960192

Odiakaosa, O. J., Dlodlo, N., & Jere, N. (2017). Teacher and learner perceptions on mobile learning technology: A case of Namibian high schools from the hardap region. *Higher Educator-An International Journal*, *1*(1), 13–41. doi:10.26762/he.2017.30000002

OECD. (2020). *Education responses to COVID-19: Embracing digital learning and online collaboration.* Retrieved November 23, 2022, from https://www.oecd.org/coronavirus/policy-responses/education-responses-to-covid-19-embracing-digital-learning-and-online-collaboration-d75eb0e8/

OkoliC.SchabramK. (2010). Una guía para realizar una revisión sistemática de la literatura sobre la investigación de sistemas de información. *Revista Electrónica* SSRN, 10 (26) doi:10.2139/ssrn.1954824

Olson, K. E., O'Brien, M. A., Rogers, W. A., & Charness, N. (2011). Diffusion of technology: Frequency of use for younger and older adults. *Ageing International*, *36*(1), 123–145. doi:10.100712126-010-9077-9 PMID:22685360

Önal, N., & Önal, N. T. (2019). Adaptation of English mobile learning attitude scale into Turkish: Validity and reliability study. *Journal of Social Sciences of Mus Alparslan University*, *7*(1), 283–289.

ONU. (2019). *Página principal de la Naciones Unidas.* https://www.un.org/sustainabledevelopment/es/objetivosde-desarrollo-sostenible/

Organisation for Economic Co-operation and Development. (2015). *Future of Education and Skills 2030.* https://www.oecd.org/education/2030-project/

Organización Mundial de la Salud. (2020). *Coronavirus.* Consultado el 18 de enero desde https://www.who.int/docs/default-source/coronaviruse/situation-reports/20200213-sitrep-24-covid-19.pdf?sfvrsn=9a7406a4_4

Ortega Ortigoza, D., Rodríguez-Rodríguez, J. & Mateos Inchaurrondo, A. (2021). Educación superior y la COVID-19: adaptación metodológica y evaluación online en dos universidades de Barcelona. *Revista digital de investigación en docencia universitaria, 15*(1), 1-13. http://dx.doi.org/. doi:10.19083/ridu.2021.1275

Ortíz-ColónA.-M.JordánJ.AgredalM. (2018). Gamification in education: an overview on the state of the art. *Educ. Pesqui.* 44. https://doi.org/ doi:10.1590/S1678-4634201844173773

Özcan, M. (2022). Evaluation of prospective teachers' digital literacy levels and mobile learning attitudes. *Journal of Educational Technology & Online Learning*, *5*(2), 367–378. doi:10.31681/jetol.1020586

Özdoğan, A. Ç., & Berkant, H. G. (2020). Covid-19 pandemi dönemindeki uzaktan eğitime ilişkin paydaş görüşlerinin incelenmesi. *Milli Eğitim Dergisi*, *49*(1), 13–43. doi:10.37669/milliegitim.788118

Özel, A. G., & Arikan, A. (2015). The use of the internet and web 2.0 tools among EFL instructors. *Mediterranean Journal of Humanities*, *5*(1), 313–325. doi:10.13114/MJH.2015111386

Özer, O., & Kılıç, F. (2018). The effect of mobile-assisted language learning environment on EFL students' academic achievement, cognitive load and acceptance of mobile learning tools. *Eurasia Journal of Mathematics, Science and Technology Education*, *14*(7), 2915–2928. doi:10.29333/ejmste/90992

Paakkari, L., & Okan, O. (2020). COVID-19: Health literacy is an underestimated problem. *The Lancet. Public Health*, *5*(5), e249–e250. Advance online publication. doi:10.1016/S2468-2667(20)30086-4 PMID:32302535

Pachler, N. (2010). The socio-cultural ecological approach to mobile learning: an overview. In B. Bachmair (Ed.), *Medienbildung in neuen Kulturräumen. Die deutschsprachige und britische Diskussion* (pp. 153–167). Wiesbaden VS Verlag für Sozialwissenschaften. doi:10.1007/978-3-531-92133-4_11

Padilla-Carmona, T., González-Monteagudo, J., & Tenorio Rodríguez, M. (2022). Lights and shadows: The inclusion of invisible students in Spanish universities. *Autonomy and Responsibility Journal of Educational Sciences*, *7*(1), 37–44. doi:10.15170/AR.2022.7.1.3.

Padmanathan, Y., & Jogulu, L. N. (2018). Mobile learning readiness among Malaysian Polytechnic students. *Journal of Information Systems and Technology Management*, *3*, 113–125.

Page, M. J., McKenzie, J. E., Bossuyt, P. M., Boutron, I., Hoffmann, T. C., Mulrow, C. D., Shamseer, L., Tetzlaff, J. M., Akl, E. A., Brennan, S. E., Chou, R., Glanville, J., Grimshaw, J. M., Hróbjartsson, A., Lalu, M. M., Li, T., Loder, E. W., Mayo-Wilson, E., McDonald, S., ... Moher, D. (2021). The PRISMA 2020 statement: An updated guideline for reporting systematic reviews. *British Medical Journal*, *372*(71), n71. Advance online publication. doi:10.1136/bmj.n71 PMID:33782057

Palacios-Hidalgo, F. J., Huertas-Abril, C. A. y Gómez-Parra, M. E. (2022). Competencia digital docente percibida del profesorado bilingüe en formación: Estudio comparativo entre enseñanza presencial y enseñanza a distancia de emergencia. *Revista de educación a distancia (RED), 22*(69), 1-15. . doi:10.6018/red.466771

Pan American Health Organization. (2020). *Communication on safe vaccination. Guidelines to support health personnel in communication with mothers, fathers, caregivers and patients.* . doi:10.37774/9789275322819

Papadopoulos, I. (2011). *Courage, compassion and cultural competence. In the 13th Anna Reynvaan Lecture.* De Stadsschouwburg - Amsterdam City Theatre, Academic medical Centre, University of Amsterdam. https://www.researchgate.net/publication/303885059_Courage_Compassion_and_Cultural_Competence_The_Anna_Reynvaan_Lecture_2011

Papadopoulos, I. (2006). *Transcultural health and social care: development of culturally competent practitioners.* Churchill Livingstone Elsevier.

Pappas, C. (2013). *Gamify the classroom.* Retrieved on February 12, 2015 from https://elearningindustry.com/gamify-the-classroom

Páramo Rengifo, C. A. (2020). Gamificación e innovación educativa: apuntes para la práctica. *Revista Conrado, 16*(S1), 241-248. Retrieved from https://conrado.ucf.edu.cu/index.php/conrado/article/view/1547

Pariser, E. (2011). *The Filter Bubble: What the Internet is hiding from you.* The Penguin Press.

Park, C. N., & Son, J. B. (2009). Implementing computer-assisted language learning in EFL classroom: Teachers' perceptions and perspectives. *International Journal of Pedagogies and Learning*, *5*(2), 80–101. doi:10.5172/ijpl.5.2.80

Paternotte, E., Scheele, F., Seeleman, C. M., Bank, L., Scherpbier, A. J. J. A., & van Dulmen, S. (2016). Intercultural doctor-patient communication in daily outpatient care: Relevant communication skills. *Perspectives on Medical Education, 5*(5), 268–275. doi:10.1007/S40037-016-0288-Y PMID:27638395

Paterson, P., Meurice, F., Stanberry, L. R., Glismann, S., Rosenthal, S. L., & Larson, H. J. (2016). Vaccine hesitancy and healthcare providers. *Vaccine, 34*(52), 6700–6706. doi:10.1016/j.vaccine.2016.10.042 PMID:27810314

Pegalajar Palomino, M. C. (2021). Implicaciones de la gamificación en Educación Superior: Una revisión sistemática sobre la percepción del estudiante. *Revista de Investigación Educacional, 39*(1), 169–188. doi:10.6018/rie.419481

Pegrum, M. (2014). *Mobile learning: languages, literacies and cultures*. Palgrave Macmillan. doi:10.1057/9781137309815

Pegrum, M., Oakley, G., & Faulkner, R. (2013). Schools going mobile: A study of the adoption of mobile handheld technologies in western Australian independent schools. *Australasian Journal of Educational Technology, 29*(1). Advance online publication. doi:10.14742/ajet.64

Peña-Acuña, B. (2022). Indagación evaluativa de una intervención con metodologías activas para estudiantes universitarios. *Revista de estilos de aprendizaje, 15*(29), 5-18. . doi:10.55777/rea.v15i29.4054

Pérez Vázquez, E., Gilabert Cerdá, A., & Lledó Carreres, A. (2019). Gamificación en la educación universitaria: El uso del escape room como estrategia de aprendizaje. In R. Roig-Vila (Ed.), *Investigación e innovación en la Enseñanza Superior. Nuevos contextos, nuevas ideas*. Octaedro.

Perraton, H. (2008). *Open and distance learning in the developing world.* Routledge.

Peterson, D., Pere, L., & Sheehan, N. (2004). *Respect cost nothing: a survey of discrimination faced by people with experience of mental illness in Aotearona.* Mental Health Foundation of New Zealand.

Pfattheicher, S., Petersen, M. B., & Böhm, R. (2022). Information about herd immunity through vaccination and empathy promote COVID-19 vaccination intentions. *Health Psychology, 41*(2), 85–93. doi:10.1037/hea0001096 PMID:34570535

Pham, H., & Ho, T. (2020). Toward a 'new normal' with e-learning in Vietnamese higher education during the post COVID-19 pandemic. *Higher Education Research & Development, 39*(7), 1327–1331. doi:10.1080/07294360.2020.1823945

Pinaka, O., Spanou, I., Papadouli, V., Papanikolaou, E., Gioulekas, F., & Mouchtouri, V. A. (2021). The role of local primary healthcare units in increasing immunization uptake among children in vulnerable social groups through vaccination campaigns. *Public Health in Practice (Oxford, England), 2*, 100185. doi:10.1016/j.puhip.2021.100185 PMID:36101591

Pinto-Foltz, M. D., Hickman, R., Logsdon, M. C., & Burant, C. H. (2012). Psychometric evaluation of the revised attribution questionnaire (r-AQ) to measure metal illness stigma in adolescents. *Journal of Nursing Measurement, 20*(1), 47–58. doi:10.1891/1061-3749.20.1.47 PMID:22679709

Plass, J. L., Homer, B. D., & Kinzer, C. K. (2015). Foundations of Game-Based Learning. *Educational Psychologist, 50*(4), 258–283. Advance online publication. doi:10.1080/00461520.2015.1122533

Poveda-Pineda, D. F., Cifuentes-Medina, J. E., Poveda-Pineda, D. F., & Cifuentes-Medina, J. E. (2020). Incorporation of information and communication technologies (ICT) during the learning process in higher education. *Formación Universitaria, 13*(6), 95–104. doi:10.4067/S0718-50062020000600095

Povey, E. (2019). Storytelling in young learner EFL education: From theory to practice. *International. Journal of Issues in Language and Teacher Education, 4*, 1–14.

Powers, K. L., Brooks, P. J., Aldrich, N. J., Palladino, M. A., & Alfieri, L. (2013). Effects of video-game play on information processing: A meta-analytic investigation. *Psychonomic Bulletin and Review, 20*(6). doi:10.3758/s13423-013-0418-z

Prensky, M. (2001). Digital natives, digital immigrants. *On the Horizon*, *9*(5), 1–6. doi:10.1108/10748120110424816

Prieto, L. (2006). Aprendizaje activo en el aula universitaria: El caso del aprendizaje basado en problemas. *Miscelánea Comillas. Revista de Ciencias Humanas y Sociales*, *64*(124), 173–196.

Puebla, C., Fievet, T., Tsopanidi, M., & Clahsen, H. (2022). Mobile-assisted language learning in older adults: Chances and challenges. *ReCALL*, *34*(2), 169–184. doi:10.1017/S0958344021000276

Purgina, M., Mozgovoy, M., & Blake, J. (2020). WordBricks: Mobile technology and visual grammar formalism for gamification of natural language grammar acquisition. *Journal of Educational Computing Research*, *58*(1), 126–159. doi:10.1177/0735633119833010

Puri, N., Coomes, E. A., Haghbayan, H., & Gunaratne, K. (2020). Social media and vaccine hesitancy: New updates for the era of COVID-19 and globalized infectious diseases. *Human Vaccines & Immunotherapeutics*, *16*(11), 2586–2593. Advance online publication. doi:10.1080/21645515.2020.1780846 PMID:32693678

Quinn, D. M., Williams, M. K., & Weisz, B. M. (2015). From discrimination to internalized mental illness stigma: The mediating roles of anticipated discrimination and anticipated stigma. *Psychiatric Rehabilitation Journal*, *38*(2), 103–108. doi:10.1037/prj0000136 PMID:25844910

Quinteros, C., & Cárdenas, N. (2021). Flipped classroom and role play: Implementation in the agricultural technical baccalaureate. *Interdisciplinary Koinonia*, *6*(3), 106–127.

Rad, F. A., Otaki, F., Baqain, Z., Zary, N., & Al-Halabi, M. (2021). Rapid transition to distance learning due to COVID-19: Perceptions of postgraduate dental learners and instructors. *PLoS One*, *16*(2), 1–22. doi:10.1371/journal.pone.0246584 PMID:33556131

Radu, I. (2014). Augmented reality in education: A meta-review and cross-media analysis. *Personal and Ubiquitous Computing*, *18*(6), 1533–1543. doi:10.100700779-013-0747-y

Ramírez, J. L. (2014). *Gamificación. Mecánicas de juegos en tu vida personal y profesional.* SC Libro.

Ramos Navas-Parejo, M., Rodríguez García, A. M., & Gómez García, G. (2019). La gamificación en el aula universitaria. Una metodología activa e inclusiva. Percepciones por parte del alumnado universitario. In R. Roig Vila (Ed.), *Investigación e Innovación en la Enseñanza Superior* (pp. 669–678). Octaedro.

Ranilla, M. (2018). Gamificación de la alfabetización digital en mayores según los estilos de aprendizaje y actividades polifásicas. *Revista de estilos de aprendizaje, 11*(22). . doi:10.55777/rea.v11i22.1085

REACU. (2020a). *Acuerdo de REACU de 3 de abril de 2020, ante la situación de excepción provocada por el COVID19.* https://bit.ly/2wuHhqK

Read, T., Kukulska-Hulme, A., & Barcena, E. (2021). Supporting listening comprehension by social network-based interaction in mobile assisted language learning. *Porta Linguarum Revista Interuniversitaria De Didáctica De Las Lenguas Extranjeras*, *35*(35), 295–309. doi:10.30827/portalin.v0i35.15341

Ree, M. E. B. van der, & Lozano, J. A. M. (2019). Competencias claves de los estudiantes universitarios para el uso de las TIC. *Revista de Comunicación de la SEECI*, 43-72. doi:10.15198/seeci.2019.50.43-72

René, A. (2022). *Profesiones para nuevas generaciones postpandemia.* https://reneaguirrephd.com/2022/06/18/profesiones-para-nuevas-generaciones-postpandemia/

Reynolds, E. D., & Taylor, B. (2020). Kahoot!: EFL instructors' implementation experiences and impacts on students' vocabulary knowledge. *Comput. Assist. Lang. Learn. Electron. J.*, *2020*(21), 70–92.

Reynosa Navarro, E., Serrano Polo, E. A., Ortega-Parra, A. J., Navarro Silva, O., Cruz-Montero, J.M., & Salazar Montoya, E. (2019). Teaching strategies for scientific research: relevance in the training of researchers. *University and Society, 12*(1), 259-266. http://scielo.sld.cu/scielo.php?script=sci_arttext&pid=S2218-36202020000100259

Richards, J. C., & Farrell, T. S. C. (2005). *Professional development for language teachers: Strategies for teacher learning*. Cambridge University Press. doi:10.1017/CBO9780511667237

Ridda, I., Motbey, C., Lam, L., Lindley, I. R., McIntyre, P. B., & Macintyre, C. R. (2008). Factors associated with pneumococcal immunisation among hospitalised elderly persons: A survey of patient's perception, attitude, and knowledge. *Vaccine*, *26*(2), 234–240. doi:10.1016/j.vaccine.2007.10.067 PMID:18054818

Riggs, N. R., Jahromi, L. B., Razza, R. P., Dillworth-Bart, J. E., & Mueller, U. (2006). Executive function and the promotion of social-emotional competence. *Journal of Applied Developmental Psychology*, *27*(4), 300–309. Advance online publication. doi:10.1016/j.appdev.2006.04.002

Rixon, S. (1981). *How to use games in language teaching?* The Macmillan Press Ltd.

Roblyer, M. D. (2016). *Integrating Educational Technology into Teaching*. Pearson.

Robson, K., Plangger, K., Kietzmann, J. H., McCarthy, I., & Pitt, L. (2015). Is it all a game? Understanding the principles of gamification. *Business Horizons*, *58*(4), 411–420. doi:10.1016/j.bushor.2015.03.006

Rocamora-Pérez, P., López-Liria, R., Padilla-Góngora, D., Morales-Montoya, A., Díaz-López, P., & Aguilar-Parra, J. M. (2018). Training needs for immigrant caregivers of older people in the home in Europe: IENE7 Project. *Revista INFAD de Psicología. International Journal of Developmental and Educational Psychology.*, *4*(1), 1. Advance online publication. doi:10.17060/ijodaep.2018.n1.v4.1307

Rodicio-García, M. L., Ríos-de-Deus, M. P., Mosquera-González, M. J. y Penado Abilleira, M. (2020). La brecha digital en estudiantes españoles ante la Crisis de la Covid-19. *Revista internacional de educación para la justicia social, 9*(3), 103-125. . doi:10.15366/riejs2020.9.3.006

Rodrigo, M. (n.d.). *La comunicación intercultural.* Portal de comunicación. https://www.educacion.navarra.es/documents/57308/57732/2011_comintercultural.pdf/8bcd2117-0621-4ad8-973b-8d13f83ce0f0

Rodríguez Ferrer, J. M., Manzano León, A., Fernández Jiménez, C., Luque de la Rosa, A., Fernández Campoy, J.M., & Aguilar Parra, J. M. (2023) Shall we play together? Game-based learning for engagement and classroom climate in Spanish socially deprived communities. *Frontiers in Psychology*. doi:10.3389/fpsyg.2023.1163441

Rodríguez, A. M., Cáceres, M. P., & Alonso, S. (2018). La competencia digital del futuro docente: Análisis bibliométrico de la productividad científica indexada en Scopus. *International Journal of Educational Research and Innovation. IJERI, 10*, 317–333. doi:10.37954.e.v6i1.164

Roffey, S. (2009). Promoting social and emotional learning with games: "It's fun and we learn things.". *Simulation & Gaming*, *40*(5), 626–644. Advance online publication. doi:10.1177/1046878109333793

Roig-Vila, R., & Álvarez Herrera, J. (2019). Repercussion on Twitter of the active methodologies ABP, Flipped Classroom and Gamification. *RIED. Iberoamerican Journal of Distance Education, 22*(2). https://www.redalyc.org/articulo.oa?id=331460297005

Ross, D., & Magris, M. (2012). The role of communication and knowledge management as evidenced by HCP vaccination programs in the Netherlands, Germany and Italy: possible suggestions for medical translators. *Linguistica Antverpiensia, New Series – Themes in Translation Studies, 11*. doi:10.52034/lanstts.v11i.300

Roulston, K. (2010). *Reflective interviewing: A guide to theory and practice*. Sage Publications. doi:10.4135/9781446288009

Royal Decree 822/2021, of September 28, establishing the organization of university education and the procedure for quality assurance.

Rubin, B., Fernandes, R., Avgerinou, M. D., & Moore, J. (2010). The effect of learning management systems on student and faculty outcomes. *The Internet and Higher Education*, *13*(1-2), 82–83. doi:10.1016/j.iheduc.2009.10.008

Rugh, M. S., Beyette, D. J., Capraro, M. M., & Capraro, R. M. (2021). Using DIME maps and STEM project-based learning to teach physics. *Interactive Technology and Smart Education*, *18*(4), 553–573. Advance online publication. doi:10.1108/ITSE-07-2020-0109

Rüsch, N., Angermeyer, M., & Corrigan, P. (2005). Mental illness stigma: Concepts, consequences and initiatives to reduce stigma. *European Psychiatry*, *20*(8), 529–539. doi:10.1016/j.eurpsy.2005.04.004 PMID:16171984

Ryan, R. M., & Deci, E. L. (2000). Self-determination theory and the facilitation of intrinsic motivation, social development, and well-being. *The American Psychologist*, *55*(1), 68–78. doi:10.1037/0003-066X.55.1.68 PMID:11392867

Rysbayeva, G., Berdaliyeva, A., Kuralbayeva, A., Baiseitova, N., Uspabayeva, A., Zhapparbergenova, E., & Poshayeva, G. (2022). Students' attitudes towards mobile learning. *International Journal of Engineering Pedagogy*, *12*(2), 129–140. doi:10.3991/ijep.v12i2.29325

Saad, A. M., Mohamad, M. B., & Tsong, C. K. (2021). Behavioural intention of lecturers towards mobile learning and the moderating effect of digital literacy in Saudi Arabian universities. *International Transaction Journal of Engineering Management & Applied Sciences & Technologies*, *12*(1), 1–9.

Şahin Kızıl, A. (2017). EFL learners in the digital age: An investigation into personal and educational digital engagement. *RELC Journal*, *48*(3), 373–388. doi:10.1177/0033688216684285

Sailer, M., Hense, J., Mandl, H., & Klevers, M. (2013). Psychological perspectives on motivation through gamification. *Ixd&a, 19*(1), 28-37.

Sakarneh, M. A., & Al-Swelmyeen, M. B. (2020). The extent to which the Jordanian inclusive basic school teachers use the constructivism theory in teaching. *Journal of Educational and Social Research*, *10*(1), 182. Advance online publication. doi:10.36941/jesr-2020-0017

Salavaría, R., Buslón, N., López-Pan, F., León, B., López-Goñi, I., & Erviti, M. (2020). Disinformation in times of pandemic: Typology of Covid-19 hoaxes. *El Profesional de la Información*, *29*(3), 1–15. doi:10.3145/epi.2020.may.15

Saldaña, J. (2015). *The coding manual for qualitative researchers* (3rd ed.). Sage Publications.

Salomoni, M. G., Di Valerio, Z., Gabrielli, E., Montalti, M., Tedesco, D., Guaraldi, F., & Gori, D. (2021). Hesitant or not hesitant? A systematic review on global COVID-19 vaccine acceptance in different populations. *Vaccines*, *9*(8), 873. doi:10.3390/vaccines9080873 PMID:34452000

Samad, L., Butler, N., Peckham, C., & Bedford, H. (2006). Incomplete immunisation uptake in infancy: Maternal reasons. *Vaccine*, *24*(47), 6823–6829. doi:10.1016/j.vaccine.2006.06.039 PMID:17050048

Sanatullova-Allison, E. (2018). Contextualized culture integration through project-based learning in EFL classrooms in Vietnam. *Journal of NELTA*, *20*(1-2), 5–15. doi:10.3126/nelta.v20i1-2.19771

Sánchez i Peris, F. J. (2015). Gamification. *Education in the Knowledge Society*, *16*(2), 13–15.

Sánchez-Cruzado, C., Santiago, R., & Sánchez-Compaña, M. T. (2021). Teacher digital literacy: The indisputable challenge after covid-19. *Sustainability (Basel)*, *13*(4), 1858. Advance online publication. doi:10.3390u13041858

Sánchez-Prieto, J. C., Hernández-García, A., García-Peñalvoa, F. J., Chaparro-Peláez, J., & Olmos-Miguelañez, S. (2019). Break the walls! Second-order barriers and the acceptance of mLearning by first-year pre-service teachers. *Computers in Human Behavior, 95*, 158–167. doi:10.1016/j.chb.2019.01.019

Sánchez-Rodríguez, J., Ruiz-Palmero, J., & Sánchez-Vega, E. (2017). Flipped Classroom. Claves para su puesta en práctica. *EDMETIC, Revista de educación mediática y TIC, 6*(2), 336-358. . doi:10.21071/edmetic.v6i2.5832

Sandoval, V., Marín, B., & Barrios, T. (2021). The flipped classroom as a didactic strategy for the generation of competences: A systematic review. *RIED Ibero-American Journal of Distance Education, 24*(2), 285–308. doi:10.5944/ried.24.2.29027

Santamaría, A., & Alcalde, E. (2020). Una experiencia universitaria de gamificación en línea o en el aula presencial: ¿es este recurso de aprendizaje posible en ambos entornos? *Revista Brasileira de Lingüística Aplicada, 20*(4), 761–786. doi:10.1590/1984-6398202016390

Santos, M. J., Leiva, J. J., Ramos, M., & Benítez-Márquez, M. D. (2020). Higher education students' assessments towards gamification and sustainability: A case study. *Sustainability (Basel), 12*(20), 8513. doi:10.3390u12208513

Sanz, R., & Lopez, E. (2022). Educational learning after the COVID-19 pandemic. What role should the school play in the new world scenario? *Complutense Journal of Education, 33*(2), 215–223. doi:10.5209/rced.73928

Sarkar, A., & Biswas, P. (2021). Blended Learning as an Approach in Post Pandemic Era in India. In D. K. Kundu (Ed.), *E-Learning-Teaching Strategies and Teachers' Stress in Post Covid-19* (pp. 161–168). Kanishka Publishers.

Sarrab, M., Shibli, I. A., & Badursha, N. (2016). An empirical study of factors driving the adoption of mobile learning in Omani higher education. *International Review of Research in Open and Distance Learning, 17*(4), 331–349. doi:10.19173/irrodl.v17i4.2614

Savignon, S. J., & Sysoyev, P. V. (2005). Cultures and comparisons: Strategies for learners. *Foreign Language Annals, 38*(3), 357–365. doi:10.1111/j.1944-9720.2005.tb02222.x

Saz, I., & Ramo, R.M. (2015). Aproximación a los impactos y beneficios del aprendizaje servicio en la Universidad de Zaragoza. *RIDAS, Revista iberoamericana de aprendizaje y servicio, 1*, 9-27. . doi:10.1344/RIDAS2015.1.2

Schachter, H. M., Girardi, A., Ly, M., Lacroix, D., Lumb, A. B., van Berkom, J., & Gill, R. (2008). Effects of school based interventions on mental health stigmatization: A systematic review. *Child and Adolescent Psychiatry and Mental Health, 2*(1), 18–32. doi:10.1186/1753-2000-2-18 PMID:18644150

Scheerder, A., van Deursen, A., & van Dijk, J. (2017). Determinants of Internet skills, uses and outcomes: A systematic review of the second-and third-level digital divide. *Telematics and Informatics, 34*(8), 1607–1624. doi:10.1016/j.tele.2017.07.007

Schleicher, A. (2020). *The Impact of Covid-19 on Education Insights From Education at a Glance 2020*. OECD Publishing. https://www.oecd.org/education/the-impact-of-covid-19-on-education-insights-education-at-a-glance-2020.pdf

Schreier, M. (2012). *Qualitative content analysis in practice*. Sage Publications.

Schülze, B., Richter-Werling, M., Matschinger, H., & Angermeyer, M. (2003). Crazy? So what! Effects of a school project on students' attitudes towards people with schizophrenia. *Acta Psychiatrica Scandinavica, 107*(2), 142–150. doi:10.1034/j.1600-0447.2003.02444.x PMID:12534440

Schulz, R. A. (2007). The challenge of assessing cultural understanding in the context of foreign language instruction. *Foreign Language Annals, 40*(1), 9–26. doi:10.1111/j.1944-9720.2007.tb02851.x

Scolari, C. A. (2015). *Media ecology: environments, evolutions and interpretations*. Gedisa.

Seaman, J. E., Allen, I. E., & Seaman, J. (2018). *Grade Increase: Tracking Distance Education in the United States*. Babson Survey Research Group. https://files.eric.ed.gov/fulltext/ED580852.pdf

Selim, H. M. (2007). Critical success factors for e-learning acceptance: Confirmatory factor models. *Computers & Education*, *49*(2), 396–413. doi:10.1016/j.compedu.2005.09.004

Semante, D., & Gomez, V. (2021). Active didactic strategies to improve academic performance in the subject of Social Studies. *Episteme Koinonia*, *4*(8). Advance online publication. doi:10.35381/ekv4i8.1386

Sentell, T., Vamos, S., & Okan, O. (2020). Interdisciplinary Perspectives on Health Literacy Research Around the World: More Important Than Ever in a Time of COVID-19. *International Journal of Environmental Research and Public Health*, *17*(9), 3010. doi:10.3390/ijerph17093010 PMID:32357457

Seppala, P., & Alamaki, H. (2003). Mobile Learning in teacher training. *Journal of Computer Assisted Learning*, *19*(3), 330–335. doi:10.1046/j.0266-4909.2003.00034.x

Serçemeli, M., & Kurnaz, E. (2020). Covid-19 pandemi döneminde öğrencilerin uzaktan eğitim ve uzaktan muhasebe eğitimine yönelik bakış açıları üzerine bir araştırma. *International Journal of Social Sciences Academic Researches*, *4*(1), 40–53.

Sert, O., & Li, L. (2019). A qualitative study on CALL knowledge and materials design: Insights from pre-service EFL teachers. In M. Khosrow (Ed.), *Computer-Assisted language learning: concepts, methodologies, tools, and applications* (pp. 2036–2051). IGI Global.

Seven, M. A. (2012). Uzaktan ve örgün eğitime devam eden öğrencilerin İngilizce dersindeki başarı düzeylerinin karşılaştırılması. *Ekev Akademi Dergisi*, *6*(50), 215–228.

Severengiz, M., Roeder, I., Schindler, K., & Seliger, G. (2018). Influence of gaming elements on summative assessment in engineering education for sustainable manufacturing. In Procedia Manufacturing (pp. 429–437). doi:10.1016/j.promfg.2018.02.141

Shadiev, R., Hwang, W., & Liu, T. (2018). Investigating the effectiveness of a learning activity supported by a mobile multimedia learning system to enhance autonomous EFL learning in authentic contexts. *Educational Technology Research and Development*, *66*(4), 893–912. doi:10.100711423-018-9590-1

Sharpe, R., & Benfield, G. (2016). The student experience of e-learning in higher education: A review of the literature. *Brookes EJournal of Learning and Teaching*, 1.

Sharples, M., Inmaculada, A. S., Marcelo, M., & Giasemi, V. (2009). Mobile learning: small devices, big issues. In N. Balacheff, S. Ludvigsen, T. de Jong, & S. Barnes (Eds.), *Technology enhanced learning: principles and products* (pp. 233–249). Springer. doi:10.1007/978-1-4020-9827-7_14

Sheehy, K., Rix, J., Collins, J., Hall, K., Nind, M., & Wearmouth, J. (2009). *A systematic review of whole class, subject based, pedagogies with reported outcomes for the academic and social inclusion of pupils with special educational needs in mainstream classrooms. EPPI-Centre, Social Science Research Unit, Institute of Education*. University of London.

Siciliani, L., Wild, C., McKee, M., Kringos, D., Barry, M. M., Barros, P. P., De Maeseneer, J., Murauskiene, L., & Ricciardi, W. (2020). Strengthening vaccination programmes and health systems in the European Union: A framework for action. *Health Policy (Amsterdam)*, *124*(5), 511–518. doi:10.1016/j.healthpol.2020.02.015 PMID:32276852

Siemens, G., Gašević, D., & Dawson, S. (2015). *Preparing for the digital university: a review of the history and current state of distance, blended and online learning*. Athabasca University Press.

Silke, C., Swords, L., & Heary, C. (2017). The predictive effect of empathy and social norms on adolescents' implicit and explicit stigma responses. *Psychiatry Research*, *257*, 118–125. doi:10.1016/j.psychres.2017.07.033 PMID:28753461

Slater, P., Hasson, F., Guillen, P., Gallen, A., & Parlour, R. (2019). Virtual simulation training: Imaged experience of dementia. *International Journal of Older People Nursing*, *14*(3), 1–11. doi:10.1111/opn.12243 PMID:31070851

Slavin, R. E. (2014). Cooperative learning and academic achievement: Why does groupwork work? *Anales de Psicología*, *30*(3). Advance online publication. doi:10.6018/analesps.30.3.201201

Smith, P. J., Chu, S. Y., & Barker, L. E. (2004). Children who have received no vaccines: Who are they and where do they live? *Pediatrics*, *114*(1), 187–195. doi:10.1542/peds.114.1.187 PMID:15231927

Söderqvist, S., Nutley, S. B., Ottersen, J., Grill, K. M., & Klingberg, T. (2012). Computerized training of non-verbal reasoning and working memory in children with intellectual disability. *Frontiers in Human Neuroscience*, *6*. Advance online publication. doi:10.3389/fnhum.2012.00271 PMID:23060775

Son, J.-B. (2004). Teacher development in e-learning environments. In J.-B. Son (Ed.), Computer-assisted language learning: concepts, contexts and practices (pp. 107-122). iUniverse.

Son, J. B. (2018). *Teacher development in technology-enhanced language teaching*. Springer International Publishing. doi:10.1007/978-3-319-75711-7

Son, J.-B. (2006). Using online discussion groups in a CALL teacher training course. *RELC Journal*, *37*(1), 123–135. doi:10.1177/0033688206063478

Son, J.-B. (2014). Learning about computer-assisted language learning: Online tools and professional development. In J.-B. Son (Ed.), APACALL Book Series: Vol. 3. *Computer-assisted language learning: Learners, teachers and tools* (pp. 173–186). Cambridge Scholars Publishing.

Sophonhiranrak, S. (2021). Features, barriers, and influencing factors of mobile learning in higher education: A systematic review. *Heliyon*, *7*(4), 1–10. doi:10.1016/j.heliyon.2021.e06696 PMID:33869873

Spitzer, A., & Cameron, C. (1995). School-age children's perceptions of mental illness. *Western Journal of Nursing Research*, *17*(4), 398–415. doi:10.1177/019394599501700405 PMID:7660662

St Clair-Thompson, H. L., & Gathercole, S. E. (2006). Executive functions and achievements in school: Shifting, updating, inhibition, and working memory. *Quarterly Journal of Experimental Psychology*, *59*(4), 745–759. Advance online publication. doi:10.1080/17470210500162854 PMID:16707360

State Service of Education Quality of Ukraine. (2020). *Report on the Results of the Survey on the State of Use of Distance Learning Technologies in Higher Education Institutions of Ukraine*. State Service of Education Quality of Ukraine. https://sqe.gov.ua/images/materials/опитування/зво/Опитування_дистанційне навчання_ЗВО.pdf

Statista. (n.d.). *Salud e Industria farmacéutica*. https://es.statista.com/sectores/1174/salud-e-industria-farmaceutica/

Strakos, J. K., Douglas, M. A., McCormick, B., & Wright, M. (2023). A learning management system-based approach to assess learning outcomes in operations management courses. *International Journal of Management Education*, *21*(2), 100802. doi:10.1016/j.ijme.2023.100802

Suárez-Guerrero C., Lloret-Catalá C. y Lizandra J. (2022). Imagen educativa de la tecnología en la COVID-19. La mirada del alumnado de Magisterio con mención TIC. *Teknokultura. Revista de cultura digital y movimientos sociales*, *19*(2), 177-188. doi:10.5209/tekn.77951

Subhash, S., & Cudney, E. (2018). Gamified learning in higher education: A systematic review of the literature. *Computers in Human Behavior, 87*, 192–206. Advance online publication. doi:10.1016/j.chb.2018.05.028

Su, C. Y., & Chao, C. M. (2022). Investigating factors influencing nurses' behavioral intention to use mobile learning: Using a modified unified theory of acceptance and use of technology model. *Frontiers in Psychology, 13*, 1–10. doi:10.3389/fpsyg.2022.673350 PMID:35651564

Sung, Y., Chang, K., & Yang, J. (2015). How effective are mobile devices for language learning? A meta-analysis. *Educational Research Review, 16*, 68–84. doi:10.1016/j.edurev.2015.09.001

Sun, J. C.-Y., & Hsieh, P.-H. (2018). Application of a gamified interactive response system to enhance the intrinsic and extrinsic motivation, student engagement, and attention of English learners. *Journal of Educational Technology & Society, 21*(3), 104–116.

Suzie, B., & Krauss, J. (2014). *Reinventing project-based learning: Your field guide to real-world projects in the digital age*. Corwin, Sage Publishing.

Swords, L., Heary, C., & Hennessy, E. (2011). Factors associated with acceptance of peers with mental health problems in childhood and adolescence. *Journal of Child Psychology and Psychiatry, and Allied Disciplines, 52*(9), 933–941. doi:10.1111/j.1469-7610.2010.02351.x PMID:21223258

Tadesse, S., & Muluye, W. (2020). The impact of COVID-19 pandemic on education system in developing countries: A review. *Open Journal of Social Sciences, 8*(10), 159–170. doi:10.4236/jss.2020.810011

Taj, I., Sulan, N., Sipra, M., & Ahmad, W. (2016). Impact of mobile assisted language learning (MALL) on EFL: A meta-analysis. *Advances in Language and Literary Studies, 7*(2), 76–83.

Taylor Tavares, J., Clark, L., Cannon, D. M., Erickson, K., Drevets, W. C., & Sahakian, B. J. (2007). Distinct Profiles of Neurocognitive Function in Unmedicated Unipolar Depression and Bipolar II Depression. *Biological Psychiatry, 62*(8), 917–924. Advance online publication. doi:10.1016/j.biopsych.2007.05.034 PMID:17825802

Teixes, F. (2014). *Gamificación: fundamentos y aplicaciones*. Editorial UOC.

Tejedor, S., Cervi, L., Pérez-Escoda, A., & Jumbo, F. T. (2020). Digital literacy and higher education during COVID-19 lockdown: Spain, Italy, and Ecuador. *Publications / MDPI, 8*(4), 48. Advance online publication. doi:10.3390/publications8040048

Tello, J. & Aguaded, J. I. (2009). Desarrollo profesional docente ante los nuevos retos de las tecnologías de la información y la comunicación en los centros educativos. *Pixel Bit. Revista de medios y educación, 34*, 31-47.

Tenorio-Rodríguez, M. A., González-Monteagudo, J., & Padilla-Carmona, T. (2021). Employability and inclusion of non-traditional university students: Limitations and challenges. *IAFOR Journal of Education: Inclusive Education, 9*(1), 133–151. doi:10.22492/ije.9.1.08

Terras, M. M., & Ramsay, J. (2012). The five central psychological challenges facing effective mobile learning. *British Journal of Educational Technology, 43*(5), 820–832. doi:10.1111/j.1467-8535.2012.01362.x

Terzi, L. (2005). Beyond the dilemma of difference: The capability approach to disability and Special Edu-cational Needs. *Journal of Philosophy of Education, 39*(3), 443–459. Advance online publication. doi:10.1111/j.1467-9752.2005.00447.x

Tham, C. M., & Werner, J. M. (2005). Designing and evaluating e-learning in higher education: A review and recommendations. *Journal of Leadership & Organizational Studies, 11*(2), 15–25. doi:10.1177/107179190501100203

Thomas, J. (2000). *A review of research on project-based learning*. Retrieved from http://www. bobpearlman.org/BestPractices/PBL_Research.pdf

Todri, A., Papajorgji, P., Moskowitz, H., & Scalera, F. (2021). Perceptions regarding distance learning in higher education, smoothing the transition. *Contemp. Educ. Technol., 13*. doi:10.30935/cedtech/9274

Toledo, P., & Sánchez, J.M. (2018). Aprendizaje basado en proyectos: una experiencia universitaria. *Revista de currículum y formación de profesorado, 22*(2), 471-491. . doi:10.30827/profesorado.v22i2.7733

Tolosa, R., Vera-Rodríguez, R., Fierrez, J., Morales, A., & Ortega-García, J. (2020). Deepfakes and beyond: A Survey of face manipulation and fake detection. *Information Fusion, 64*, 131–148. doi:10.1016/j.inffus.2020.06.014

Tommerdahl, J. M., Dragonflame, C. S., & Olsen, A. A. (2022). Online first). A systematic review examining the efficacy of commercially available foreign language learning mobile apps. *Computer Assisted Language Learning*, 1–30. doi:10.1080/09588221.2022.2035401

Tondeur, J., van Braak, J., Ertmer, P. A., & Ottenbreit-Leftwich, A. (2017). Understanding the relationship between teachers' pedagogical beliefs and technology use in education: A systematic review of qualitative evidence. *Educational Technology Research and Development, 65*(3), 555–575. doi:10.100711423-016-9481-2

Toquero, C. M. (2020). Challenges and Opportunities for Higher Education amid the COVID-19 Pandemic: The Philippine Context. *Pedagogical Research, 5*(4), 1-5. https://doi.org/https://doi.org/10.29333/pr/7947

Toring, H., Legaspi, G., Omolon, J., Amadeo, R., Amadeo, E., Opolentisima, Q., Barina, V., Cacho, T., Illustrimo, F., & Cortes, S. (2022). *Evaluation of students' satisfaction toward an adopted learning management system at Indiana Aerospace University: A structural equation modelling approach*. Asia Pacific Management Review., doi:10.1016/j.apmrv.2022.12.002

Torres- López. Díaz-López y Méndez, (2020). Educación a distancia: e-learning, m-learning y b-learning, conceptos, características y necesidades actuales ante el COVID.19. *XIV International Conference on Education and Innovation.*

Torsani, S. (2016). *CALL teacher education: language teachers and technology integration*. Springer. doi:10.1007/978-94-6300-477-0

Trigueros, G., & María, I. (2017). Propuestas transmedia y gamificación aplicadas al EEES: Nuevas metodologías activas para implementar las competencias en Ciencias Sociales y en Literatura. In Memorias del Programa de Redes-I3CE de calidad, innovación e investigación en docencia universitaria. Convocatoria 2016-17 = Memòries del Programa de Xarxes-I3CE de qualitat, innovació i investigació en docència universitària. Convocatòria 2016-2017. Alicante: Universidad de Alicante, Instituto de Ciencias de la Educación (ICE).

Trigueros, R., Aguilar-Parra, J. M., Cangas, A. J., Bermejo, R., Ferrandiz, C., & López-Liria, R. (2019). Influence of emotional intelligence, motivation and resilience on academic performance and the adoption of healthy lifestyle habits among adolescents. *International Journal of Environmental Research and Public Health, 16*(16), 2810. doi:10.3390/ijerph16162810 PMID:31394722

Trujillo, F. (2017). Project-based learning: lines of progress for an innovation. *Language and Literature Didactics Texts, 78,* 42-48. https://www.researchgate.net/publication/320961737_Aprendizaje_basado_en_proyectos_Lineas_de_avance_para_una_innovacion_centenaria_Textos_Didactica_de_la_Lengua_y_de_la_Literatura_num_78_pp_42-48_octubre_2017

Tschannen-Moran, M., Hoy, A. W., & Hoy, W. K. (1998). Teacher efficacy: Its meaning and measure. *Review of Educational Research, 68*(2), 202–248. doi:10.3102/00346543068002202

Tsuei, M., & Hsu, Y. Y. (2019). Parents' acceptance of participation in the integration of technology into children's instruction. *The Asia-Pacific Education Researcher*, *28*(5), 457–467. doi:10.100740299-019-00447-3

Tuncer, M., & Taşpınar, M. (2008). Sanal ortamda eğitim ve öğretimin geleceği ve olası sorunlar. *Sosyal Bilimler Dergisi*, *10*(1), 125–144.

Turgut, K., & Akdemir, E. (2019). The attitude of prospective teachers towards mobile learning. *International Journal of Society Researchers*, *13*(19), 500–526.

Turnbull, D., Chugh, R., & Luck, J. (2020). Learning management systems, an overview. Encyclopedia of Education and Information Technologies, 1052-1058.

Tversky, A., & Kahneman, D. (1981). The framing of decisions and the psychology of choice. *Science*, *211*(4481), 453–458. doi:10.1126cience.7455683 PMID:7455683

UNESCO & IESALC. (2020). *COVID-19 and higher education: Today and tomorrow. Impact analysis, policy responses and recommendations*. https://bit.ly/34TOSvu

UNESCO. (1994). The Salamanca statement and framework for action on special needs education. In *The Salamanca statement and framework for action on special needs education*. UNESCO.

UNESCO. (2005). *Towards knowledge societies*. UNESCO.

UNESCO. (2011). *UNESCO ICT competency framework for teachers*. UNESCO. https://bit.ly/3g2Fj2s

UNESCO. (2013). *Global Media and Information Literacy Assessment Framework: country readiness and competencies*. UNESCO.

UNESCO. (2020). *Distance learning strategies. What do we know about effectiveness?* https://unesdoc.unesco.org/ark:/48223/pf0000373350

UNESCO. (2022). *Global standards on guidelines for the development of media and information literacy curricula*. UNESCO. https://www.unesco.org/sites/default/files/medias/files/2022/02/Global%20Standards%20for%20Media%20and%20Information%20Literacy%20Curricula%20Development%20Guidelines_ES.pdf

Uni. (2020). *Flipped Classroom, las claves de una metodología rompedora*. https://www.unir.net/educacion/revista/flipped-classroom/

UNICEF. (2019). *United Nations children's fund; participants' manual: interpersonal communication for immunization package*. https://ipc.unicef.org/sites/ipcfi/files/2019-06/UNICEF_Participant_Manual_FINAL_2.pdf

UNICEF. (n.d.) *Interpersonal Communication for Immunization Initiative*. https://ipc.unicef.org/

United Nations Educational, Scientific and Cultural Organization. (2022). *What you need to know about digital learning and education transformation*. https://www.unesco.org/es/digital-education/need-kno

Valverde-Martínez, M. Á., López-Liria, R., Martínez-Cal, J., Benzo-Iglesias, M. J., Torres-Álamo, L., & Rocamora-Pérez, P. (2023). Telerehabilitation, a viable option in patients with persistent post-COVID syndrome: A systematic review. *Health Care*, *11*(2), 2. Advance online publication. doi:10.3390/healthcare11020187 PMID:36673555

van Deursen, A. J., Helsper, E. J., & Eynon, R. (2016). Development and validation of the Internet Skills Scale (ISS). *Information Communication and Society*, *19*(6), 804–823. doi:10.1080/1369118X.2015.1078834

van Deursen, A. J., & van Dijk, J. A. (2019). The first-level digital divide shifts from inequalities in physical access to inequalities in material access. *New Media & Society*, *21*(2), 354–375. doi:10.1177/1461444818797082 PMID:30886536 van Dijk, J. A. (2006). Digital divide research, achievements and shortcomings. *Poetics, 34*(4-5), 221–235. doi:10.1016/j.poetic.2006.05.004

Van Manen, M. (1990). *Researching lived experience: human science for an action sensitive pedagogy.* Althouse Press.

Van Volkom, M., Stapley, J. C., & Amaturo, V. (2014). Revisiting the digital divide: Generational differences in technology use in everyday life. *North American Journal of Psychology, 16*(3), 557–574.

Varea, R. (2021, May 17). Aulas virtuales y contenidos personalizados: Así será la Universidad en unos años. *El País.* Retrieved April 8, 2023, from https://elpais.com/sociedad/educacion-online/2021-09-21/aulas-virtuales-y-contenidos-personalizados-asi-sera-la-universidad-en-unos-anos.html

Vasilevska, D., Rivza, B., Pivac, T., Alekneviciene, V., & Parlińska, A. (2017). Analysis of the Demand for Distance Education at Eastern and Central European Higher Education Institutions. *Journal of Teacher Education for Sustainability, 19*(1), 106–116. doi:10.1515/jtes-2017-0007

Vasiliou, A., & Economides, A. A. (2007). Mobile collaborative learning using multicast MANETs. *International Journal of Mobile Communications, 5*(4), 423–444. doi:10.1504/IJMC.2007.012789

Vega Pardo, M. L., & Guerrero Castillo, G. S. (2023). La gamificación y su incidencia en la calidad educativa, en la escuela Manuel Reinaldo Benalcázar Castillo. *Revista social fronteriza, 3*(2), 38–62. . doi:10.5281/zenodo.7686135

Venegas-Ramos, L., Luzardo-Martínez, H. J., & Pereira-Santana, A. (2020). Conocimiento, formación y uso de herramientas TIC aplicadas a la Educación Superior por el profesorado de la Universidad Miguel de Cervantes. *Edutec. revista electrónica de tecnología educativa, 71*, 35-52. . doi:10.21556/edutec.2020.71.1405

Vergara, D., Mezquita, J.M. & Gómez Vallecillo, A.I. (2019). Metodología innovadora basada en la gamificación educativa: evaluación tipo test con la herramienta Quizizz. *Revista de currículum y formación de profesorado, 23*(3), 363-387. doi: .v23i3.11232 doi:10.30827/profesorado

Vergara, D., Mezquita, J. M., & Gómez Vallecillo, A. I. (2019). Metodología innovadora basada en la gamificación educativa: Evaluación tipo test con la herramienta Quizizz. *Profesorado. Revista de Currículum y Formación de Profesorado, 23*(3), 363–387. doi:10.30827/profesorado.v23i3.11232

Vesa, M., Hamari, J., Harviainen, J. T., & Warmelink, H. (2017). Computer games and organization studies. *Organization Studies, 38*(2), 273–284. doi:10.1177/0170840616663242

Vesga Aldana, I., Herrera Montero, M., & Horta Guette, M. (2021). *Estrategia pedagógica mediada por un recurso de gamificación para el fortalecimiento del componente gramatical del inglés en estudiantes de quinto grado.* Universidad de Cartagena.

Viberg, O., & Grönlund, Å. (2012, October). Mobile assisted language learning: A literature review. *11th world conference on mobile and contextual learning.*

Vilardaga, R. (2012). Estigma y prejuicios. In M. Páez-Blarrina & O. Gutiérrez-Martínez (Eds.), *Múltiples Aplicaciones de la Terapia de Aceptación y Compromiso* (pp. 251–267). Pirámide.

Villalustre, L., & Del Moral, M. E. (2017). Juegos perceptivos con realidad aumentada para trabajar contenido científico. *Educação. Formação & Tecnologias, 10*(1), 36–46.

Villaverde-Caramés, E. J., Fernández-Villarino, M. Á., Toja-Reboredo, B., & González Valeiro, M. Á. (2021). Revisión de la literatura sobre las características que definen a un buen docente de Educación Física: Consideraciones desde la formación del profesorado [A literature review of the characteristics that define a good physical education teacher considera-tions]. *Retos*, (41), 471–479. doi:10.47197/retos.v0i41.84421

Viñals, A., & Cuenca, J. (2016). El rol del docente en la era digital. *Revista Interuniversitaria de Formación del Profesorado, 30*(2), 103–114.

Voloshinov, S., Kruglyk, V., Osadchyi, V., Osadcha, K., & Symonenko, S. (2020). Realities and prospects of distance learning at higher education institutions of Ukraine. *Ukrainian Journal of Educational Studies and Information Technology, 8*(1), 1–16. doi:10.32919/uesit.2020.01.01

Voudoukis, N., & Pagiatakis, G. (2022). Massive Open Online Courses (MOOCs): Practices, trends, and challenges for the higher education. *European Journal of Education and Pedagogy, 3*(3), 288–295. doi:10.24018/ejedu.2022.3.3.365

Vraga, E. K., Tully, M., & Bode, L. (2020). Empowering Users to Respond to Misinformation about Covid-19. *Media and Communication, 8*(2), 475–479. doi:10.17645/mac.v8i2.3200

Wan, Z. H., So, W. M., & Zhan, Y. (2020). Developing and validating a scale of STEM project-based learning experience. *Research in Science Education, 17*. Advance online publication. doi:10.100711165-020-09965-3

Warin, B., Talbi, O., Kolski, C., & Hoogstoel, F. (2016). Multi-role project (MRP): A new Project-Based Learning method for STEM. *IEEE Transactions on Education, 59*(2), 137–146. doi:10.1109/TE.2015.2462809

Warschauer, M., & Meskill, C. (2000). Technology and second language learning. In J. Rosenthal (Ed.), *Handbook of undergraduate second language education* (pp. 303–318). Lawrence Erlbaum.

Washington State Department of Health. (n.d.). *Immunization Action Coalition Washington.* Immunity Community. https://immunitycommunitywa.org/iacw/

Watson, A. C., Ottati, V. C., Morabito, M. S., Draine, J., Kerr, A. N., & Angell, B. (2010). Outcomes of pólice contacts with persons with mental illness: The impact of CIT. *Administration and Policy in Mental Health, 37*(4), 302–317. doi:10.100710488-009-0236-9 PMID:19705277

Watson, R., & Watson, S. (2012). An argument for clarity: What are learning management systems, what are they not, & what should they become? *TechTrends, 51*(2), 28–34.

Watts, L. (2016). Synchronous and asynchronous communication in distance learning: A review of the literature. *Quarterly Review of Distance Education, 17*(1), 23–32. doi:10.1007/BF02763377

Weiss, A. P., Alwan, A., Garcia, E. P., & García, J. (2020). Surveying fake news: Assessing university faculty's fragmented definition of fake news and its impact on teaching critical thinking. *International Journal for Educational Integrity, 16*(1), 1–30. doi:10.100740979-019-0049-x

Wei, X., Lenz, K. B., & Blackorby, J. (2013). Math Growth Trajectories of Students with Disabilities: Disa-bility category, gender, racial, and Socioeconomic Status Differences From Ages 7 to 17. *Remedial and Special Education, 34*(3), 154–165. Advance online publication. doi:10.1177/0741932512448253

Werbach, K., & Hunter, D. (2012). *For the win: how game thinking can revolutionize your business.* Wharton Digital Press.

Werbach, K., & Hunter, D. (2015). *The gamification toolkit: Dynamics, mechanics, and components for the win.* Wharton Digital Press.

WHO. (2020b). *Coronavirus disease (Covid-19) outbreak.* https://www.who.int/es/emergencies/diseases/novel-coronavirus-2019/advice-for-public

WHO. (2022). *Implementation guide for vaccination of health workers.* https://apps.who.int/iris/rest/bitstreams/1453340/retrieve

Will Crescioni, A., Ehrlinger, J., Alquist, J. L., Conlon, K. E., Baumeister, R. F., Schatschneider, C., & Dutton, G. R. (2011). High trait self-control predicts positive health behaviors and success in weight loss. *Journal of Health Psychology, 16*(5), 750–759. Advance online publication. doi:10.1177/1359105310390247 PMID:21421645

Williamson, B., Eynon, R., & Potter, J. (2020). Pandemic politics, pedagogies and practices: Digital technologies and distance education during the coronavirus emergency. *Learning, Media and Technology, 45*(2), 107–114. doi:10.1080/17439884.2020.1761641

Willig, J. H., Croker, J., McCormick, L., Nabavi, M., Walker, J., Wingo, N. P., Roche, C. C., Jones, C., Hartmann, K. E., & Redden, D. (2021). Gamification and education: A pragmatic approach with two examples of implementation. *Journal of Clinical and Translational Science, 5*(1), e181. doi:10.1017/cts.2021.806 PMID:34849256

Winke, P., & Goertler, S. (2008). Did we forget someone? Students' computer access and literacy for CALL. *CALICO Journal, 25*(3), 482–509. doi:10.1558/cj.v25i3.482-509

Wlodkowski, R., & Ginsberg, M. (2017). *Enhancing adult motivation to learn: A comprehensive guide for teaching all adults* (4th ed.). Jossey-Bass.

Woolfolk Hoy, A. (2013). A reflection on the place of emotion in teaching and teacher education. In Emotion and School: Understanding how the Hidden Curriculum Influences Relationships, Leadership, Teaching, and Learning (pp. 255-270). Emerald Group Publishing Limited. doi:10.1108/S1479-3687(2013)0000018017

Word Economic Forum. (2020) *Report on the future of employment 2020.* https://www-weforum-org.translate.goog/reports/the-future-of-jobs-report-2020/?_x_tr_sl=en&_x_tr_tl=es&_x_tr_hl=es&_x_tr_pto=sc

World Health Organization WHO. (2020a). *Global Vaccine Action Plan 2011-2020.* https://www.who.int/immunization/global_vaccine_action_plan/en/

Writers, S. (2022, December 20). *Vaccine myths debunked.* https://www.publichealth.org/public-awareness/understanding-vaccines/vaccine-myths-debunked/

Wu, Q. (2014). Learning ESL vocabulary with smartphones. *Procedia: Social and Behavioral Sciences, 143*, 302–307. doi:10.1016/j.sbspro.2014.07.409

Wu, W. H., Wu, Y. C. J., Chen, C. Y., Kao, H. Y., Lin, C. H., & Huang, S. H. (2012). Review of trends from mobile learning studies: A meta-analysis. *Computers & Education, 59*(2), 817–827. doi:10.1016/j.compedu.2012.03.016

Yağan, S. A. (2023). Mobile learning in higher education: A mixed research synthesis. Erzincan University Journal of Education Faculty, 25(1), 108-125.

Yağcı, M. (2012). Çevrimiçi Sınav Ortamlarının Öğrencilerin Akademik Başarılarına Etkisi. *E-Journal of New World Sciences Academy, 7*(1), 331–339.

Yang, J., Cervera, R. L., Tye, S. J., Ekker, S. C., & Pierret, C. (2018). Adolescent mental health education InSciEd Out: A case study of an alternative middle school population. *Journal of Translational Medicine, 16*(1), 84. doi:10.118612967-018-1459-x PMID:29615090

Yang, X. Z., Lin, L., Cheng, P. Y., Yang, X., & Ren, Y. (2019). Which EEG feedback works better for creativity performance in immersive virtual reality: The reminder or encouraging feedback? *Computers in Human Behavior, 99*, 345–351. doi:10.1016/j.chb.2019.06.002

Yavuz-Konokman, G., Yanpar-Yelken, T., & Sancar-Tokmak, H. (2013). An investigation of primary school pre-service teachers' perception on their Tpack in terms of a variety factors: Mersin university case. *Kastamonu Eğitim Dergisi., 21*(2), 665–684.

Yeh, C. Y. (2018). *Investigating technology integration in United States TESOL teacher preparation programs* [Doctoral dissertation]. University of Delaware.

Yılmaz İnce, E., Kabul, A., & Diler, İ. (2020). Distance Education in Higher Education in the COVID-19 Pandemic Process: A Case of Isparta Applied Sciences University. *International Journal of Technology in Education and Science*, *4*(4), 343–351. doi:10.46328/ijtes.v4i4.112

Yılmaz, M. (2007). Instructional technology in training primary school teacher. *Gazi Eğitim Fakültesi Dergisi.*, *27*(1), 155–167.

Yılmaz, N., & Çakır, R. (2022). Preservice teachers' digital literacy and attitudes towards mobile learning. *Turkish Journal of Educational Studies*, *9*(2), 291–325.

Yokuş, G. (2016). *Eğitim fakültesi öğrencilerinin mobil öğrenmeye ilişkin görüşlerinin incelenmesi ve eğitim bilimleri alanına yönelik mobil uygulama geliştirme çalışması: mobil akademi* [Unpublished master's thesis]. Mersin University.

Yuan, R., & Lee, I. (2014). Pre-service teachers' changing beliefs in the teaching practicum: Three cases in an EFL context. *System*, *44*, 1–12. doi:10.1016/j.system.2014.02.002

Yu, Z., Xu, W., & Sukjairungwattana, P. (2023). Motivation, learning strategies, and outcomes in mobile English language learning. *The Asia-Pacific Education Researcher*, *32*(4), 545–560. doi:10.100740299-022-00675-0

Zamecnik, A., Kovanović, V., Joksimović, S., & Liu, L. (2022). Exploring non-traditional learner motivations and characteristics in online learning: A learner profile study. *Computers and Education: Artificial Intelligence*, *3*, 100051. doi:10.1016/j.caeai.2022.100051

Zaremohzzabieh, Z., Roslan, S., Mohamad, Z., Ismail, I. A., Ab Jalil, H., & Ahrari, S. (2022). Influencing factors in MOOCs adoption in higher education: A meta-analytic path analysis. *Sustainability (Basel)*, *14*(14), 14. Advance online publication. doi:10.3390u14148268

Zarzycka-Piskorz, E. (2016). Kahoot it or not? Can games be motivating in learning grammar? *Teaching English with Technology*, *16*(3), 17–36.

Zelazo, P. D. (2006). The Dimensional Change Card Sort (DCCS): A method of assessing executive function in children. *Nature Protocols*, *1*(1), 297–301. Advance online publication. doi:10.1038/nprot.2006.46 PMID:17406248

Zhaohui, W. (2020). *How a top Chinese university is responding to coronavirus.* World Economic Forum. Retrieved December 10, 2022, from https://www.weforum.org/agenda/2020/03/coronavirus-china-the-challenges-of-online-learning-for-universities/

Zonneveld, M., Patomella, A.-H., Asaba, E., & Guidetti, S. (2020). The use of information and communication technology in healthcare to improve participation in everyday life: A scoping review. *Disability and Rehabilitation*, *42*(23), 3416–3423. doi:10.1080/09638288.2019.1592246 PMID:30966833

Zou'bi, R. A. (2022). The impact of media and information literacy on students' acquisition of the skills needed to detect fake news. *The Journal of Media Literacy Education*, *28*(2), 58–71. doi:10.23860/JMLE-2022-14-2-5

About the Contributors

Madalina Armie earned a Master's Degree in English language and literature (2014) from the University of Almería. She completed her PhD on the contemporary Irish short story at the turn of the twenty-first century at the University of Almería (2019), for which she obtained the EIDUAL Dissertation Award 2019 for Best Doctoral Dissertation (2021) and the honorary second prize for the Best Doctoral Dissertation in Studies for Equality and the Fight against Gender Violence of the University of Almeria (2022). Her current areas of research include the contemporary Irish short story and Irish women's writing. She has published articles and reviews in international journals, such as *Irish Studies Review*, *Estudios Irlandeses*, *Review of Irish Studies in Europe (RISE)* and *Studi Irlandesi*. Armie is the author of the monograph *The Irish Short Story at the Turn of the Twenty-first Century: Tradition, Society and Modernity* and one of the editors of *Trauma, Memory and Silence of the Irish Woman in Contemporary Literature Wounds of the Body and the Soul* published both by Routledge. She is currently teaching at the University of Almeria, Spain.

Germán Asensio obtained both his Bachelor's degree and his Master's degree in English Studies at the University of Almería (Spain), both with honorary distinctions. He completed his Ph.D. dissertation at the University of Almeria (2020) on Irish writer Brian O'Nolan (1911-1966), also known as Flann O'Brien and Myles na gCopaleen, and mid-twentieth century Irish politics in his *Cruiskeen Lawn* column (1940-1966). He is currently a Research Fellow at the University of Almería (Department of Philology/ English Division) and has published articles and delivered papers on O'Nolan's novels, short stories and columns. His research focuses on the field of Irish literature with a particular interest.

María Enriqueta Cortés de los Ríos is a Senior Lecturer of English Language in the Departament of Philology at the University of Almeria. Dr. Cortés obtained a licenciatura and her Ph.D. in English Philology at the University of Granada. She teaches English Language for Specific Purposes and English Language in different degrees at the University of Almería. She has published articles in both national and international journals primarily in business English and the advertising language. She is a member of the research group HUM-859.

Ángeles Jordán Soriano obtained her bachelor's in English Studies from the University of Almería in 2019 and earned her master's in English Studies at the same university in 2020. In 2021, she obtained the Patricia Shaw Research Award granted by The Spanish Association for Anglo-American Studies (AEDEAN), thanks to which she conducted a research visit to the University of Kent. She has also been a visiting researcher at Cardiff Metropolitan University for 3 months. As a predoctoral fellow funded by the Spanish Ministry of Universities, she is currently teaching at the University of Almería in the Department of Philology and working on her Ph.D. dissertation, which examines the influence of politics in culture during the 1960s in Great Britain. Her research and publications primarily focus on artistic production during Harold Wilson's first government (1964–1970).

* * *

Patricia Bárcena Toyos has a Ph.D. in English and Applied Linguistics from the University of Memphis. She has extensive experience teaching foreign languages at secondary schools and universities in the USA and Spain. She teaches at graduate and postgraduate levels at Universidad Internacional de La Rioja. Her research focuses on the use of active methodologies in bilingual and English as a foreign language classrooms, Content and Language Integrated Learning (CLIL) and the use of Sheltered Instruction Observation Protocol (SIOP) in CLIL settings. She has published research studies on these areas, such as Barcena-Toyos, P. (2022). La gamificación como herramienta para dinamizar la evaluación continua en un máster universitario. *Revista De Estilos De Aprendizaje*, *15*(30), 109–119. , and Bárcena-Toyos, P. (2023) CLIL and SIOP: An effective partnership? *International Multilingual Research Journal*, *17*(1), 1-14. 10.1080/19313152.2022.2075191.

María Jesús Benzo-Iglesias graduated in Physiotherapy. Her final degree project received the first Alejandro Plaza Prize, awarded by the Professional College of Physiotherapists of Andalusia. It has also been published in a magazine (UAL, 2021). She has a master's in Neurological Physiotherapy from Universidad de Cádiz (UCA), which she received in 2022, and a master's in Physical Activity and Health from Universidad Internacional de Andalucía (UNIA), which she completed in 2023. She is now a doctoral student in Physiotherapy at the University of Almería (16/12/2022–present) in the research group HUMANITIES-98, where she is studying "Intervention Psychology in Development, Education, and Guidance."

Rubén Camacho Sánchez is a doctoral student at GIAM INEFC Lleida, specializing in higher education. He has a deep passion for the effects of playful learning and is dedicated to advancing knowledge and contributing to the scientific community. Rubén Camacho Sánchez completed his master's in education at Universidad Camilo José Cela (UCJC) in 2018. He has a strong background in educational technology, with a particular focus on game-based learning and gamification. He is currently a Ph.D. student at the University of Lleida, Spain. Throughout his academic career, Rubén Camacho Sánchez has actively participated in several research projects, collaborating with researchers and institutions to address critical challenges in motivational and learning effects on students. Alongside his research work, Rubén Camacho Sánchez is committed to knowledge exchange. He has participated in conferences and seminars, presenting his research results and participating in interdisciplinary discussions. Rubén Camacho Sánchez continues to actively contribute to the scientific community through his involvement in professional organizations and conferences.

Edgar Oliver Cardoso Espinosa is a professor and researcher who has been working in the Division of Postgraduate Studies and Research of the Superior School of Commerce and Administration (ESCA) at the Instituto Politécnico Nacional (IPN) in Mexico City since 2009. He conducts research projects focused on managing and developing education in educational programs at a senior level. Cardoso teaches courses like Methodological Seminar I, II, III, and Statistics for Educational Research to master's students. He also teaches Research Seminars I, II, and III and Research Visits I, II, III, and IV to Ph.D. students. Cardoso Espinosa earned his doctorate at IPN. He has an MSc in the Management and Development of Education from the ESCA-IPN. His research evaluates educational programs and executive education. He belongs to the National System of Researchers and has a level I distinction in the CONAHCyT of Mexico. Cardoso has directed several research projects in the area of educational management and conducted thesis addresses at master's and doctoral levels in the educational field.

Anabel Corral Granados is a postdoctoral research fellow involved in the "Childhood, School and Inequality in the Nordic Countries (Unequal Childhood)" project at the Department of Teacher Education. In her Ph.D. thesis at Anglia Ruskin University, she explored the conceptualizations of inclusion and understanding and the implementation of inclusive ways of working with children with disabilities. She has in-depth competence in organizational theories, policy analysis, and inclusion teaching. She also has experience from the European Agency for Special Needs and inclusive education in Brussels and Odense. Corral has previously worked as an associate professor at the State University of Alicante, where she coordinated three different units on educational psychology and taught about children's rights, educational psychology, and international educational policy.

Jésica Alhelí Cortés Ruiz has a Ph.D. in Administrative Sciences from the Higher School of Commerce and Administration and an MSc in Systems Engineering from the Higher School of Mechanical and Electrical Engineering. She is a specialist in transdisciplinary systems and has authored chapters of dissemination books and various articles in research journals.

Muhammet Demirbilek is a professor who studies educational technology. He is a diverse thinker with over 20 years of academic achievements and internationally funded projects and research. His experience includes developing various computer games and simulations, artificial intelligence, learning analytics, social networks, and virtual worlds. He earned his doctorate and master's in Educational Technology from the University of Florida. He was previously a postdoctoral researcher at the University of Wisconsin-Madison and the University of Florida. Demirbilek also holds BS and MS degrees in Electronics Engineering from Istanbul University.

María del Pilar Díaz-López is a graduate who focuses on child teaching. She received the Prize for the Best Academic Record from 2004 to 2007. She is a graduate of Educational Psychology and holds a master's in Special Education and a Ph.D. in Health, Psychology, and Psychiatry (Cum Laude). Díaz-López is a Professor of the Junta de Andalucía and a member of the research group HUM-498. She has authored several scientific articles Q1 and Q2 and book chapters Q1. Her main lines of research include physical activity and sport for quality of life; the development of cognitive processes; teaching-learning methods and strategies in curricular knowledge areas; models and programs of psychoeducational intervention that improve, optimize, and accelerate processes of development; schools' learning and orientation in their different educational levels; regulated and special needs; affective and emotional processes; aging processes; and orientation and employment programs for active life. Díaz-López has participated in and coordinated several national and international research and teaching innovation projects.

Patricia Fidalgo is an associate professor who joined the Emirates College for Advanced Education (ECAE) in 2014. She has taught in undergraduate and graduate programs. She holds a Ph.D. in Sciences of Education from Nova's University of Lisbon. Fidalgo is an expert in technology, networks, and multimedia in education and training, with over 24 years of teaching experience in Europe, Africa, and the Middle East. She was Chair of the Emerging Researchers' Group of the European Educational Research Association from 2011 to 2016. In this role, she managed and organized the group's activities, annual international conference, and summer schools. Her research interests include the use of technological devices by higher education students, the usefulness of learning information literacy skills to enhance 21st-century skills, and the integration of 3D printing in educational settings. Fidalgo is currently researching distance education's potential as a strategy for higher education institutions. She also has experience using blended and distance learning models. Fidalgo is the Head of the Curriculum and Instruction Division and the Chair of the Research Grants Committee at ECAE.

Maria João Figueiras is a clinical health psychologist (specialist). She has a Ph.D. in Health Psychology from Kings College at the University of London. She also has a master's in Health Psychology from United Medical and Dental Schools, University of London, and a bachelor's in Clinical Psychology from Instituição de referência nas áreas da Psicologia, Biociências e Educação (ISPA-IU). She specializes in Psychotherapy and is a registered EuroPsy psychologist. She is a full professor of clinical health psychology (with Habilitation) and a researcher. Figueiras currently teaches at Zayed University in the Department of Psychology, College of Health and Natural Sciences, where she is an active research community member. She develops social and human sciences research with an emphasis on clinical health psychology. Her work has significantly impacted the lives of individuals adjusting to chronic illnesses. She has received four research awards. She has collaborated with several higher education institutions. She was President of the Portuguese Association of Health and Behavioral Sciences (2015–2019) and has been a full member of the European Health Psychology Society since 1994. Between 2001 and 2014, she was awarded four research grants funded by the Fundação para a Ciência e a Tecnologia and coordinated these research projects alongside other entities. She has published in scientific journals and written several book chapters. Figuerias has also been a consultant for international research projects, participated in several scientific committees at national and international conferences, and was an invited speaker in several seminars, workshops, and conferences.

Anastasio García-Roca has a Ph.D. in Human and Social Sciences from the University of Almería. His doctoral thesis, entitled *Reading and Writing Practices in Spaces of Affinity: New Keys for Literary Education*, obtained Outstanding Cum Laude with international mention. He has carried out research stays at the University of Seville, the Georgia Tech Institute, and the University of Silesia. The lines of research he develops are related to language and literature didactics, digital reading, the reading habits and behaviors of young people and adolescents, and media and information literacy in media contexts of disinformation. His studies have been published in high-impact international journals and publications. In addition, he has presented at international conferences in different countries (Poland, Portugal, Italy, Chile, and the United States) and universities in Spain (Almería, Alicante, Badajoz, Madrid, Málaga, Cuenca, Seville, and Granada). Many of these experiences have involved publishing chapters with prestigious publishers. Finally, he has participated as a working group member in three national research, development, and innovation projects with references EDU201569924-R, PGC2018-101457-B-I00, and TED2021-129984A-I00. These were financed by the Ministry of Economy and Competitiveness and the Ministry of Science, Innovation and Universities, respectively. Currently, he is conducting a research stay at the University of Murcia financed through a Margarita Salas postdoctoral contract from the teacher requalification program for 2021–2023. This is convened by the University of Almería (ref. RR_A_2021_13) and financed by the Spanish Ministry of Universities and the European Union.

Tuğra Karademir Coşkun completed her master's in 2012 and her doctorate in 2018 in Computer and Instructional Technologies Education at Ankara University. From 2015 to 2017, she worked as a researcher at the University of Oklahoma with the support of The Scientific and Technological Research Council of Turkey. In 2022, she was awarded the title of associate professor in the field of instructional technologies. Coşkun continues researching instructional technologies, robotics education, distance education, game development, and artificial intelligence.

Oleksandr Kulyk is a professor in the Philosophy Department at Oles Honchar Dnipro National University. He holds a Ph.D. in philosophy and is the author of five books and 40 academic articles. His research focuses on teaching methods and philosophy. Since 2008, Kulyk has worked at Oles Honchar Dnipro National University, delivering lectures in undergraduate and graduate programs. As part of his commitment to improving his teaching, Kulyk has conducted several sociological studies to evaluate the effectiveness of various teaching methods on students' perceptions and published their results in peer-reviewed journals.

José Alberto Lencastre is an assistant professor of educational technology at the Institute of Education at the University of Minho. His academic qualifications include a degree in Visual Literacy (1991), a master's in Educational Technology (2002), a Ph.D. in Educational Technology (2009), and a Postdoc in Educational Technology (2019). His teaching expertise relates to exploring innovative pedagogical practices focusing on active methodologies, blended learning models, and digital technologies. Lecastre's interests surround the design of innovative pedagogy with new technologies to enhance teaching and learning processes.

Remedios López Liria has a diploma in Physiotherapy (UGR, 1998). She received the Award for the Best Academic Record for her grade in psychology. She received a Ph.D. in Physiotherapy from the University of Almeria (UAL) in 2009. She was awarded an extraordinary doctorate award in health for her work from 2008 to 2009. Since 2006, she has worked as a professor in the Department of Nursing, Physiotherapy, and Medicine. López received a mention for teaching excellence in the Docentia Program at UAL in 2013 and 2018. She is a director in the research group HUM-498: "Psychological Intervention in Development, Education and Guidance." She has conducted multiple teaching innovation projects and European projects as a principal researcher. She focuses on disability, physiotherapy, education, and telerehabilitation.

Ana María Martínez-Martínez has a degree in Educational Psychology from the University of Almería and received an Extraordinary Prize for her degree record. Martínez-Martínez has complemented her training with two master's: Intervention in School Coexistence (2011) and the International Master's Degree in Spanish as a Foreign Language (2014). These were both carried out at the University of Almería (UAL). She received her doctorate in the Educational Innovation Program in 2020 with Outstanding Cum Laude. Martínez-Martínez was involved in the Humanities-498 Research group from 2009–2022, where she participated in numerous prestigious national and international research projects as a collaborating researcher (Ministry of Education, AECID, Grundtvig, Leonardo Da Vinci, Erasmus+). She currently belongs to the International Comparative Research group at UAL, where she is involved in the project "Effects of hate speech on offline relationships among adolescents in the western area of Almería" (DIS-HATE). Her research focuses on models and programs of psychoeducational intervention that improve, optimize, and accelerate development processes and schools' learning and orientations at different levels. She has focused on educational, regulated, and special needs contexts. She has also researched bullying, cyberbullying, hate speech, and emotional intelligence. Her extensive research experience is supported by indexed articles with a good impact index in the JCR repository. She has authored books and book chapters published by renowned publishers. She has also conducted oral communications at national and international conferences.

Noelia Navarro Gómez has a Ph.D. in Psychology, specializing in personality, evaluation, and psychological treatment. She is currently a professor at the University of Málaga. Her research mainly focuses on stigma in mental health, suicide prevention, and the application of new technologies in the prevention and intervention of psychopathologies. She has just deposited her doctoral thesis in education, entitled "Applications of new technologies in the reduction of stigma towards severe mental disorder in educational contexts," which will be presented next September.

Ali Orhan earned his Ph.D. focusing on Curriculum and Instruction from Düzce University. He also holds a master's in Curriculum and Instruction from Düzce University and a bachelor's in English Language and Literature from Ege University. He is currently working at Zonguldak Bülent Ecevit University in Turkey. His research focuses on higher-order thinking skills, such as critical thinking and problem-solving, curriculum development and evaluation, meta-analysis, flipped learning, and online learning. He is the author of numerous articles published in international peer-reviewed journals like the *Journal of Educational Research*, the *Journal of Psychoeducational Assessment*, *Smart Learning Environments*, and *Learning Environments Research*.

Patricia Rocamora-Pérez has a diploma in Physiotherapy from the University of Granada (UGR), which she received in 1999. She received an award for the best academic record of her promotion. She achieved a master's in nursing sciences at the University of Almería (UAL) in 2009. She also received her Ph.D. in Physiotherapy from UAL in 2011, which earned her an extraordinary doctorate award in health. Rocamora-Pérez holds a certificate of Proficiency in English (CPE, C2) from the University of Cambridge, which she received in 2017. She has been a lecturer in the Department of Nursing, Physiotherapy and Medicine at UAL since 2007. Rocamora-Pérez received a mention of teaching excellence from the Docentia Program at UAL in both 2013 and 2019. She is involved in the research Group HUM-498L: "Psychological Intervention in Development, Education and Guidance" at UAL. She is involved in teaching innovation projects at UAL and the university's plurilingualism promotion plan. Rocamora-Pérez's research focuses on rare diseases, physiotherapy, education, and multilingualism.

Jose Miguel Rodriguez Ferrer is a graduate of Psychology. He has a master's in General Health Psychology and a doctorate in Medical Sciences from the University of Almería. Currently, he serves as an interim substitute professor at the University of Jaén. His research focuses on using playful methodologies, specifically gamification (learning through games) and escape rooms, in various contexts, primarily in education and as a sensitization tool. Within the realm of playful methodologies, his studies examine variables like motivation, learning, performance, classroom climate, and student engagement, where these learning strategies are employed rather than more passive approaches. He has conducted these investigations across multiple educational stages, particularly emphasizing secondary education and university settings. He has also made contributions to educational innovation in the university environment. His publications explore innovative approaches in education, such as flipped classrooms, service learning, and project-based learning.

Christian Roith received a Ph.D. in Educational Sciences from the Complutense University of Madrid in 1999 and a degree in Pedagogy from the Bayerische Julius-Maximilians-Universität Würzburg in 1986. Since 2011, he has worked as a professor at the University of Almeria, focusing on the theory and history of education. Between 2000 and 2011, he worked as a professor at the University of Alcalá, focusing on the same topic. His academic interests are in comparative education and the history of education, particularly academic discourses in the educational sciences in Spain and Germany and their history. He is also interested in reconstructing the history of education using alternative sources, such as children's drawings or autobiographies. In 2000, Roith was accepted as a member of the International Standing Conference for the History of Education. He has participated in numerous congresses of this association. He is also a member of the Spanish Society for the History of Education.

María del Mar Sánchez Pérez is a lecturer in the Department of Philology at the University of Almería (UAL). She graduated in Translation and Interpreting from the University of Granada. Sánchez completed her postgraduate studies at UAL, where she obtained her Ph.D. in Arts and Humanities, for which she received Cum Laude and international mention and a Ph.D. Extraordinary Award. Sánchez is a sworn translator and interpreter appointed by the Spanish Ministry of Foreign Affairs. She is a member of the research group HUM-859: "English language: Applied Linguistics and Literary Studies." Her main research interests include English language acquisition/learning, English-teaching methodology, bilingual education, content and language integrated learning, English-medium instruction, integrating content and language in higher education, and translation.

Joan Thormann is a professor emeritus in the Division of Educational Technology. She taught at Lesley University from 1989 to 2016. Thormann has developed and taught courses online since 1996. Her recent research addresses distance education, online teaching strategies such as synchronous conferencing, and student moderating. She incorporates the principles of Universal Design for Learning, information literacy, and student use of technology. She has written numerous articles and has co-authored books about technology and education, including her most recent book, *The Complete Step-by-Step Guide to Designing and Teaching Online Courses* (2012), published by Teachers College Press at Columbia University.

María de los Ángeles Valverde Martínez graduated in Physiotherapy from the University of Almería (UAL) in 2017. She holds a master's in Physiotherapy and Disability from UAL, which she received in 2018, and a master's in Osteopathy from the Escuela de Osteopatía de Madrid (EOM), which she studied between 2018 and 2023. She is a doctoral student in Physiotherapy at UAL and belongs to the research group Humanities-98: "Intervention Psychology in Development, Education and Guidance."

María Elena Zepeda Hurtado is a pedagogue and teacher in administration and educational development. She is certified in teaching skills by ANUIES. She has experience in designing and updating training actions, teacher training, and curricular design, including preparing and updating study plans and programs.

Gülin Zeybek is a doctor in English Language Teaching (ELT) and is currently working in the School of Foreign Languages at Isparta University of Applied Sciences. She holds two MA degrees in Foreign Language Education and Educational Technologies. Her research interests center on pre-service teacher education, multimodal mobile-assisted language learning, and various technology integrations like augmented reality and artificial intelligence into foreign language teaching. She has worked as a foreign language teacher trainer for over 10 years. Zeybek has published various research papers nationally and internationally. Some of her recent publications include: Kuru Gönen, S. İ., & Zeybek, G. (2022). Using QR code enhanced authentic texts in EFL extensive reading: A qualitative study on student perceptions. *Education and Information Technologies*, *27*(2), 2039-2057; Kuru Gönen, S. İ., & Zeybek, G. (2022). Training on multimodal mobile-assisted language learning: a suggested model for pre-service EFL teachers. *Computer Assisted Language Learning*, 1-22; Zeybek, G., & Sayin, İ. (2022). Acceptance and Use of Smartphones: AR-enhanced EFL Reading Practices. *Language Teaching and Educational Research*, *5*(1), 16-35; and Zeybek, G., Erdemir, N., & Sayın, İ. (2023). The Effect of an Online Writing Community on ELT Students' Academic Writing Motivation. In *New Directions in Technology for Writing Instruction* (pp. 367-389). Cham: Springer International Publishing.

Index

H

I

L

M

N

P

S

T

U

V

Ensure Quality Research is Introduced to the Academic Community

Become an Evaluator for IGI Global Authored Book Projects

The overall success of an authored book project is dependent on quality and timely manuscript evaluations.

Applications and Inquiries may be sent to:
development@igi-global.com

Applicants must have a doctorate (or equivalent degree) as well as publishing, research, and reviewing experience. Authored Book Evaluators are appointed for one-year terms and are expected to complete at least three evaluations per term. Upon successful completion of this term, evaluators can be considered for an additional term.

If you have a colleague that may be interested in this opportunity, we encourage you to share this information with them.

www.ingramcontent.com/pod-product-compliance
Lightning Source LLC
LaVergne TN
LVHW060421300926
849162LV00006B/1488
* 9 7 8 1 6 6 8 4 8 8 6 1 4 *